GAUGE FIELD THEORIES

AN INTRODUCTION

Other Pergamon Titles of Interest

GIBSON
The Physics of Nuclear Reactions

GORBACHEV et al
Nuclear Reactions in Heavy Elements

HODGSON
Growth Points in Nuclear Physics

Volume 1: Nuclear Physics

Volume 2: Nuclear Forces and Nuclear Reactions

Volume 3: Nuclear Structures, Potentials and Reactions

Journals

Progress in Particle and Nuclear Physics
Recent special issues include:
WILKINSON: Heavy Ion Interactions at High Energies
WILKINSON: Nuclear Astrophysics
Nuclear Tracks and Radiation Measurements

Full details of all Pergamon publications and a free specimen copy of any Pergamon journal available on request from your nearest Pergamon office

GAUGE FIELD THEORIES
AN INTRODUCTION

by

J. LEITE LOPES
Centre de Recherches Nucléaires
Université Louis Pasteur, Strasbourg

PERGAMON PRESS
OXFORD · NEW YORK · TORONTO · SYDNEY · PARIS · FRANKFURT

U.K.	Pergamon Press Ltd., Headington Hill Hall, Oxford OX3 0BW, England
U.S.A.	Pergamon Press Inc., Maxwell House, Fairview Park, Elmsford, New York 10523, U.S.A.
CANADA	Pergamon Press Canada Ltd., Suite 104, 150 Consumers Road, Willowdale, Ontario M2J 1P9, Canada
AUSTRALIA	Pergamon Press (Aust.) Pty. Ltd., P.O. Box 544, Potts Point, N.S.W. 2011, Australia
FRANCE	Pergamon Press SARL, 24 rue des Ecoles, 75240 Paris, Cedex 05, France
FEDERAL REPUBLIC OF GERMANY	Pergamon Press GmbH, Hammerweg 6, D-6242 Kronberg-Taunus, Federal Republic of Germany

First edition 1981
Reprinted 1983

British Library Cataloguing in Publication Data

Lopes, J. Leite
Gauge field theories.
1. Gauge fields (Physics)
I. Title
539.7'2 QC793.3.F5.
ISBN 0-08-026501-4 (Hardcover)
ISBN 0-08-026500-6 (Flexicover)

In order to make this volume available as economically and as rapidly as possible the author's typescript has been reproduced in its original form. This method unfortunately has its typographical limitations but it is hoped that they in no way distract the reader.

Printed in Great Britain by A. Wheaton & Co. Ltd., Exeter

Sind wir vielleicht hier, um zu sagen : Haus,
Brücke, Brunnen, Tor, Krug, Obstbaum, Fenster,
höchstens : Säule, Turm ... aber zu sagen,
verstehs, oh zu sagen so, wie selber die Dinge
niemals innig meinten zu sein.

Rainer Maria Rilke, Die neunte Elegie
Duineser Elegien
Im Insel-Verlag, Frankfurt am Main, 1962

Preface

In the last fifteen years the attempts at a unified description of the fundamental physical interactions by gauge field theories have given rise to exciting developments in particle physics.

In this book, which grew out of lectures I gave in the last few years in several places, at Strasbourg University, at the 1980 XIV Curso Centro Americano de Fisica, held at the University of Panama, at the Federal University of Rio de Janeiro and the Centro Brasileiro de Pesquisas Fisicas, I try to explain in an elementary way the basic notions and principles of gauge theories. In particular, the Salam-Weinberg model of electro-weak interactions is developed in some detail including its verification in the study of neutrino-lepton scattering and the parton model. This model is at present the most successful attempt at a unified theory of physical interactions.

The aim of this book is to give a self-contained introduction to these theories.

The reader will be assumed to know basic quantum mechanics and special relativity together with the elements of group theory needed for these disciplines ; a knowledge of the qualitative description of elementary particles and their quantum numbers will also be required, as well as the elements of the Feynman diagrams technique.

The first Chapter contains the basic notions of classical field theory and the all important Noether's theorem. An introduction is also given to solitons and instantons and the topological quantum numbers, subjects which arose from the study of the non-linear field equations in gauge theories and which have been developed in the recent few years.

Besides the study of the electromagnetic and the Yang-Mills gauge fields, a chapter on the gravitational field is included. We think that this chapter is of interest for two reasons : firstly, it may be suggestive for the graduate students to learn that there are several common features between this and the Yang-Mills field-non-linear equations, similar covariance behaviour of certain quantities such as the gauge field and its source, under the corresponding groups. Secondly, it is precisely the unification of gravitational with strong, electromagnetic and weak interactions, which present the greatest challenge to theoreticians nowadays. It would be stimulating that the young readers acquire a basic knowledge of the situation for each gauge field, gravity included.

Perturbation calculations, renormalization and path-integral quantization are not studied in this book. Two excellent books on this subject which were recently published, one by C. Nash, the other by J.C. Taylor, are indicated in the bibliography ; they fully develop the basic ideas and techniques in this domain. The reader is invited to consult excellent reports and review articles mentioned in the bibliography.

A section in Chapter IX deals with very recent speculations on possible lepton and quark structures, for which there is so far no experimental evidence. An introduction to the SU(5) model of grand unification is presented in Chapter X. Problems are given for each chapter and solutions are collected at the end of the book.

I am most grateful to Abdus Salam, Director of the International Centre for Theoretical Physics, for sponsoring my lectures in Panama and to Mario Bunge for his support and encouragement ; my best thanks are also due to B. Fernandez and his colleagues of the University of Panama, to R. Lobo, E. Lerner and their colleagues of the Centro Brasileiro de Pesquisas Fisicas and of the University of Rio de Janeiro, respectively for the humanly warm and kind hospitality. The author greatly profited from conversations with J.J. Giambiagi, Ch. Ragiadakos, C.A. Savoy, J.A. Martins Simões and D. Spehler on topics of this book.

I am deeply grateful to the authors and to the Nobel Foundation for their kind permission to reprint the lectures given by the 1979 Nobel Laureates Sheldon L. Glashow, Abdus Salam and Steven Weinberg. I am equally grateful to the Physical Society of Japan and to the author for kindly permitting the reproduction of the Table IX from the article by C. Baltay in the Proceedings of the 19th International Conference on High Energy Physics, 885-903, Tokyo (1978).

Madame Erice North prepared the typescript with great ability and patience, my warmest thanks go to her.

J. LEITE LOPES

Strasbourg, January 1981

Contents

Introduction

1. - You all know that the philosophical dream of physicists has always been to reduce (and thus "explain") the enormous varieties of material bodies and events in nature to configurations of a small number of basic constituents and their interactions -the Greek atoms, the atoms and molecules of the chemistry and physics of the XVII th century, the elementary particles of the last fifty years, the quarks, leptons and fundamental bosons of today.

The ninety two elements of the Mendelejev table were explained in terms of three particles, the electron and the proton and neutron ; these, together with the photons, responsible for the electromagnetic interaction among electrons and nuclei, were the primordial objects of the physicists around 1934.

The later discovery of pions, postulated by Yukawa in 1935, to describe the nucleon interactions, and then of muons and neutrinos, of strange particles and ressonances, seemed to suggest that the underlying reality of fundamental particles was perhaps too rich to be possibly reduced to a small number of objects.The number of supposed elementary particles soon became at least as large as the number of elements in the Mendelejev table

2. - On the other hand the idea that physical forces propagate in space with a finite velocity through the action of a field was introduced by Maxwell and Lorentz in electrodynamics. This idea was further developed by Einstein and in his relativistic theory of gravitation -perhaps the most beautiful achievement in theoretical physics up to our days- the unifying power of the description by the field concept was greatly enhanced, the gravitational field being identified with the metric tensor in a Riemannian space-time.

GFT - B

With the development of quantum mechanics and of the principles of quantum field theory, physicists were led to associate a field to each particle. However, the large number of elementary particles which were discovered in the fifties discouraged many physicists in their belief of the unifying rôle of field theory. The efforts developed by Einstein to find a unitary theory of the gravitational and the electromagnetic field, seemed meaningless to quantum and particle physicists since many other fields would have to be taken into account in such a unifying theory. It was mainly in the domain of strong interaction physics that the notion of field seemed useless.

The developments in the last ten years which culminated with the Salam-Weinberg model of gauge fields which unify electromagnetic and weak interactions and, more recently, the discovery of quantum chromodynamics, restored the full theoretical value of field theory. It is believed that the unification which we must seek is rather that of the basic forces of nature, rather than of the bodies and their constituents. The elementary particles are now reduced to leptons and quarks but the number of these admitted basic objects seems to be increasing. Instead, the Salam-Weinberg model opened up a new style and a new aim, in the spirit of the great unification of physical fields as dreamed of by Einstein. Strong interactions are now assumed to be described by massless vector gauge fields associated with the colour degrees of freedom of quarks. And this theory is expected to reproduce the strong interactions between hadrons, although for the moment mathematical difficulties have been preventing an early completion of this program.

Current research actively develops efforts in the sense of a "grand" unification of all the basic interactions of nature, such as the Georgi-Glashow SU(5) model which attempts to unify the strong, weak and electromagnetic interactions.

3. - We know that the elementary particles are classified according to their spin into bosons -particles with integral spin, obeying therefore to the Bose-Einstein statistics- and fermions particles with half-integral spin and which obey the Fermi-Dirac statistics.

I will show you now a table (Table I) which indicates the basic physical interactions between particles. These are in small number : all forces in nature result from the interplay of : 1) gravitational interactions which are created by and act upon all forms of energy and matter ; 2) weak interactions, which act between leptons (electrons, muons, tauons and their neutrinos) and also hadrons ; 3) electromagnetic interactions, created by and acting upon all particles with a charge, a dipole moment ; 4) strong interactions, which act only on heavy particles called hadrons. As we said above, it is today believed that these interactions may be described in a unified way : massless vector fields -the gauge fields- are defined in association with the postulated invariance of the theory under gauge transformation and these fields give rise after a spontaneous break-down of the symmetry, to the fields of the weak and electromagnetic interactions. The strong interactions are assumed to be governed by another gauge theory with unbroken symmetry, the colour SU(3) gauge group, which introduces eight massless gauge fields, the gluons.

TABLE I - Basic interactions

Interactions	strength of coupling constant	Transmitted by	Gauge fields	
Gravitation	$\mathcal{G}\frac{m_e^2}{e^2} \sim 0.2\times10^{-42}$	spin 2 massless field quantum : graviton	general coordin. transform gauge field	
Weak	$G_F\frac{(m_p c)^2}{\hbar^3} \sim 1.01\times10^{-5}$	spin 1 massive fields quanta : W^+, W^-, Z^0	$SU(2) \otimes U(1)$ gauge fields	The SU(5) model defines 24 gauge fields
Electromagnetic	$\alpha \equiv \frac{e^2}{4\pi\hbar c} \sim \frac{1}{137} \sim 10^{-2}$	spin 1 massless field quantum : photon		
Strong	$\frac{g^2}{4\pi\hbar c} \sim 10$ for hadronic. matter Momentum transfer dependent $\alpha_s(Q^2)$ for quark interaction	spin 1 massless fields gluons	colour SU(3) gauge fields	

Supergravity postulates a massless spin 3/2 quantum -the gravitino- in addition to the graviton

TABLE II - Observed fermions

	Display	observed
Leptons	weak and electromagnetic forces	$e^-, \nu_e; \mu^-, \nu_\mu; \tau^-, \nu_\tau$ and their antiparticles (spin 1/2)
Baryons	weak, electromagnetic and strong forces	nucleons; hyperons; baryonic resonances (spin 1/2, 3/2,...)

TABLE III - Observed bosons (1980)

Photons	
Hadronic Mesons	π, ρ, κ, ϕ, D, ψ, Υ, etc.

TABLE IV - Quark quantum numbers

Flavours	Q	I	I_3	B	Y	S	C	b	t(?)
u_a	2/3	1/2	1/2	1/3	1/3	0	0	0	0
d_a	- 1/3	1/2	- 1/2	1/3	1/3	0	0	0	0
s_a	- 1/3	0	0	1/3	- 2/3	- 1	0	0	0
c_a	2/3	0	0	1/3	- 2/3	0	1	0	0
t_a ?	2/3	0	0	1/3	- 2/3	0	0	0	1
b_a	- 1/3	0	0	1/3	- 2/3	0	0	- 1	0
⋮(?)									

Q = charge
I = isospin
I_3 = third component of isospin

B = baryon number
Y = hypercharge
S = strangeness
C = charm
b = beauty (bottom-ology ; bottomness ?)
t = top-ology (topness ?)

$Y = B + S - C + b - t$

$Q = I_3 + \frac{1}{2}(B + S + C + b + t)$

Each quark is assumed to exist in three states which differ among themselves only by a new quantum number, the colour a = 1,2,3

TABLE V - Lepton quantum numbers

	L_e	L_μ	L_τ
ν_μ	0	1	0
μ^-	0	1	
ν_e	1	0	
e	1	0	
$\bar{\nu}_\mu$	0	-1	
μ^+	0	-1	
$\bar{\nu}e$	-1	0	
e^+	-1	0	
ν_τ	0	0	1
τ	0	0	1
τ^+	0	0	-1
$\bar{\nu}_\tau$	0	0	-1

L_a : a-onic lepton quantum number

TABLE VI - Basic fermions

Leptons	(ν_e, e) ; (ν_μ, μ) ; (ν_τ, τ) ; ... (?)
Quarks	(u_a, d_a) ; (c_a, s_a) ; (t_a, b_a) ; ... (?) a = green, yellow, blue

TABLE VII - Basic boson fields

	Gauge field	Broken field	Quanta
Gravitational	40 component gauge field	no	spin two graviton
Weak and electro-magnetic	Four vector gauge fields	Three massive vector fields W^{μ^-}, W^{μ^+}, Z^μ and one abelian gauge massless field A^μ	spin one $W^\pm$, Z^0, and photon γ
Colour fields	Eight vector massless gauge fields	no	spin one gluons

Supergravity introduces a massless spin 3/2 field in addition to the spin two graviton.

The spontaneous breakdown of gauge symmetry to generate boson masses, is carried out by certain scalar fields called Higgs fields. Higgs bosons are therefore assumed to be also basic particles but they, as well as the vector bosons W and Z, have not been observed so far (as of November 1980).

TABLE VIII - Quark structure of hadrons

1) Hadrons are colourless	
2) Hadronic mesons are assumed to be formed of a pair quark-antiquark (summed over colours)	meson field = $\sum_c \bar{q}_c \, q'_c$ thus $\pi^+ = \bar{d}_1 u_1 + \bar{d}_2 u_2 + \bar{d}_3 u_3$ $K^+ = \bar{s}_1 u_1 + \bar{s}_2 u_2 + \bar{s}_3 u_3$ where c = 1,2,3 or g,y,b
3) Baryons are constructed out of three quarks, their field wave functions being antisymmetric	baryon field = $= N \begin{vmatrix} q_{1g} & q_{1y} & q_{1b} \\ q_{2g} & q_{2y} & q_{2b} \\ q_{3g} & q_{3y} & q_{3b} \end{vmatrix}$ thus $\Lambda \simeq u_g d_y s_b + u_b d_g s_y + u_y d_b s_g -$ $- u_d d_b s_y - u_y d_g s_b - u_b d_y s_g$

TABLE IX - Questions

Are leptons to be associated to quarks as in Table VI or are they to be associated to hadrons, observable particles like them ?

Are new sub-leptonic particles to be associated to quarks to give a unified description of leptons and hadrons ? Are therefore leptons point-like or do they have a subtle structure ? Do leptons with spin 3/2 exist ?

Are the fundamental bosons without structure ? And quarks ?
Is the SU(5) model enough for grand unification ?

The classification of observed fermions and observed bosons is shown in Tables II and III.

It is today assumed that the binding of certain basic particles called quarks give rise to hadronic matter. Quarks, with their defining quantum numbers, are shown in Table IV. As quarks have not been observed, it is assumed that their mutual interaction increases strongly with their distances and they are in this way, confined, not to be found in a free state. Observed would be only states with zero colour and quarks exist in three different colour states. Besides quarks, there exist the leptons, namely, the electrons and electronic-neutrinos, the muons and muonic neutrinos and the tauons and the tauonic neutrinos. Both quarks and leptons are supposed to be the basic pieces of all matter (Tables V-VI). Table VIII indicates the quark structure of hadronic matter.

4. - The aim of this book is to give a self-contained introduction to the theory of gauge fields, which plays a basic role in the description and unification of the basic physical interactions.

The reader will be assumed to know the foundations of quantum mechanics and of the special theory of relativity. For the sake of completeness, we dedicate the Chapter I to the establishment of the basic equations in particle physics, namely the equations for free particles described

a) by scalar fields ; this is the case of spin zero bosons

b) vector fields, which describe particles with spin one such as the vector bosons (and the photon for massless real fields)

c) Dirac spinor fields, which apply to particles with spin 1/2,

d) Rarita-Schwinger vector spinor fields, which describe fermions with spin 3/2 ;

e) Symmetric second rank tensor fields, which describe particles with spin 2 ; the case of such massless tensor field applies to the description of the gravitational field as will be seen in Chapter V.

In addition, dual or pseudofields are defined and this allows us to introduce the currents constructed with Dirac fields and which play a central role in the theory of weak interactions.

The establishment of these equations can be made in a very simple way by postulating in each case, the existence of only one such independent field, and of no other field of lower tensor or spinor-rank. The establisment of some usual non-linear field equations is also made by a similar method.

Still in Chapter I , we recall the lagrangeans from which these equations may be derived and the all-important Noether-theorem. A detailed derivation of the conserved Noether tensors is given since Noether conserved quantities, such as the energy and momentum, the angular momentum, the electromagnetic current, the isospin currents, play a special role in the theory.

We thought it would be helpful to the reader to have the expression of some of these physical quantities, deduced in a detailed fashion from the lagrangeans.

An introduction to the study of the soliton and instanton solutions of non-linear field equations which appear in gauge theories, is given ; these solutions introduce the important notion of conserved topological numbers, a notion which does not follow from Noether's theorem at all, but is rather associated to topological properties of those solutions.

In Chapter II, these notions and equations are applied to the study of the electromagnetic interactions. The electromagnetic field is the first example of a gauge field and the presentation of Maxwell's theory under the gauge field view point is beautiful and instructive. The generalization of this idea, by Yang and Mills, to the so-called non-abelian gauge fields, was an important step taken in 1954, and which ultimately made possible the recent developments in particle theory in the last ten years.

The Yang-Mills gauge field is studied in Chapter IV. Its origins are recalled, the field equations and the corresponding lagrangeans are given and the important case of the colour gauge field is described.

We thought that a study of the gravitational field as a gauge field would be important for two reasons : 1) it would be suggestive for the young readers to see that there are many common features between the gravitation field and the Yang-Mills field -non-linearity of the equations, covariant properties of certain quantities such as the gauge field itsef and its source, similar behaviour, from the point of view of covariance, of corresponding equations. Secondly it is precisely the unification of the gravitational interactions with the other ones -strong, electromagnetic and weak- which presents the greatest challenge nowadays. I believe it would be instructive that this treatment of the gravitational field be included in a book dedicated more to the study of the other interactions, even at the price of having this Chapter somehow disconnected from the following chapters.

The remaining chapters are dedicated to introduce the ideas and techniques needed to a detailed study of the Salam-Weinberg model, which describes in a unified form the electromagnetic and the weak interactions.

Finally, in Chapter X, we mention the recent attempts at a "grand" unification of the electromagnetic and weak forces with the strong interactions and present an introduction to the study of the SU(5) model, the simplest such model.

It is our hope that this book, even in its form not completely homogeneous, due of course to the author's fault but also due to the presentation of recent theories still under intensive investigation, it is our hope that it will help young graduate students to follow a hopefully clear path which will lead them further in the study of particle physics.

CHAPTER I

Field Equations, Conserved Tensors and Topological Quantum Numbers

I. 1 - FREE FIELD EQUATIONS

Elementary particles are described by fields which obey certain equations, the so-called relativistic wave equations. These fields have basic properties which follow from a postulated invariance of the wave equations under certain groups of transformations, the symmetry groups. The latter are suggested by experiment. Examples of these symmetries are the independence of the laws of physics from the origin of time and from the position and orientation of laboratories in ordinary space. Invariance of the wave equations under the groups of time and space translations and rotations leads to the important principles of conservation of energy and momentum and angular momentum.

The most important invariance principle in field theory is the principle of relativity. It states that the laws of physics do not depend on the choice of an origin for a coordinate system and for the counting of time ; nor do they depend on the spatial orientation of this coordinate system nor on the state of rectilinear and uniform motion of the laboratory. Mathematically, this principle imposes that the equations of motion be invariant under the proper orthochronous Poincaré group. Wave fields are assumed to belong to representation spaces of this group[1-12] ; their space-time geometrical nature is thus determined and the fields can only be scalars, spinors, vectors and higher order spinors and tensors under the Poincaré group.

Experiment has shown, after the discovery of a number of elementary particles, that there exist fields which, besides their Poincaré geometrical nature, have internal degrees of freedom. They may be scalars, two-component spinors, three-dimensional vectors with respect to certain additional symmetry groups related to internal quantum numbers such as isospin, flavor, colour and so on. This internal degree of freedom results from the invariance of the equations of motion under certain groups of transformations acting on the field regarded as an entity in a new internal space - such as the group $U(1)$ of phase transformations of complex fields or the group $SU(2)$ of phase operator transformations of isospinors and so on.

Whereas the Poincaré invariance of the equations of field theory is universal the invariance under internal symmetry groups is obeyed by specific field models as in chromodynamics ; there are also internal symmetries which are only approximate, as in flavordynamics. And the recent discovery of a mechanism of mass generation by spontaneous break down of certain subgroups of an internal symmetry group has been of the utmost importance for the formulation of the current unification models of physical interactions.

In this paragraph, we shall establish the equations of free fields with respectively spin zero, spin one half, spin one, spin three-halves and spin two by a simple method which consists in postulating, in each case, the existence of only one such independent field and of no other field of lower tensor or spinor rank.

a) Scalar field : let us assume that we are given a scalar field $\varphi(x)$, i.e, such that under a proper, orthochronous Poincaré transformation of the geometrical frame of reference

$$\begin{aligned} & x^{\mu} \rightarrow x'^{\mu} = a^{\mu} + \ell^{\mu}_{\ \nu}\, x^{\nu} \\ & \ell^{\mu}_{\ \alpha}\, g_{\mu\nu}\, \ell^{\nu}_{\ \beta} = g_{\alpha\beta} \qquad\qquad (I.\ 0) \\ & \det \ell = 1,\ \ell^{0}_{\ 0} \geq 1. \end{aligned}$$

transforms like :

$$\varphi'(x') = \varphi(x)$$

We are told then that there must exist no other scalar independent from $\varphi(x)$. Now the equipement of space-time analysis gives us the differential operator

$\partial_{\mu} \equiv \dfrac{\partial}{\partial x^{\mu}}$ and the metric tensor :

$$\begin{aligned} & g_{\mu\nu} = 0 \quad \text{for} \quad \mu \neq \nu \\ & g_{00} = - g_{11} = - g_{22} = - g_{33} = 1. \end{aligned}$$

Thus we may form a four-vector $\partial_{\mu}\varphi$ and a tensor $\partial_{\mu}\partial_{\nu}\varphi$. With the latter and

$g_{\mu\nu}$ we obtain a new scalar* :

$$\Box \varphi = g_{\mu\nu} \partial^{\mu} \partial^{\nu} \varphi$$

In view of our postulate which forbids another scalar, built out from $\varphi(x)$, independent from $\varphi(x)$, we shall necessarily have :

$$a \Box \varphi + b \varphi = 0$$

The ratio $\frac{b}{a}$ will be called m^2 and this is the Klein-Gordon equation for a <u>free</u> field

$$\Box \varphi + m^2 \varphi = 0 \qquad \text{(I. 1)}$$

m is found out to be the mass of the particles described by the field. Note that we have elected a system of units in which

$$\hbar = 1, \quad c = 1$$

and the constant m^2 above has the dimension ℓ^{-2} since the Compton wavelength associated to these particles is

$$\lambda = \frac{\hbar}{mc}$$

If besides $\varphi(x)$, we are given another scalar, $\rho(x)$ independent from $\varphi(x)$ and forbid the existence of a third independent scalar we are led then to write

$$\Box \varphi = a \varphi + g \rho$$

and the requirement that when $\rho(x)$ is not present we get equation (I. 1) gives :

$$(\Box + m^2) \varphi(x) = g \rho(x) \qquad \text{(I. 2)}$$

g is the coupling constant, $\rho(x)$ is the source of the scalar field $\varphi(x)$.

Now we need another assumption, which is rather a quantum-mechanical result in the theory of representation of the rotation group and of the Lorentz group : a particle with spin $s = 0,1,2,\ldots$, <u>is described by a field with $2s+1$</u>

---

*We stop at second derivatives. A more general equation than (I. 2) is :

$$F(\Box)(\Box + m^2) \phi(x) = g \rho(x)$$

where F is a function of the operator $\Box$; this is the Pais-Uhlenbeck equation. In general, locality and causality properties are lost by such higher-order equations.

independent components, whereas if $s = 1/2, 3/2, \ldots$ the number of independent components is $2(2s + 1)$.

b) Vector field : let us now assume we are given a vector field $\phi^\mu(x)$ and forbid the existence of any scalar function built out of ϕ^μ (which then would describe a spin 0 particle) and of any other vector independent from $\phi^\mu(x)$; under the transformation (I. 0) the vector field transforms in the following way :

$$\phi'^\mu(x') = \ell^\mu{}_\nu \, \phi^\nu(x)$$

With $\phi^\mu(x)$ we construct a tensor $\partial^\alpha\phi^\mu$ and another $\partial^\alpha\partial^\beta\phi^\mu$. And with this we get a vector :

$$\Box \, \phi^\mu = g_{\alpha\beta} \, \partial^\alpha \, \partial^\beta \, \phi^\mu$$

By our assumption we must have then :

$$\begin{cases} (\Box + m^2) \, \phi^\mu = 0 \\ \\ \partial_\mu \, \phi^\mu = 0 \end{cases} \qquad \text{(I. 3)}$$

and

since $\partial_\mu\phi^\mu$ is a scalar made with ϕ^μ. This gives us three independent components for ϕ^μ which thus describes spin 1 particles.

If we are given two independent vectors, $\phi^\mu(x)$ and $j^\mu(x)$ and forbid the existence of any other independent vector and rule out the existence of any scalar then we get

$$(\Box + m^2) \, \phi^\mu = g \, j^\mu$$

$$\partial_\mu \, \phi^\mu = \partial_\mu \, j^\mu = 0$$

We could also have another postulate : there exists only a four-vector $A^\mu(x)$ and no other four-vector can be different from zero, as well as no scalar. This gives us the massless free field equations

$$\begin{cases} \Box \, A^\mu(x) = 0 \\ \\ \partial_\mu \, A^\mu(x) = 0 \end{cases} \qquad \text{(I. 4)}$$

Massless free fields are described by only two independent field components.

c) Dirac's equation for spin 1/2 field

If we are given a Dirac spinor $\psi(x)$ (together with the matrix machinery) which transforms in the following way, under the group (I. 0) :

$$\psi'(x') = D(\varepsilon)\ \psi(x) \quad ; \quad D(\varepsilon) = \exp\left(-\frac{i}{4}\sigma^{\mu\nu}\varepsilon_{\mu\nu}\right)$$

$$\frac{1}{2}(\gamma^{\mu}\gamma^{\nu} + \gamma^{\nu}\gamma^{\mu}) = g^{\mu\nu} \quad ; \quad \frac{i}{2}(\gamma^{\mu}\gamma^{\nu} - \gamma^{\nu}\gamma^{\mu}) = \sigma^{\mu\nu}$$

we see that

$$\partial_{\mu}\psi$$

is a vector-spinor and hence $i\,\gamma^{\mu}\,\partial_{\mu}\,\psi$ is another spinor

$$\varphi_a = i\,\gamma^{\mu}{}_{ab}\,\partial_{\mu}\,\psi_b$$

If we forbid the existence of any other spinor independent from ψ we are then led to the equation :

$$(i\,\gamma^{\mu}\,\partial_{\mu}\,\psi - m\,\psi) = 0 \tag{I. 5}$$

for a free spinor field.

In the presence of a source one obtains the equation

$$(i\,\gamma^{\mu}\,\partial_{\mu}\,\psi - m\,\psi) = \rho$$

where ρ is a spinor linearly dependent on ψ. Thus :

$$\rho = \begin{cases} g\,\varphi\,\psi, \text{ or} \\ g\,\gamma_{\mu}\,\phi^{\mu}\,\psi \\ \text{etc} \end{cases}$$

The number of components is here four, due to the existence of negative-energy states.

d) Rarita-Schwinger equation for spin 3/2 fields

Suppose we are given a spinor-vector $\psi_a{}^{\mu}(x)$ and say that this is the only independent field of this nature ; suppose further that there exists no pure spinor, which would describe spin 1/2. With $\psi_a{}^{\mu}$ we can form two spinors, namely :

$$\varphi \equiv \gamma \cdot \psi \equiv \gamma_{\mu} \psi^{\mu}$$

and

$$\chi \equiv \partial \cdot \psi \equiv \partial_{\mu} \psi^{\mu}$$

We see then that the equations :

$$\begin{aligned} &(i \gamma^{\mu} \partial_{\mu} - m) \psi^{\alpha} = 0 \\ &\gamma_{\alpha} \psi^{\alpha} = 0 \quad ; \quad \partial_{\alpha} \psi^{\alpha} = 0 \\ &\psi'^{\alpha}{}_{a}(x') = [D(\varepsilon)]_{aa'} \ell^{\alpha}{}_{\beta} \psi^{\beta}{}_{a'}(x) \end{aligned} \tag{I. 6}$$

describe a field with eight independent components corresponding to spin 3/2. The particles with spin 1/2 contained in ψ^{α} are transformed away by the subsidiary conditions.

e) Equation for spin 2-field.

A tensor of second rank $\phi^{\mu\nu}(x)$ has 16 components. A particle with spin 2 must be described by a field with five independent components.

For an antisymmetric tensor it is not possible to describe a spin 2 field since it has six independent components and no single condition can be imposed to it (the scalar formed with $\partial_{\mu} \partial_{\nu} \phi^{\mu\nu}$ would vanish identically with $\phi^{\mu\nu}$ antisymmetric).

Consider then a symmetric tensor

$$\phi^{\mu\nu} = \phi^{\nu\mu}$$

and forbid the existence of any other independent symmetric tensor (except the source if it is the case of interaction with other fields) as well as of any vector or scalar formed with $\phi^{\mu\nu}$. We then obtain the following equations :

$$(\Box + m^2)\, \phi^{\mu\nu} = 0 \quad , \phi^{\mu\nu} = \phi^{\nu\mu}$$

$$\partial_\mu \phi^{\mu\nu} = 0 \qquad (I.\ 7)$$

$$g_{\mu\nu} \phi^{\mu\nu} = 0$$

which reduce the number of independent components to five and these therefore describe spin 2 particles.

f) Equation for spin 2 massless field

An alternative way of establishing these equations will be illustrated for the case of a spin 2 massless field $h^{\mu\nu}$ generated by a source $T^{\mu\nu}$ where :

$$h^{\mu\nu}(x) = h^{\nu\mu}(x) \quad , \quad T^{\mu\nu}(x) = T^{\nu\mu}(x)$$

With the field $h^{\mu\nu}$ we may construct the following tensors :

$$\Box\, h^{\mu\nu} \;;\; \partial_\lambda \partial^\mu h^{\nu\lambda} + \partial_\lambda \partial^\nu h^{\mu\lambda} \;;\; \partial^\mu \partial^\nu g_{\alpha\beta} h^{\alpha\beta} \;;$$

$$g^{\mu\nu} \Box\, g_{\alpha\beta} h^{\alpha\beta} \;;\; g^{\mu\nu} \partial_\lambda \partial_\eta h^{\lambda\eta} \;;\; g^{\mu\nu} g_{\alpha\beta} h^{\alpha\beta} .$$

As for massless fields no terms in $h^{\mu\nu}$ and $g^{\mu\nu}h$ must occur in its equation we must have :

$$\Box\, h^{\mu\nu} - a\,(\partial_\lambda \partial^\mu h^{\nu\lambda} + \partial_\lambda \partial^\nu h^{\mu\lambda}) + b\, \partial^\mu \partial^\nu h + c\, g^{\mu\nu} \Box\, h +$$

$$+ d\, g^{\mu\nu} \partial_\alpha \partial_\beta h^{\alpha\beta} = - \kappa\, T^{\mu\nu}$$

where

$$h \equiv g_{\alpha\beta} h^{\alpha\beta}$$

is the trace of the field. By differentiation with respect to x^μ we get

$$\Box\, \partial_\mu h^{\mu\nu} (1 - a) + (d - a)\, \partial^\nu \partial_\alpha \partial_\beta h^{\alpha\beta} + (b + c)\, \Box\, \partial^\nu h =$$

$$= - \kappa\, \partial_\mu T^{\mu\nu}$$

Conservation of the source tensor gives

$$\partial_\mu T^{\mu\nu} = 0$$

therefore the left-hand side of the above equation must vanish identically which leads to

$$a = 1, \quad d = a = 1, \quad b = -c$$

So the equation will be :

$$\Box h^{\mu\nu} - (\partial_\lambda \partial^\mu h^{\mu\lambda} + \partial_\lambda \partial^\nu h^{\mu\lambda}) + b \partial^\mu \partial^\nu h - b g^{\mu\nu} \Box h +$$

$$+ g^{\mu\nu} \partial_\alpha \partial_\beta h^{\alpha\beta} = - \kappa T^{\mu\nu}$$

or

$$\frac{1}{b} \Box h^{\mu\nu} - \frac{1}{b} (\partial_\lambda \partial^\mu h^{\nu\lambda} + \partial_\lambda \partial^\nu h^{\mu\lambda}) + \partial^\mu \partial^\nu h - g^{\mu\nu} \Box h +$$

$$+ \frac{1}{b} g^{\mu\nu} \partial_\alpha \partial_\beta h^{\alpha\beta} = - \frac{\kappa}{b} T^{\mu\nu}$$

Now we choose a field scale so that the coefficient of $\Box h^{\mu\nu}$ be the unity hence $b = 1$ and :

$$\Box h^{\mu\nu} - (\partial_\lambda \partial^\mu h^{\nu\lambda} + \partial_\lambda \partial^\nu h^{\mu\lambda}) + \partial^\mu \partial^\nu h - g^{\mu\nu} \Box h +$$

$$+ g^{\mu\nu} \partial_\alpha \partial_\beta h^{\alpha\beta} = - \kappa T^{\mu\nu} \qquad \text{(I. 8)}$$

This equation is invariant under the gauge transformation

$$h^{\mu\nu} \rightarrow h'^{\mu\nu} = h^{\mu\nu} - \partial^\nu \Lambda^\mu - \partial^\mu \Lambda^\nu$$

If we then impose a gauge-fixing condition

$$\partial_\mu (h^{\mu\nu} - \frac{1}{2} g^{\mu\nu} h) = 0$$

then the equation (I. 8) reduces to

$$\Box(h^{\mu\nu} - \frac{1}{2} g^{\mu\nu} h) = - \kappa T^{\mu\nu}$$

We then introduce a new field variable

$$\phi^{\mu\nu} = h^{\mu\nu} - \frac{1}{2} g^{\mu\nu} h$$

which then leads to the equations

$$\Box \phi^{\mu\nu} = - \kappa T^{\mu\nu}$$

$$\partial_\mu \phi^{\mu\nu} = 0 \tag{I. 8a}$$

We shall see that the Einstein's equation for the gravitational field reduce to an equation to the type (I. 8) in the weak-field approximation (Chapter V).

g) Pseudotensor fields and currents

There is an important instrument of the space-time tensor machinery, namely the Levi-Civita totally antisymmetric tensor $\varepsilon_{\alpha\beta\gamma\delta}$ thus defined :

$$\varepsilon_{\alpha\beta\gamma\delta} = 0 \quad \text{for any two equal indices,}$$

$$\varepsilon_{0123} = 1,$$

$\varepsilon_{\alpha\beta\gamma\delta}$ changes sign under interchange of two consecutive indices. It allows us to define dual or pseudotensor fields and currents which transform under space reflection with an opposite sign as compared to that of the associated tensors.

In view of the expression for the determinant of a Lorentz transformation

$$\varepsilon_{\alpha\beta\gamma\delta}\, \ell^{\alpha}{}_{\mu}\, \ell^{\beta}{}_{\nu}\, \ell^{\gamma}{}_{\lambda}\, \ell^{\delta}{}_{\eta} = (\det \ell)\, \varepsilon_{\mu\nu\lambda\eta}$$

we define the following dual or pseudotensor fields, the transformation of which under a Lorentz transformation gets $(\det \ell)$ as a factor :

I) Pseudoscalar field $\varphi(x)$, the dual of a totally antisymmetric 4th rank tensor $\varphi^{\mu\nu\lambda\eta}(x)$:

$$\varphi(x) = \frac{1}{4!}\, \varepsilon_{\mu\nu\lambda\eta}\, \varphi^{\mu\nu\lambda\eta}(x)$$

$$\varphi'(x') = (\det \ell)\, \varphi(x)$$

Thus $\varphi(x)$ changes sign under improper transformations.

An example is the pseudoscalar

$$P(x) = i\, \bar{\psi}(x)\, \gamma^5\, \eta(x)$$

constructed with two Dirac spinors $\psi(x)$, $\eta(x)$ and their adjoints

$$\bar{\psi}(x) = \psi^+(x)\, \gamma^0,\ \bar{\eta}(x) = \eta^+(x)\, \gamma^0$$

and the definition :

$$\gamma^5 = \frac{i}{4!}\, \varepsilon_{\mu\nu\lambda\eta}\, \gamma^\mu\, \gamma^\nu\, \gamma^\lambda\, \gamma^\eta$$

where γ^μ are the Dirac matrices given in c). The transformation properties of the spinors :

$$\psi'(x') = D(\ell)\, \psi(x)$$
$$\bar{\psi}'(x') = \bar{\psi}(x)\, D^{-1}(\ell),$$

(and similarly for $\eta(x)$) and

$$D^{-1}\, \gamma^\mu\, D = \ell^\mu{}_\nu\, \gamma^\nu$$

show that $P(x)$ is a pseudoscalar.

II) Axial vector field $\phi_\mu(x)$, equivalent to a totally antisymmetric 3rd rank tensor $\phi^{\nu\lambda\eta}(x)$ by the relationship :

$$\phi_\mu(x) = \frac{1}{3!}\, \varepsilon_{\mu\nu\lambda\eta}\, \phi^{\nu\lambda\eta}(x)$$

so that :

$$\phi'_\mu(x')\, \ell^\mu{}_\alpha = (\det \ell)\, \phi_\alpha(x)$$

An important example is the axial vector current constructed with two spinors $\psi(x)$, $\eta(x)$:

$$A^\mu(x) = \bar{\psi}(x)\, \gamma^5\, \gamma^\mu\, \eta(x)$$

III) <u>Dual tensor field</u> $\tilde{F}_{\mu}(x)$, equivalent to an antisymmetric second rank tensor $F^{\lambda\eta}(x)$:

$$\tilde{F}_{\mu\nu}(x) = \frac{1}{2}\,\varepsilon_{\mu\nu\lambda\eta}\,F^{\lambda\eta}(x)$$

such as the one associated to the Maxwell field tensor.

An example is the pseudotensor current :

$$\Sigma^{\mu\nu}(x) = \bar{\psi}(x)\,\gamma^5\,\sigma^{\mu\nu}\,\eta(x)$$

Besides the currents $P(x)$, $A^{\mu}(x)$ and $\Sigma^{\mu\nu}(x)$, the proper tensor currents :

$$S(x) = \bar{\psi}(x)\,\eta(x), \text{ scalar}$$
$$V^{\mu}(x) = \bar{\psi}(x)\,\gamma^{\mu}\eta(x), \text{ vector}$$
$$S^{\mu\nu}(x) = \bar{\psi}(x)\,\sigma^{\mu\nu}\,\eta(x), \text{ tensor}$$

are the basic tools for the construction of interaction lagrangeans between spin $\frac{1}{2}$ particles among themselves and with given tensor fields.

I. 2 - NON-LINEAR FIELD EQUATIONS FOR A SINGLE SCALAR FIELD

The equations established in the preceding paragraph are the basis of the simplest field theories.

The recent development of gauge theories has given rise to classical non-linear field equations, which have been the subject of much investigation in view of the fact that they admit of non-singular solutions of finite total energy which are non-dissipative.

It is well known in quantum mechanics that as time goes to infinity a free wave packet spreads out indefinitely. Only for sufficiently small time intervals is the spreading of the wave packet negligible and only then may the latter represent a free particle.

In general, a solution of a classical field equation is called dissipative if it gives rise to an energy density $T_{00}(\vec{x}, t)$ which vanishes after an infinitely long time :

$$\lim_{t \to \infty} T_{00}(\vec{x}, t) = 0$$

for all $\vec{x}$.

This notion of dissipation is a generalization of the following simple physical process, in the words of Sidney Coleman : "A stone thrown into a still body of water makes ripples that spread out and eventually die away. The stone disturbs the water, gives it energy, but, even if we ignore friction, this energy tends in the course of time to spread out over the water. If we imagine the water to be infinite in extent, then, if we wait long enough, at no point is the water appreciably different from its state before the stone was cast. The disturbance dissipates".

The non-singular solutions with finite energy of the linear free-field equations have the property of being dissipative.

There exist, however, some non-linear classical field equations that have non-singular non-dissipative solutions. The importance of these solutions resides in the fact that the simplest of them are time-independent lumps of energy which do not spread out with time. These solutions are frequently called solitons in the literature. The lumps provide a description of extended objects with a finite energy such as might be the classical limit of hadrons. The non-linearity of the field equation implies a self-interaction which is responsible for the concentration in space of these lumps of energy.

Let us examine the simplest non-linear field equations which are the subject of interest to particle physics.

a) Scalar field equation with quartic self-interaction : in the preceding paragraph I. 1, item a), the equation (I. 1) was obtained from the postulate that the term in $\Box\varphi$ was not linearly independent from the field itself. This equation may be generalized if we assume that $\Box\varphi$ and some arbitrary function of φ, $F(\varphi)$, (but not of its derivatives) are also not linearly independent :

$$\Box\varphi + F(\varphi) = 0 \qquad \text{(I. 9)}$$

$F(\varphi)$ may be either a polynomial or a series in φ. In the first case we write :

$$\Box \varphi + \sum_{n \geq 1}^{N} \lambda_n \phi^n = 0 \qquad \text{(I. 9a)}$$

The particular case in which :

$$\lambda_1 = m^2, \lambda_n = 0 \quad \text{for} \quad n > 1$$

leads to the linear equation (I. 1).

An important equation is obtained when we take :

$$\lambda_1 = \mu^2, \lambda_2 = 0, \quad \lambda_3 = \frac{\lambda}{3!}, \lambda_n = 0 \quad \text{for} \quad n > 3$$

where λ is a constant and φ is a real scalar function.

It is the equation of the so called φ^4 - theory

$$(\Box + \mu^2)\, \varphi + \frac{\lambda}{3!} \varphi^3 = 0 \qquad \text{(I. 9b)}$$

which plays an important role in gauge theories (see Chapter VII). The name φ^4 - theories results from the occurence of a term with the 4th power of φ in the lagrangean which gives rise to equation (I. 9b). The non inclusion of a term in φ^2 in this equation means that we impose a symmetry namely that the equation be invariant under the transformation $\varphi \to -\varphi$, an important requirement in the study of spontaneous symmetry breakdown (Chapter VII).

b) The sine-Gordon equation : this equation is obtained from equation (I. 9) by choosing the function $F(\varphi)$ as a special power series, namely :

$$F(\varphi) = \frac{\alpha}{\beta} \sin (\beta\varphi) \qquad \text{(I. 10)}$$

where α and β are positive constants :

$$\Box \varphi + \frac{\alpha}{\beta} \sin (\beta\varphi) = 0 \qquad \text{(I. 10a)}$$

The field is a regarded as a function of t and of only one spacial dimension.

Expansion in powers of $\beta \varphi$ gives :

$$\Box \varphi + \alpha \varphi - \frac{\alpha \beta^2}{3!} \varphi^3 + \ldots = 0 \qquad \text{(I. 10b)}$$

the constants α and $\alpha\beta^2$ correspond to μ^2 and $-\lambda$ in equation (I. 9b).

I. 3 - NON-LINEAR VECTOR FIELD EQUATIONS

a) The linear Maxwell equations and electromagnetic gauge invariance

Maxwell's equations for the free electromagnetic field are :

$$\partial_\nu F^{\mu\nu}(x) = 0 \qquad \text{(I. 11)}$$

where $F^{\mu\nu}(x)$ is the field tensor, expressed as the curl of the potential field $A^\mu(x)$:

$$F^{\mu\nu}(x) = \partial^\nu A^\mu(x) - \partial^\mu A^\nu(x) \qquad \text{(I. 12)}$$

These are the simplest linear equations for a vector field.

These equations are gauge-invariant, i e , they do not change if the gradient of an arbitrary function $\Lambda(x)$, $\partial^\mu \Lambda(x)$, is added to the field $A^\mu(x)$. Therefore, $A'^\mu(x)$, a new field obtained from $A^\mu(x)$ by an equation of the type :

$$A'^\mu(x) = A^\mu(x) - \partial^\mu \Lambda(x) \qquad \text{(I. 13)}$$

gives rise to the same field tensor :

$$F'^{\mu\nu} = F^{\mu\nu}$$

The requirement of gauge invariance of the electromagnetic theory (which will be discussed in Chapter II) imposes the tensor $F^{\mu\nu}$ as defined in equation (I. 12) as the only second rank gauge-invariant tensor obtained by differentiation of the field $A_\mu(x)$.

In general, given a vector field $\phi^\mu(x)$ which describes a spin 1 field with mass m, it obeys, in the free-field case, the equations (I. 3). These equations are equivalent to the following ones :

$$\partial_\nu \mathcal{G}^{\mu\nu} + m^2 \phi^\mu = 0 \qquad \text{(I. 14)}$$

where

$$\mathcal{G}^{\mu\nu} = \partial^\nu \phi^\mu - \partial^\mu \phi^\nu \qquad \text{(I. 14a)}$$

In view of the mass term, which is proportional to the field ϕ^μ the equation (I. 14) is not gauge-invariant if ϕ^μ undergoes the transformation (I. 13). Maxwell's equations are therefore the only possible second order differential equations in $A^\mu(x)$ which are gauge invariant. And a comparison of the equations (I. 11) and (I. 14) shows that the electromagnetic field is massless.

However, the free field ϕ^μ is divergence-less as a consequence of the equation (I. 14) with $m \neq 0$, whereas the Lorentz condition $\partial_\mu A^\mu = 0$ has to be explicitly and independently assumed in the Maxwell case :

$$\Box A^\mu - \partial^\mu (\partial_\nu A^\nu) = 0$$

in order to reduce it to the pair of equations (I. 4).

b) Internal degrees of freedom. The Yang-Mills fields

The classical electromagnetic field equations are linear. Examples of non-linear equations for vector fields are provided by the Yang-Mills gauge fields. These are special fields which have a certain internal degree of freedom.

The simplest example of field is a real scalar field : it has zero spin (scalar nature of field) and is electrically neutral (the field is a real function in classical theory, a hermitian operator in quantum theory). It has no internal degree of freedom*. If we consider two real functions $\phi_a(x)$, $a = 1, 2$, they are equivalent to the pair of complex functions :

$$\phi(x) = \frac{1}{\sqrt{2}} (\phi_1(x) + i\, \phi_2(x)), \quad \phi^* (x) = \frac{1}{\sqrt{2}} (\phi_1(x) - i\, \phi_2(x)) \qquad (I.\ 15)$$

The fields ϕ, ϕ^*, that obey, like ϕ_a, the Klein-Gordon equation, give rise, as we shall see in § I. 6, to an electromagnetic current. The corresponding electric charge is a quantum number which is the first example of an internal degree of freedom. The latter results from the fact that the theory with these fields is assumed to be invariant under transformations of the phases of the function $\phi(x)$. Physically this means that the observables do not depend on the phase of the field ; the physics constructed with $\phi(x)$, $\phi^*(x)$ is the same as the physics constructed with the fields $\phi'(x)$, $\phi^{*\prime}(x)$ where :

$$\begin{aligned} \phi'(x) &= e^{i\lambda}\, \phi(x), \\ \phi'^*(x) &= e^{-i\lambda}\, \phi^*(x) \end{aligned} \qquad (I.\ 16)$$

λ being and arbitrary real number. These are the so-called global phase transformations.

The energy, the momentum and the angular momentum of a field have

---

* However, lagrangeans with such a field such as the one in equ. (VII. 1) may have a symmetry ($\varphi \to -\varphi$) which defines an internal degree of freedom for the corresponding system.

eigenvalues which are related to the geometrical degrees of freedom of the field (see § I. 5, 1), 2)). The momentum four-vector is the generator of space-time translations, angular momentum is that of Lorentz transformations. They are so to say geometrical quantum numbers.

The electric charge of a field $\psi^\alpha(x)$ is associated with the reality character of the function which describes it and arises from the invariance properties of its equations of motion (or lagrangean) under the one-parameter group U(1) of phase transformations (I. 16) :

$$\psi^\alpha(x) \to \psi'^\alpha(x) = e^{i\lambda}\, \psi^\alpha(x) \qquad \text{(I. 16a)}$$

The field $\psi^\alpha(x)$ is equivalently described by the pair of real (or self-charge conjugate) functions $\psi^\alpha{}_a(x)$, $a = 1, 2$ related to ψ^α, $\psi^{\alpha *}$ by equations of the type (I. 15).

There exist fields which have other types of internal degree of freedom. The simplest example is given by an isospin doublet $\psi(x)$, that is, a pair of complex functions $\psi_1(x)$, $\psi_2(x)$ represented as a one-column, two lines matrix :

$$\psi(x) = \begin{pmatrix} \psi_1(x) \\ \psi_2(x) \end{pmatrix} \qquad \text{(I. 17)}$$

similar to the Pauli spinor for the spinning electron. We may think of ψ_1 and ψ_2 as representing two possible states of a given particle such as the proton and neutron considered as different charge states of the nucleon. In the limiting case of equal mass for these two particles, experiment indicates that one must have the same nuclear physics for the neutron and for the proton, if one makes abstraction of the charge of the latter and of the corresponding electromagnetic forces. Therefore, in this approximation, the physics in a laboratory which describes the nucleon by $\psi(x)$ as given in (I. 17) is the same as that in another laboratory which describes the nucleon by $\psi'(x)$ where :

$$\psi'(x) = U(\vec{\alpha})\ \psi(x) \qquad \text{(I. 18)}$$

$U(\vec{\alpha})$ being a general unitary unimodular matrix which mixes the components ψ_1, ψ_2 into the new components ψ'_1, ψ'_2 of $\psi'(x)$. As $U(\vec{\alpha})$ must be a 2×2 matrix which depends on three real parameters $\alpha_1, \alpha_2, \alpha_3$ it will be expressed in terms of the identity and the three Pauli spin matrices τ_k (which form a basis in the space of the 2×2 matrices :

$$I = \begin{pmatrix} 1 & 0 \\ 0 & 1 \end{pmatrix}, \quad \tau_1 = \begin{pmatrix} 0 & 1 \\ 1 & 0 \end{pmatrix}, \quad \tau_2 = \begin{pmatrix} 0 & -i \\ i & 0 \end{pmatrix}, \quad \tau_3 = \begin{pmatrix} 1 & 0 \\ 0 & -1 \end{pmatrix} \qquad \text{(I. 18a)}$$

For infinitesimal values of the three parameters $\alpha_1, \alpha_2, \alpha_3$ we have :

$$U(\vec{\alpha}) = I + i\,\vec{\alpha} \cdot \frac{\vec{\tau}}{2}\,, \quad \text{infinitésimal } \alpha_k \qquad \text{(I. 18b)}$$

For finite values of α_k, $U(\vec{\alpha})$ will be :

$$U(\vec{\alpha}) = \exp\left(i\,\vec{\alpha} \cdot \frac{\vec{\tau}}{2}\right). \qquad \text{(I. 18c)}$$

If we fix the three parameters, knowledge of the three operators $\frac{\tau_k}{2}$ determines the transformation corresponding to the values of the α_k's. These three operators are the generators of the group SU(2) (the identity is clearly the generator of the group U(1), (I. 16a)).

As will be seen in Chapter IV, the generalization of the transformations of the group SU(2) to the case when the parameters $\vec{\alpha}$ depend on the point of space-time,(the group of local SU(2) transformations) leads us to introduce, if the theory is to be invariant under this local group, certain vector fields which are traceless 2×2 matrices $\mathscr{A}^{\mu}(x)$, expressed in terms of Pauli matrices in the following way :

$$\mathscr{A}^{\mu}(x) = \sum_{k=1}^{3} A^{\mu}{}_{k}(x) \frac{\tau_k}{2} \qquad \text{(I. 19)}$$

The set of three vector fields thus defined, $A^{\mu}{}_{k}(x)$, k = 1, 2, 3, constitutes the SU(2) Yang-Mills fields.

Similarly and more simply, the electromagnetic vector field is defined in connection with the postulate of invariance of complex field theories under the local group of transformations U(1), the parameter λ in (I. 16) being now space-time dependent as the function $\Lambda(x)$ in (I. 13).

Another example of Yang-Mills vector fields is provided by vector fields which are 3 x 3 matrices and which are expressed as a function of the generators of the SU(3) group, $\frac{\lambda_a}{2}$:

$$\boldsymbol{a}^{\mu}(x) = \sum_{a=1}^{8} \boldsymbol{a}^{\mu}{}_{a}(x) \frac{\lambda_a}{2} \qquad \text{(I. 20)}$$

The group SU(3) of 3 x 3 unitary unimodular matrices transforms the space of triplets :

$$\psi(x) = \begin{pmatrix} \psi_1(x) \\ \psi_2(x) \\ \psi_3(x) \end{pmatrix} \qquad \text{(I. 20a)}$$

into itself (see Chapter IV. 9). These triplets describe, for instance, quarks which are assumed to exist, for each flavour, in three different colour states with the same mass. It is assumed that the physics of quarks is invariant under the colour SU(3) group, that is, it will be the same for $\psi(x)$ and for $\psi'(x)$ such that

$$\psi'(x) = U(\omega)\,\psi(x) \qquad \text{(I. 21)}$$

where now $\psi(x)$ is the triplet (I. 20a) and $U(\omega)$ depends on eight parameters ω_a, a = 1, ...8 and has the form :

$$U(\omega) = I + i \sum_{a=1}^{8} \omega_a \frac{\lambda_a}{2} \qquad \text{(I. 21a)}$$

for infinitesimal ω_a, a = 1, ... 8 and :

$$U(\omega) = e^{i\omega_a \frac{\lambda_a}{2}} \qquad \text{(I. 21b)}$$

for finite values of the parameters.

The colour Yang-Mills vector field (I. 20) appears in association with the construction of a theory which is invariant under the group of local SU(3) transformations, the parameters of which are functions of point in space-time, $\omega_a = \omega_a(x)$.

The generators, $\frac{\tau_a}{2}$ for SU(2), $\frac{\lambda_a}{2}$ for SU(3), obey certain commutation rules which characterise their respective algebra. The structure constants of these groups, namely ε_{abc} for SU(2) :

$$\left[\frac{\tau_a}{2}, \frac{\tau_b}{2} \right] = i\,\varepsilon_{abc} \frac{\tau_c}{2}, \quad a, b = 1, 2, 3 \qquad \text{(I. 22)}$$

$$\varepsilon_{123} = \varepsilon_{312} = \varepsilon_{231} = 1,$$

$$\varepsilon_{abc} = - \varepsilon_{acb} = - \varepsilon_{bac}$$

and $f_{k\ell n}$ for SU(3) group (as given in Chapter IV, § IV. 9)

$$\left[\frac{\lambda_k}{2}, \frac{\lambda_\ell}{2}\right] = i\, f_{k\ell n}\, \frac{\lambda_n}{2}\ ;\ k, \ell = 1 \ldots 8 \qquad \text{(I. 22a)}$$

are tools to be used in the theories which admit one of these groups as a symmetry.

c) Examples of non-linear field equations involving Yang-Mills fields

Let us consider a Yang-Mills field $A^{\mu}{}_{a}(x)$ associated to a local SU(n) group so that :

$$A^{\mu}(x) = \sum_{a=1}^{N} A^{\mu}{}_{a}(x)\, T_a, \qquad N = n^2 - 1$$

is a $n \times n$ matrix. The $N^2 - 1$ operators T_a are the generators of this group and obey the commutation rules :

$$\left[T_a, T_b\right] = i\, C_{abc}\, T_c$$

where the constants C_{abc} are the structure constants of the group.

Clearly, in order to construct an antisymmetric tensor field $F^{\mu\nu}{}_{a}(x)$ which will be the generalization of Maxwell's tensor $F^{\mu\nu}$, we dispose not only of the curl of $A^{\mu}{}_{a}(x)$ but also of another tensor constructed with the help of the structure constants, namely, $C_{abc}\, A^{\mu}{}_{b}(x)\, A^{\nu}{}_{c}(x)$. Therefore, we may write in general :

$$F^{\mu\nu}{}_{a}(x) = \partial^{\nu} A^{\mu}{}_{a}(x) - \partial^{\mu} A^{\nu}{}_{a}(x) + g\, C_{abc}\, A^{\mu}{}_{b}(x)\, A^{\nu}{}_{c}(x) \qquad \text{(I. 23)}$$

where g is a constant.

Similarly, the generalization of Maxwell's free-field equation to the Yang-Mills case will have to take into account that besides $\partial_\nu F^{\mu\nu}{}_a(x)$ there is another vector which can be constructed with $A^\mu{}_a(x)$, $F^{\mu\nu}{}_a(x)$ and the structure constants, namely $C_{abc}\ A_{\nu,b}(x)\ F^{\mu\nu}{}_c(x)$. Therefore the Yang-Mills field equations in the absence of external sources are of the form :

$$\partial_\nu F^{\mu\nu}{}_a(x) + g\, C_{abc}\, A_{\nu,b}(x)\, F^{\mu\nu}{}_c(x) = 0 \qquad \text{(I. 23a)}$$

These are non-linear equations for the field $A^\mu{}_a(x)$. The non-linear terms express a self-interaction of this field with itself.

As we shall see in Chapter IV and in Chapter V, § 1, the construction of a field theory invariant under a local group of transformations with generators T_a requires the generalization of the differential operator into one which is a matrix in the space in which the T_a's are defined. The generalized differential operator, the so-called covariant derivative, has the form :

$$D_\mu = \partial_\mu I + i\, g\, A_{\mu,a}\, T_a \qquad \text{(I. 24)}$$

where I is the unity operator and the T_a's are the generators of the symmetry group under consideration.

Depending on the choice of the representation space of the group we shall have appropriate matrix representations for these generators.

In the case of SU(2), the principal representation space is the space of two-component isospinors (I. 17) and the generators are one-half the Pauli-matrices :

$$T_a = \frac{\tau_a}{2} \quad \text{for two-dimensional representation of SU(2)}$$

If the representation space of SU(2) has three dimensions, the space of isovectors, the generators are the usual 3 x 3 angular momentum matrices :

$$(T_a)_{bc} = - i \, \varepsilon_{abc} \quad \text{for three-dimensional representation of SU(2)}$$

In the case of the SU(3) group, the principal representation space is the space of complex three-dimensional vectors and the generators are the Gell-Mann[10-33] matrices $\frac{\lambda_a}{2}$:

$$T_a = \frac{\lambda_a}{2} \quad \text{for the three-dimensional representation of SU(3).}$$

Another example of non-linear equations is obtained if one considers a complex scalar field $\phi(x)$ with a quartic self-interaction and in interaction with the electromagnetic field $A_\mu(x)$. The equations must be invariant under the electromagnetic gauge group of transformations namely under the transformations

$$\varphi(x) \to \varphi'(x) = e^{ie\Lambda(x)} \varphi(x)$$

$$A_\mu(x) \to A'_\mu(x) = A_\mu(x) - \partial_\mu \Lambda(x) \qquad \text{(I. 25)}$$

where $\Lambda(x)$ is the point-dependent parameter of the transformation. We see that the gauge group for fields in interaction with the electromagnetic field interconnects the phase transformations of the complex field mentioned in (I. 16a) (but now with space-time dependent parameter λ as given in (I. 25)) with the gauge transformation of the electromagnetic field mentioned in (I. 13).

A comparison of the infinitesimal phase transformation

$$\varphi'(x) = (I + i e \Lambda(x)) \varphi(x) \qquad \text{(I. 26)}$$

in (I. 25) with similar transformations for isospinors as given in (I. 18b) with $\vec{\alpha}$ dependent on space-time points, shows that the generator of the electromagnetic gauge group is the identity.

Therefore, the electromagnetic-gauge generalized differential operator obtain from (I. 24) is :

$$D_\mu = \partial_\mu + i\, e\, A_\mu(x) \qquad \text{(I. 27)}$$

The equation for the field ϕ with quartic self-interaction and interacting with $A_\mu(x)$ is obtained from equation (I. 9b) by replacing the invariant differential operator $\square$ by the gauge-invariant differential operator

$$\square = \partial^\alpha \partial_\alpha \to D_\alpha D^\alpha \qquad \text{(I. 28)}$$

where D_α is given in (I. 27)

We may therefore write the equation

$$(D_\alpha D^\alpha + \mu^2)\, \phi + \eta(\phi^* \phi)\, \phi = 0 \qquad \text{(I. 29)}$$

as corresponding to equation (I. 9b) for a real scalar field φ. The constant η corresponds to the coupling constant λ in (9b).

The associated Maxwell equations for the electromagnetic field in interaction with the scalar field ϕ are of the form :

$$\partial_\nu F^{\mu\nu}(x) = e\, j^\mu(x) \qquad \text{(I. 30)}$$

where

$$j^\mu = i\, \{ \phi^* D^\mu \phi - \phi(D^\mu \phi)^* \}$$

is the gauge-invariant current.

Other examples of non linear equations, such as those for 't Hoft's monopole and for the instanton, are more easily established starting from their lagrangians.

I. 4 - FIELD EQUATIONS AND ACTION PRINCIPLE

As is well known, the field equations are assumed to be derived from an action principle. A relativistically invariant function L of the field variables $\psi^\alpha(x)$ and its first derivatives $\partial^\mu \psi^\alpha(x)$ is assumed -possibly also an explicit function of the coordinates- out of which a functional is constructed, the action S :

$$S = \int L\,(\psi^\alpha(x), \partial^\mu \psi^\alpha(x))\, d^4x \qquad (I.\ 31)$$

the integration being taken over the whole space. The action principle postulates that the variation of S vanishes when one varies the fields $\psi^\alpha(x)$ in such a way that $\delta\,\psi^\alpha(x)$ vanishes at infinity

$$\delta\, S = 0 \quad \text{for} \quad \delta\,\psi^\alpha(x) = 0 \quad \text{at infinity} \qquad (I.\ 32)$$

The variation of the field $\psi^\alpha(x)$ means that we consider a family of fields $\psi^\alpha(x\ ;\ \lambda)$ characterised by a parameter λ and then :

$$\delta\,\psi^\alpha = \left[\frac{\partial\,\psi^\alpha}{\partial\,\lambda}\right]_{\lambda=0} \delta\lambda$$

We have :

$$\delta\, S = \int \delta\, L\, d^4x = \int d^4x \left\{ \frac{\partial\, L}{\partial\,\psi^\alpha(x)}\, \delta\,\psi^\alpha(x) + \frac{\partial\, L}{\partial(\partial^\mu\,\psi^\alpha(x))}\, \delta(\partial^\mu\,\psi^\alpha) \right\}$$

$$= \int d^4x \left\{ \frac{\partial L}{\partial \psi^\alpha(x)} - \partial^\mu \frac{\partial L}{\partial(\partial^\mu \psi^\alpha(x))} \right\} \delta \psi^\alpha(x) +$$

$$+ \int d^4x\, \partial^\mu \left(\frac{\partial L}{\partial(\partial^\mu \psi^\alpha)} \delta \psi^\alpha \right)$$

Now

$$\int d^4x\, \partial^\mu \left(\frac{\partial L}{\partial(\partial^\mu \psi^\alpha)} \delta \psi^\alpha \right) = \int_{\substack{\text{outer} \\ \text{surface}}} \delta \psi^\alpha \frac{\partial L}{\partial(\partial^\mu \psi^\alpha)} d\sigma^\mu = 0$$

because $\left[\delta \psi^\alpha\right]_{\text{outer surface}} = 0.$

So for $\delta \psi^\alpha$ otherwise arbitrary, the postulate (I. 32) implies the field equations :

$$\frac{\partial L}{\partial \psi^\alpha(x)} - \partial^\mu \frac{\partial L}{\partial(\partial^\mu \psi^\alpha(x))} = 0 \qquad \text{(I. 33)}$$

I. 5 - EXAMPLES OF LAGRANGEANS

a) Scalar complex field $\varphi(x), \varphi^*(x)$ with mass m :

$$L = \partial^\mu \varphi^* \partial_\mu \varphi - m^2 \varphi^* \varphi \;; \qquad \text{(I. 34)}$$

$$\frac{\partial L}{\partial \varphi^*} - \partial^\mu \frac{\partial L}{\partial(\partial^\mu \varphi^*)} \equiv (\Box + m^2)\, \varphi(x) = 0 \qquad \text{(I. 35)}$$

b) Vector complex field $\varphi^\mu(x), \varphi^{\mu *}(x)$ with mass m :

$$L = \frac{1}{2} \mathcal{G}^*_{\mu\nu} \mathcal{G}^{\mu\nu} - \frac{1}{2} \mathcal{G}^*_{\mu\nu} (\partial^\nu \varphi^\mu - \partial^\mu \varphi^\nu) - \frac{1}{2} (\partial^\nu \varphi^{\mu *} - \partial^\mu \varphi^{\nu *}) \mathcal{G}_{\mu\nu} +$$

$$+ m^2 \varphi^*_\mu \varphi^\mu \;; \qquad \text{(I. 36)}$$

$$\frac{\partial L}{\partial \mathcal{G}^*_{\mu\nu}} - \partial^\alpha \frac{\partial L}{\partial(\partial^\alpha \mathcal{G}^*_{\mu\nu})} = 0 \Rightarrow \quad \mathcal{G}^{\mu\nu} = \partial^\nu \varphi^\mu - \partial^\mu \varphi^\nu$$

$$\frac{\partial L}{\partial \varphi^*_\mu} - \partial^\alpha \frac{\partial L}{\partial(\partial^\alpha \varphi^*_\mu)} = 0 \Rightarrow \partial_\nu \mathcal{G}^{\mu\nu} + m^2 \varphi^\mu = 0 \tag{I. 37}$$

These are the Proca equations for the vector field. These equations are equivalent to the following ones :

$$(\Box + m^2)\, \varphi^\mu = 0$$
$$\partial_\mu \varphi^\mu = 0 \tag{I. 38}$$

There are thus only three independent components of $\varphi^\mu(x)$ as required if this field is to describe a spin 1 field.

c) <u>Spinor field</u> $\psi(x)$, $\bar{\psi}(x)$ with mass m where :

$$\bar{\psi}(x) = \psi^+(x)\, \gamma^0$$

or

$$\bar{\psi}_a(x) = \psi^*{}_b(x)(\gamma^0)_{ba}$$

the matrices γ^μ are such that :

$$\frac{1}{2}(\gamma^\mu\gamma^\nu + \gamma^\nu\gamma^\mu) = g^{\mu\nu}$$

$$L = \bar{\psi}(x)\, (i\, \gamma^\alpha\, \partial_\alpha - m)\, \psi(x), \tag{I. 39}$$

$$\frac{\partial L}{\partial \bar{\psi}(x)} - \partial^\alpha \frac{\partial L}{\partial(\partial^\alpha \bar{\psi}(x))} = 0 \Rightarrow (i\, \gamma^\alpha\, \partial_\alpha - m)\, \psi(x) = 0 \tag{I. 40}$$

Lagrangeans which differ by the divergence of a four-vector which vanishes at infinity are equivalent since they give the same action. Then the following Dirac's lagrangean is equivalent to the above one

$$L = \frac{i}{2}\, (\bar{\psi}\, \gamma^\alpha\, \partial_\alpha\, \psi - (\partial_\alpha\, \bar{\psi})\, \gamma^\alpha\, \psi) - m\, \bar{\psi}\, \psi \tag{I. 41}$$

d) Scalar field with quartic self-interaction

The equation for a real scalar field φ with a polynomial interaction (containing no derivatives of φ) such as equation (I. 9) is deduced from a lagrangean :

$$L = \frac{1}{2} \partial^{\mu} \varphi \, \partial_{\mu} \varphi - U(\varphi) \qquad (I.\ 42)$$

One obtains

$$\Box \varphi + F(\varphi) = 0$$

where

$$F(\varphi) = U'(\varphi)$$

For the case of a quartic self-interaction we have :

$$U(\varphi) = \frac{1}{2} \mu^2 \varphi^2 + \frac{\lambda}{4!} \varphi^4 \qquad (I.\ 42a)$$

and equation (I. 9b) for φ.

Of particular interest is the case where the term in μ^2 is negative. As will be studied in Chapter VII § 1, the potential energy $U(\varphi)$ (that is, that part of $-L$ or of the hamiltonian which is different from zero for a constant field) will have two minima for :

$$\varphi^2 = - \frac{6\mu^2}{\lambda} \equiv a^2 > 0 , \quad \mu^2 < 0 \qquad (I.\ 42b)$$

The existence of more than one minimum for the function $U(\varphi)$ or of more than one zero for the function

$$U_1(\varphi) = \frac{\lambda}{4!} (\varphi^2 - a^2)^2 , \qquad (I.\ 42c)$$

$$a^2 = - \frac{6\mu^2}{\lambda} > 0, \ \mu^2 < 0$$

in the equivalent lagrangian

$$L_1 = \frac{1}{2}\ \partial^\mu \varphi\ \partial_\mu \varphi - U_1(\varphi) \sim L \qquad \text{(I. 42d)}$$

is important in that it makes possible the existence of non-trivial time-independent solutions of the equation (I. 9b) with finite energy, for the case in which the function $\varphi(y, t)$ depends on t and only one spatial dimension y.

e) The sine-Gordon equation

This is derived from the lagrangean (I. 42) by assuming :

$$U(\varphi) = -\frac{\alpha}{\beta^2}\ (\cos \beta\varphi - 1)$$

If we want to compare the constants α and β with those of (I. 42), (I. 42a) we may introduce m^2 and η such that :

$$\frac{\alpha}{\beta^2} = \frac{m^4}{\eta}\ , \quad \beta = \frac{\sqrt{\eta}}{m}$$

so that the sine-Gordon lagrangean is :

$$L = \frac{1}{2}\ (\partial_\mu \varphi)(\partial^\mu \varphi) + \frac{m^4}{\eta}\left[\cos\left(\frac{\sqrt{\eta}}{m}\varphi\right) - 1\right]$$

and $\varphi = \varphi(y, t)$ depends on only one spatial dimension.

f) Yang-Mills field

The lagrangean for a Yang-Mills vector field $A^\mu{}_a(x)$ in the absence of sources from other fields, which will lead to equation (I. 23a), has the form :

$$\mathscr{L} = -\frac{1}{4}\ F^{\mu\nu}{}_a\ F_{\mu\nu,a} \qquad \text{(I. 43)}$$

where $F^{\mu\nu}{}_a$ is the tensor field given by equation (I. 23)

It will be seen in Chapter IV that this tensor is generated by the algebra of the covariant derivatives (such as given in equation (IV. 16)).

We give here, as two additional examples, the lagrangean for

g) Scalar field in interaction with the electromagnetic field, as defined by equations (I. 27), (I. 29), (I. 30), (I. 30a), namely :

$$L = -\frac{1}{4} F^{\mu\nu} F_{\mu\nu} + (D^{\mu}\phi)^*(D_{\mu}\phi) - \frac{1}{2}\mu^2\phi^*\phi + \frac{\eta}{2}(\phi^*\phi)^2 \tag{I. 44}$$

and h) the lagrangean of the 't Hooft-Polyakov monopole, described by a Yang-Mills vector field $A^{\mu}{}_a(x)$ and a set of scalar fields $\phi_a(x)$ where the index $a = 1,2,3$ means that these fields are isovectors under the group of three-dimensional rotations SO(3) as an internal degree of freedom. It is :

$$L = -\frac{1}{4} F^{\mu\nu}_a F_{\mu\nu a} + \frac{1}{2}(D_{\mu}\phi_a)(D^{\mu}\phi_a) - \frac{\eta}{4}(\phi_a\phi_a - \frac{m^2}{\eta})^2 \tag{I. 45}$$

where

$$F_{\mu\nu a} = \partial_{\nu} A_{\mu a} - \partial_{\mu} A_{\nu a} + e\,\varepsilon_{abc} A_{\mu b} A_{\mu c}$$

$$D_{\mu}\phi_a = \partial_{\mu}\phi_a + e\,\varepsilon_{abc} A_{\mu b}\phi_c$$

This gives rise to the equations

$$D_{\nu} F^{\mu\nu}_a + e\,\varepsilon_{abc}\phi_b D^{\mu}\phi_c = 0$$

$$D^{\mu} D_{\mu}\phi_a + \lambda\phi_a(\phi_a\phi_a - \frac{m^2}{\lambda}) = 0$$

Note that the term in m^2 ($m^2 > 0$) corresponds to the case $\mu^2 < 0$ in (I. 42a), (I. 42b), (I. 42c).

I. 6 - NOETHER'S CONSERVED TENSORS

The lagrangean formalism of classical field theory allows the construction of physical quantities which are conserved, of observables which do not change in time.

Noether's theorem states that if an action is invariant under a continuous group of transformations of the fields the corresponding lagrangean determines a conserved tensor and an associated time-independent observable.

If the field transformation law corresponds to a geometrical transformation of the space-time coordinates, the Noether conserved objects are a) the energy-momentum tensor density, which is divergenceless, and the associated time-independent energy-momentum vector ; b) the divergenceless angular-momentum tensor density and the time-independent total angular momentum.

Other conserved objects, such as currents and charges (electric or baryonic or isospin, etc) result from invariance of the lagrangean density under continuous groups of transformation of the fields, corresponding to internal degrees of freedom.

In general we assume the action :

$$S = \int L\ (\psi^{\alpha}(x),\ \partial^{\mu}\ \psi^{\alpha}(x)\ ;\ x)\ d^4x$$

the lagrangean may depend explicitly on the coordinates x.

Suppose an infinitesimal coordinate transformation(an element of a continuous transformation group)is carried out

$$x^{\mu} \rightarrow x'^{\mu} = x^{\mu} + \delta\ x^{\mu} \qquad \text{(I. 46)}$$

and is defined by the parameters $\delta \omega^k$, so that :

$$\delta x^\mu = f^\mu{}_k(x)\, \delta \omega^k \qquad (I.\ 47)$$

This transformation induces a transformation of the field variables

$$\psi^\alpha(x) \rightarrow \psi'^\alpha(x') = \psi^\alpha(x) + \bar{\delta}\, \psi^\alpha(x) \qquad (I.\ 48)$$

where we shall set :

$$\bar{\delta}\, \psi^\alpha(x) = F^\alpha{}_k(x)\, \delta \omega^k \qquad (I.\ 49)$$

We see that the transformation law for the fields involve their total variation

$$\bar{\delta}\, \psi^\alpha(x) \equiv \psi'^\alpha(x') - \psi^\alpha(x)$$

whereas in (I.9) we used the variation in form only of the field, $\delta \psi^\alpha$:

$$\delta\, \psi^\alpha(x) = \psi'^\alpha(x) - \psi^\alpha(x) \qquad (I.\ 50)$$

i.e relative to the same point x.

We have the following identity :

$$\bar{\delta}\, \psi^\alpha(x) = \psi'^\alpha(x') - \psi'^\alpha(x) + \psi'^\alpha(x) - \psi^\alpha(x)$$

or

$$\bar{\delta}\, \psi^\alpha = \delta\, \psi + (\partial_\mu\, \psi)\, \delta x^\mu \qquad (I.\ 51)$$

This is seen more clearly if we make use of the parameters $\delta\,\omega^k$ and identify $\psi'^{\alpha}(x')$ with $\psi^{\alpha}(x'\,;\,\delta\,\omega^k)$ and $\psi^{\alpha}(x)$ with $\psi^{\alpha}(x\,;\,0)$. Then :

$$\bar{\delta}\,\psi^{\alpha}(x) = \frac{\partial\,\psi^{\alpha}}{\partial\,\omega^k}\,\delta\,\omega^k + \frac{\partial\,\psi^{\alpha}}{\partial\,x^{\mu}}\quad\delta\,x^{\mu}$$

since the variation $\bar{\delta}$ refers to a differential of $\psi^{\alpha}(x)$ with respect to the parameters $\delta\,\omega^k$ and these appear explicitly in $\psi'^{\alpha}(x') \equiv \psi^{\alpha}(x'\,;\,\delta\,\omega^k)$ and implicitly through x'.

We then see that the variation of the action will be :

$$\delta\,S = \int \delta\,L\,d^4x + \int L\,\delta(d^4x) + \int \left(\frac{\partial L}{\partial\,x^{\mu}}\,\delta\,x^{\mu}\right)d^4x$$

or

$$\delta\,S = \int\left[\frac{\partial L}{\partial\,\psi^{\alpha}}\,\delta\,\psi^{\alpha} + \frac{\partial\,L}{\partial(\partial^{\mu}\,\psi^{\alpha})}\,\delta(\partial^{\mu}\,\psi^{\alpha})\right]\,d^4x +$$

$$+ \int \frac{\partial\,L}{\partial\,x^{\mu}}\,\delta\,x^{\mu}\,d^4x + \int L\,\delta\,(d^4x)$$

Indeed we have, assuming that as $x \to \pm\,\infty$, $x' \to \pm\,\infty$:

$$\delta\,S = \int L(\psi^{\alpha'}(x'),\,\partial^{\mu}\,\psi'^{\alpha}(x')\,;\,x')\,d^4x' - \int L(\psi^{\alpha}(x),\,\partial^{\mu}\,\psi^{\alpha}(x)\,;\,x)\,d^4x$$

$$= \int L(\psi'^{\alpha}(x'),\,\partial^{\mu}\,\psi'^{\alpha}(x'),\,x')\,d^4x' - \int L(\psi^{\alpha}(x'),\,\partial^{\mu}\,\psi^{\alpha}(x'),\,x')\,d^4x' +$$

$$+ \int L(\psi^{\alpha}(x'),\,\partial^{\mu}\,\psi^{\alpha}(x'),\,x')\,d^4x' - \int L(\psi^{\alpha}(x),\,\partial^{\mu}\,\psi^{\alpha}(x)\,;\,x)\,d^4x$$

$$= \int \delta\,L\,d^4x' + \int L(\psi^{\alpha}(x'),\,\partial^{\mu}\,\psi^{\alpha}(x'),\,x')(1 + \partial_{\lambda}(\delta x^{\lambda}))\,d^4x\,-$$

$$- \int L(\psi^{\alpha}(x),\,\partial^{\mu}\,\psi^{\alpha}(x),\,x)\,d^4x$$

where we made use of the relationship :

$$d^4x' = \left|\frac{\partial x'}{\partial x}\right|\,d^4x = (1 + \partial_{\lambda}(\delta x^{\lambda}))\,d^4x$$

Thus in first order in $\delta\,\omega^k$:,

$$\delta S = \int \delta L\, d^4x + \int \frac{\partial L}{\partial x^\lambda}\, \delta x^\lambda\, d^4x + \int L\, \partial_\lambda(\partial x^\lambda)\, d^4x$$

therefore

$$\delta S = \int\left(\frac{\partial L}{\partial \psi^\alpha}\, \delta\,\psi^\alpha + \frac{\partial L}{\partial(\partial^\mu \psi^\alpha)}\, \delta(\partial^\mu \psi^\alpha) + \frac{\partial L}{\partial x^\lambda}\, \delta x^\lambda + \right.$$

$$\left. + L\, \partial_\lambda(\delta x^\lambda)\right) d^4x \qquad (I.\ 52)$$

or since by the equations of motion :

$$\frac{\partial L}{\partial \psi^\alpha} = \partial^\mu \frac{\partial L}{\partial(\partial^\mu \psi^\alpha)}$$

we get

$$\delta S = \int \partial_\lambda \left\{ \frac{\partial L}{\partial(\partial_\lambda \psi^\alpha)}\, \delta\,\psi^\alpha + L\,\delta\, x^\lambda \right\} d^4x$$

$$= \int \partial_\lambda \left\{ \frac{\partial L}{\partial(\partial_\lambda \psi^\alpha)} \left[\bar{\delta}\,\psi^\alpha - (\partial_\nu \psi^\alpha)\, \delta x^\nu \right] + L\, \delta x^\lambda \right\} d^4x$$

As

$$\delta x^\mu = f^\mu{}_k(x)\, \delta\,\omega^k$$

$$\bar{\delta}\,\psi^\alpha = F^\alpha{}_k(x)\, \delta\,\omega^k$$

we may write

$$\delta S = \int \partial_\lambda \left\{ \frac{\partial L}{\partial(\partial_\lambda \psi^\alpha)} \left[F^\alpha{}_k - (\partial_\nu \psi^\alpha)\, f^\nu{}_k \right] + \right.$$

$$\left. + L\, f^\lambda{}_k \right\} \delta\,\omega^k\, d^4x$$

Invariance of the action under the transformations we have considered means that

$$\delta S = 0$$

and hence, for arbitrary infinitesimal $\delta\,\omega^k$:

$$\partial_\lambda\, N^\lambda{}_k = 0 \tag{I. 53}$$

where

$$N^\lambda{}_k = \frac{\partial\, L}{\partial(\partial_\lambda\, \psi^\alpha)} \left\{ (\partial_\nu\, \psi^\alpha)\, f^\nu{}_k - F^\alpha{}_k \right\} - L\, f^\lambda{}_k \tag{I. 54}$$

is the Noether conserved generalised current. The conserved Noether generalised charge is :

$$N_k = \int N^o{}_k\, d^3x \quad , \quad \frac{d\, N_k}{d\, x^o} = 0 \tag{I. 55}$$

I. 7 - EXAMPLES OF NOETHER TENSORS

1) Field energy-momentum tensor and energy-momentum vector

Suppose the transformation is space-time translation :

$$x'^\mu = x^\mu + a^\mu$$

then :

$$\psi'^\alpha(x') = \psi^\alpha(x)$$

$\Big\{$ this is seen by the fact that if for one variable one sets $f'(x') = f(x)$ then $f'(x + a) = f(x)$ and we see that $f'(x + a) \equiv e^{-a \frac{d}{dx}}\, f(x + a) = f(x)$, so that $f'(x) \equiv e^{-iap}\, f(x) = f(x - a)$, $p = - i\, \frac{d}{dx}$ $\Big\}$.

Therefore, in the relations :

$$x'^{\mu} = x^{\mu} + f^{\mu}_{\ k}\ \delta\,\omega^{k},$$

$$\psi'^{\alpha}(x') = \psi^{\alpha}(x) + F^{\alpha}_{\ k}(x)\ \delta\,\omega^{k}$$

one sets :

$$f^{\mu}_{\ \lambda} \equiv \delta^{\mu}_{\ \lambda}\ ,\ \delta\,\omega^{\lambda} \equiv a^{\lambda}$$

$$F^{\alpha}_{\ \lambda}(x) = 0$$

Therefore Noether's conserved current is the energy-momentum tensor and corresponds to the postulated invariance of the theory under space-time translations :

$$N^{\alpha\beta} \equiv T^{\alpha\beta} = \frac{\partial L}{\partial(\partial_{\alpha}\ \psi^{a})}\ (\partial^{\beta}\ \psi^{a}) + \text{h.c.} - L\ g^{\alpha\beta} \qquad \text{(I. 56)}$$

The energy-momentum vector is :

$$P^{\alpha} = \int T^{\alpha 0}\ d^{3}x \qquad \text{(I. 57)}$$

and satisfies the equation :

$$\frac{d}{dx^{0}}\ P^{\alpha} = 0$$

2) Field angular momentum tensor

Assume the coordinate transformation group to be the infinitesimal Lorentz propre orthochrone group :

$$x'^{\mu} = x^{\mu} + \varepsilon^{\mu\nu}\ x_{\nu},$$

$$\varepsilon^{\mu\nu} = -\ \varepsilon^{\nu\mu}$$

Compare with :

$$x'^{\mu} = x^{\mu} + \frac{1}{2} f^{\mu}{}_{\alpha\beta}\, \varepsilon^{\alpha\beta} \quad ; \quad f^{\mu}{}_{\alpha\beta} = - f^{\mu}{}_{\beta\alpha} \, ,$$

since the parameters $\delta\, \omega^{k}$ are now $\varepsilon^{\alpha\beta}$, then we have :

$$f^{\mu}{}_{\alpha\beta} = (\delta^{\mu}{}_{\alpha}\, g_{\beta\nu} - \delta^{\mu}{}_{\beta}\; g_{\alpha\nu})x^{\nu}$$

[so that :

$$f^{\mu}{}_{\alpha\beta}\, \varepsilon^{\alpha\beta} = \varepsilon^{\alpha\beta}(\delta^{\mu}{}_{\alpha}\, g_{\beta\nu} - \delta^{\mu}{}_{\beta}\, g_{\alpha\nu})x^{\nu} = (\varepsilon^{\mu\beta}\, g_{\beta\nu} - \varepsilon^{\beta\mu}\, g_{\beta\nu})x^{\nu} = 2\,\varepsilon^{\mu\nu}\, x_{\nu}$$]

Let us write for the corresponding field transformation

$$\psi'^{\alpha}(x') = \psi^{\alpha}(x) + \frac{1}{2} F^{\alpha}{}_{\mu\nu}(x)\, \varepsilon^{\mu\nu}$$

and set :

$$F^{\alpha}{}_{\mu\nu}(x) = D^{\alpha}{}_{\beta;\mu\nu}\, \psi^{\beta}(x)$$

so that

$$\psi'^{\alpha}(x') = \psi^{\alpha}(x) + \frac{1}{2} D^{\alpha}{}_{\beta;\mu\nu}\, \psi^{\beta}(x)\, \varepsilon^{\mu\nu}$$

where

$$D^{\alpha}{}_{\beta;\mu\nu} = - D^{\alpha}{}_{\beta;\nu\mu}$$

The above form for $\psi'^{\alpha}(x')$ is an imitation of the corresponding form for x'^{μ}.

Then the Noether generalised current is the angular momentum density :

$$M^{\lambda}{}_{\mu\eta} = - N^{\lambda}{}_{\mu\eta} = x_{\mu}\, T^{\lambda}{}_{\eta} - x_{\eta}\, T^{\lambda}{}_{\mu} + \frac{\partial L}{\partial(\partial_{\lambda}\, \psi^{\alpha})}\, D^{\alpha}{}_{\beta;\mu\eta}\, \psi^{\beta} \qquad \text{(I. 58)}$$

which is conserved

$$\partial_\lambda M^\lambda{}_{\mu\eta} = 0$$

This equation gives :

$$T_{\mu\eta} - T_{\eta\mu} + \partial_\lambda \left(\frac{\partial L}{\partial(\partial_\lambda \psi^\alpha)} D^\alpha{}_{\beta;\mu\eta} \psi^\beta \right) = 0 \qquad \text{(I. 59)}$$

Let us introduce a tensor $F_{\lambda\mu\eta}$ such that :

$$F_{\lambda\mu\eta} - F_{\lambda\eta\mu} = \frac{\partial L}{\partial(\partial^\lambda \psi^\alpha)} D^\alpha{}_{\beta;\mu\eta} \psi^\beta \qquad \text{(I. 60)}$$

then clearly the tensor

$$\Theta_{\mu\eta} = T_{\mu\eta} + \partial^\lambda F_{\lambda\mu\eta} \qquad \text{(I. 61)}$$

is symmetric :

$$\Theta_{\mu\eta} = \Theta_{\eta\mu} \qquad \text{(I. 62)}$$

we impose further that :

$$F_{\lambda\mu\eta} = - F_{\mu\lambda\eta} \qquad \text{(I. 63)}$$

so that

$$\partial^\lambda \partial^\mu F_{\lambda\mu\eta} = 0$$

and hence :

$$\partial^\mu \Theta_{\mu\eta} = 0 \qquad \text{(I. 64)}$$

From equation (I. 60) we get :

$$F_{\mu\lambda\eta} - F_{\mu\eta\lambda} = \frac{\partial L}{\partial(\partial^{\mu} \psi^{\alpha})} D^{\alpha}{}_{\beta;\lambda\eta} \psi^{\beta}$$

$$F_{\eta\mu\lambda} - F_{\eta\lambda\mu} = \frac{\partial L}{\partial(\partial^{\eta} \psi^{\alpha})} D^{\alpha}{}_{\beta;\mu\lambda} \psi^{\beta}$$

whence we deduce, thanks to equality (I. 63)

$$F_{\lambda\eta\mu} = \frac{1}{2} \left\{ \frac{\partial L}{\partial(\partial^{\lambda} \psi^{\alpha})} D^{\alpha}{}_{\beta;\eta\mu} \psi^{\beta} + \frac{\partial L}{\partial(\partial^{\eta} \psi^{\alpha})} D^{\alpha}{}_{\beta;\mu\lambda} \psi^{\beta} - \right.$$

$$\left. - \frac{\partial L}{\partial(\partial^{\mu} \psi^{\alpha})} D^{\alpha}{}_{\beta;\lambda\eta} \psi^{\beta} \right\} \qquad \text{(I. 65)}$$

as a result of equations (I. 61) and (I. 63) the energy momentum vector may be calculated from either $T^{\alpha\beta}$ or $\Theta^{\alpha\beta}$:

$$P^{\beta} = \int d^3x \; T^{0\beta} = \int d^3x \; \Theta^{0\beta} \qquad \text{(I. 66)}$$

Now if we replace $T_{\mu\eta}$ as given by equation (I. 61) into the angular momentum density $M_{\lambda\mu\eta}$ (I. 58) we get :

$$M_{\lambda\mu\eta} = x_{\mu} \; \Theta_{\lambda\eta} - x_{\eta} \; \Theta_{\lambda\mu} + \frac{\partial L}{\partial(\partial^{\lambda} \psi^{\alpha})} D^{\alpha}{}_{\beta;\mu\eta} \psi^{\beta} +$$

$$+ x_{\eta} \, \partial^{\alpha} F_{\alpha\lambda\mu} - x_{\mu} \, \partial^{\alpha} F_{\alpha\lambda\eta} \qquad \text{(I. 67)}$$

that is, in view of equation (I. 60):

$$M_{\lambda\mu\eta} = x_{\mu} \; \Theta_{\lambda\eta} - x_{\eta} \; \Theta_{\lambda\mu} - \partial^{\alpha}(F_{\alpha\lambda\eta} \, x_{\mu} - F_{\alpha\lambda\mu} \, x_{\eta}) \qquad \text{(I. 68)}$$

So the angular-momentum tensor as defined by :

$$J_{\mu\nu} = \int d^3x \; M_{0\mu\nu} \qquad \text{(I. 69)}$$

is equal to :

$$J_{\mu\nu} = \int d^3x \; (x_\mu \; \Theta_{0\nu} - x_\nu \; \Theta_{0\mu}) \qquad \text{(I. 70)}$$

note that the skew-symmetry of F as given in (I. 63) imposes $\alpha = 1,2,3$ for $\lambda = 0$ in (I. 61).

3) Current-vectors for internal degrees of freedom

Assume now that the geometrical coordinates are kept unchanged :

$$f^\mu{}_k = 0$$

and that the fields undergo a continuous transformation corresponding to an internal degree of freedom.

That is the case of a change of the phase of the complex field $\psi^\alpha(x)$ by a constant factor :

$$\psi^\alpha(x) \to \psi'^\alpha(x) = e^{i\omega} \psi^\alpha(x) \qquad \text{(I. 71)}$$

or, for infinitesimal ω :

$$\psi'^\alpha(x) = (1 + i\omega) \; \psi^\alpha(x) \qquad \text{(I. 72)}$$

This is the global gauge group with one parameter, U(1). Then the Noether's current is the current vector :

$$j^\mu(x) = - \frac{\partial L}{\partial(\partial_\mu \psi^\alpha)} \; F^\alpha(x)$$

and as $\delta\omega$ is here denoted ω :

$$F^\alpha(x) = i \; \psi^\alpha(x)$$

thus :

$$j^\mu(x) = - i \frac{\partial L}{\partial(\partial_\mu \psi^\alpha)} \; \psi^\alpha(x)$$

If we want that $j^\mu(x)$ be hermitian we write :

$$j^\mu(x) = i \left\{ \psi^{+\alpha}(x) \frac{\partial L}{\partial(\partial_\mu \psi^{+\alpha}(x))} - \frac{\partial L}{\partial(\partial_\mu \psi^{\alpha}(x)} \psi^\alpha(x) \right\} \qquad (I.\ 73)$$

The charge is given by :

$$Q = \int j^0\text{-}x)\ d^3x$$

and is conserved in time :

$$\partial_\mu\ j^\mu = 0 \quad ; \quad \frac{dQ}{dt} = 0$$

Examples of Noether currents correponding to invariance of the theory under the internal SU(2) and SU(3) groups respectively will be exhibited in Chapter IV.

I. 8 - CONSERVED NOETHER TENSORS FOR SPECIFIC FIELDS

1) Scalar complex field

The lagrangean is :

$$L = (\partial^\mu\ \varphi^+)(\partial_\mu\ \varphi) - m^2\ \varphi^+\ \varphi$$

Invariance under space-time translation defines the energy momentum tensor :

$$\begin{aligned} T^{\alpha\beta} &= \frac{\partial L}{\partial(\partial_\alpha\ \varphi)} (\partial^\beta \varphi) + \text{h.c.} - L\ g^{\alpha\beta} = \\ &= (\partial^\alpha\ \varphi^+)(\partial^\beta \varphi) + (\partial^\beta \varphi^+)(\partial^\alpha \varphi) - \\ &- g^{\alpha\beta} \left[(\partial^\mu \varphi^+)(\partial_\mu \varphi) - m^2\ \varphi^+\ \varphi \right] \end{aligned} \qquad (I.\ 74)$$

which is symmetric in α, β. The energy momentum vector is

$$P^\alpha = \int T^{\alpha 0}\ d^3x \qquad (I.\ 75)$$

from which we get the hamiltonian

$$H = P^0 = \int d^3x \left\{ \pi^+\pi + (\vec{\nabla}\ \varphi_{\bullet}^{\ +}\ \vec{\nabla}\ \varphi) + m^2\ \varphi^+\varphi \right\} \qquad (I.\ 76)$$

with :

$$\pi(x) = \partial^0 \varphi$$

The linear momentum is :

$$P^k = - \int d^3x \left\{ \pi^+ \partial_k \varphi + (\partial_k \varphi)^+ \pi \right\} \qquad \text{(I. 77)}$$

Invariance under the Lorentz group determines the angular momentum which is purely orbital:

$$M_{\lambda\mu\nu} = x_\mu T_{\lambda\nu} - x_\nu T_{\lambda\mu} \equiv L_{\lambda\mu\nu} \qquad \text{(I. 78)}$$

so that :

$$L_{k\ell} = \int d^3x \, (x_k T_{0\ell} - x_\ell T_{ok}) = \int d^3x \left\{ \left[x_k (\partial_\ell \varphi^+) - x_\ell (\partial_k \varphi^+) \right] \pi + \right.$$

$$\left. + \pi^+ \left[(\partial_\ell \varphi) x_k - (\partial_k \varphi) x_\ell \right] \right\} \qquad \text{(I. 79)}$$

the current will be :

$$j^\mu(x) = i \left\{ \varphi^+(x) \partial^\mu \varphi - (\partial^\mu \varphi)^+ \varphi \right\} \qquad \text{(I. 80)}$$

and the charge

$$Q = \int d^3x \, j^0(x) = \int d^3x \, i \left\{ \varphi^+ \partial^0 \varphi - (\partial^0 \varphi^+) \varphi \right\} \qquad \text{(I. 81)}$$

2) Complex vector field

Lagrangean :

$$L = \frac{1}{2} \mathcal{G}^{\mu\nu+} \mathcal{G}_{\mu\nu} - \frac{1}{2} \mathcal{G}^{\mu\nu+} (\partial_\nu \varphi_\mu - \partial_\mu \varphi_\nu) -$$

$$- \frac{1}{2} (\partial^\nu \varphi^{\mu+} - \partial^\mu \varphi^{\nu+}) \mathcal{G}_{\mu\nu} + m^2 \varphi^{\mu+} \varphi_\mu$$

The energy - momentum tensor is

$$T^{\alpha\beta} = \frac{\partial L}{\partial(\partial_\alpha \varphi_\mu)} \partial^\beta \varphi_\mu + \text{h.c.} - L g^{\alpha\beta} =$$

$$= \mathcal{G}^{\alpha\mu+} \partial^\beta \varphi_\mu + \partial^\beta \varphi_\mu^{+} \mathcal{G}^{\alpha\mu} - L g^{\alpha\beta} \qquad \text{(I. 82)}$$

It is not symmetric ; the symmetric tensor $\Theta^{\alpha\beta}$ will be according to equations (I. 61) and (I. 62)

$$\Theta_{\mu\eta} = T_{\mu\eta} + \partial^\lambda F_{\lambda\mu\eta} \qquad \text{(I. 83)}$$

where

$$F_{\lambda\mu\eta} = \frac{1}{2}\Big\{ \mathcal{G}^{+}_{\lambda\alpha} (\delta^\alpha_{\ \mu} g_{\eta\beta} - \delta^\alpha_{\ \eta} g_{\mu\beta}) \varphi^\beta +$$

$$+ \mathcal{G}^{+}_{\mu\alpha} (\delta^\alpha_{\ \eta} g_{\lambda\beta} - \delta^\alpha_{\ \lambda} g_{\eta\beta}) \varphi^\beta - \mathcal{G}^{+}_{\eta\alpha} (\delta^\alpha_{\ \lambda} g_{\mu\beta} - \delta^\alpha_{\ \mu} g_{\lambda\beta}) \varphi^\beta \Big\} +$$

$$+ \text{h.c.} = \Big\{ \frac{1}{2} \mathcal{G}^{+}_{\lambda\mu} \varphi_\eta - \mathcal{G}^{+}_{\mu\lambda} \varphi_\eta \Big\} + \text{h.c.}$$

so

$$F_{\lambda\mu\eta} = \mathcal{G}^{+}_{\lambda\mu} \varphi_\eta + \varphi_\eta^{+} \mathcal{G}_{\lambda\mu} \qquad \text{(I. 84)}$$

Therefore

$$\partial^\lambda F_{\lambda\mu\eta} = (\partial^\lambda \mathcal{G}^{+}_{\lambda\mu}) \varphi_\eta + \mathcal{G}^{+}_{\lambda\mu} \partial^\lambda \varphi_\eta + \text{h.c.}$$

but

$$\partial^\lambda \mathcal{G}_{\mu\lambda} + m^2 \varphi_\mu = 0$$

so

$$\partial^\lambda F_{\lambda\mu\eta} = m^2(\varphi^+_{\ \mu}\ \varphi_\eta + \varphi^+_{\ \eta}\ \varphi_\mu) + \mathscr{G}^+_{\ \lambda\mu}\ \partial^\lambda\ \varphi_\eta + \partial^{\lambda +}\ \varphi_\eta\ \ \mathscr{G}_{\lambda\mu}$$

thus

$$\theta_{\mu\eta} = \mathscr{G}^+_{\ \mu\alpha}\ \partial_\eta\ \varphi^\alpha + \partial_\eta\ \varphi^{\alpha +}\ \ \mathscr{G}_{\mu\alpha} - L\ g_{\mu\eta} +$$

$$+ m^2(\varphi^+_{\ \mu}\ \varphi_\eta + \varphi^+_{\ \eta}\ \varphi_\mu) + \mathscr{G}^+_{\ \lambda\mu}\ \partial^\lambda\ \varphi_\eta + \partial^\lambda\ \varphi_\eta + \mathscr{G}_{\lambda\mu}$$

or

$$\theta_{\mu\eta} = \mathscr{G}^+_{\ \mu\alpha}(\partial_\eta\ \varphi^\alpha - \partial^\alpha\ \varphi_\eta) + \mathscr{G}^+_{\ \mu\alpha}\ \partial^\alpha\ \varphi_\eta + \text{h.c.}\ -$$

$$- L\ g_{\mu\eta} + m^2(\varphi^+_{\ \mu}\ \varphi_\eta + \varphi^+_{\ \eta}\ \varphi_\mu) + \mathscr{G}^+_{\ \alpha\mu}(\partial^\alpha\ \varphi_\eta - \partial_\eta\ \varphi^\alpha) +$$

$$+ \mathscr{G}^+_{\ \alpha\mu}\ \partial_\eta\ \varphi^\alpha + \text{h.c.} = \mathscr{G}^+_{\ \mu\alpha}\mathscr{G}^\alpha_{\ \eta} + \mathscr{G}^{\alpha +}_{\ \ \eta}\ \mathscr{G}_{\mu\alpha}\ -$$

$$- L\ g_{\mu\eta} + m^2(\varphi^+_{\ \mu}\ \varphi_\eta + \varphi^+_{\ \eta}\ \varphi_\mu) + \mathscr{G}^+_{\ \alpha\mu}\ \mathscr{G}^\alpha_{\ \eta} + \mathscr{G}^+_{\ \mu\alpha}\mathscr{G}^\alpha_{\ \eta} +$$

$$+ \mathscr{G}^{\alpha +}_{\ \ \eta}\ \mathscr{G}_{\alpha\mu} + \mathscr{G}^{\alpha +}_{\ \ \eta}\ \mathscr{G}_{\mu\alpha}$$

So the symmetric energy-momentum tensor is :

$$\theta_{\mu\eta} = \mathscr{G}^+_{\ \mu\alpha}\ \mathscr{G}^\alpha_{\ \eta} + \mathscr{G}^{\alpha +}_{\ \ \eta}\ \mathscr{G}_{\mu\alpha} + m^2(\varphi^+_{\ \mu}\ \varphi_\eta + \varphi^+_{\ \eta}\ \varphi_\mu) +$$

$$+ g_{\mu\eta}(\tfrac{1}{2}\ \mathscr{G}^{\alpha\beta +}\ \ \mathscr{G}_{\alpha\beta} - m^2\ \varphi^{\mu +}\ \varphi_\mu) \qquad \text{(I. 85)}$$

where we have taken into account the equations (I. 37).

We know that it is indifferent to calculate the energy-momentum vector from $T^{\alpha\beta}$ or from $\Theta^{\alpha\beta}$, according to equation (I. 66).

We have for the hamiltonian density :

$$T^{oo} = \mathcal{G}^{o\mu+} \partial^o \varphi_\mu + \partial^o \varphi_\mu{}^+ \mathcal{G}^{o\mu} - L \tag{I. 86}$$

Define

$$\pi_\mu = \frac{\partial L}{\partial(\partial^o \varphi^{\mu+})} = -(\partial_o \varphi_\mu - \partial_\mu \varphi_o) = \mathcal{G}_{o\mu} \tag{I. 87}$$

whence

$$\pi_o = 0 \quad , \quad \pi_k = \mathcal{G}_{ok} \tag{I. 88}$$

From the equation for π_k we get :

$$\partial^o \varphi_k = \partial_k \varphi^o - \pi_k \tag{I. 89}$$

or

$$\partial^o \vec{\varphi} = - \vec{\nabla} \varphi^o - \vec{\pi} \tag{I. 90}$$

From the field equations :

$$\partial_\nu \mathcal{G}^{\mu\nu} + m^2 \varphi^\mu = 0$$

we deduce for φ^o :

$$\varphi^o = - \frac{1}{m^2} \partial_k (\partial^k \varphi^o - \partial^o \varphi^k)$$

therefore, from (I. 89) we are able to express φ^o in terms of $\vec{\pi}$:

$$\varphi^o = - \frac{1}{m^2} (\vec{\nabla} \cdot \vec{\pi}) \tag{I. 91}$$

So from (I. 89) we get :

$$\partial^{0} \varphi^{k} = \frac{1}{m^2} \partial_k (\vec{\nabla} \cdot \vec{\pi}) - \pi^{k}$$

We now proceed to replace φ^{0} and $\mathcal{G}^{ok}$ in equation (I.86) by expressions (I. 91.) and (I. 88.) respectively, which gives :

$$T^{00} = \vec{\pi}^{+} \cdot \vec{\pi} - \frac{2}{m^2} \vec{\pi}^{+} \cdot \vec{\nabla} (\vec{\nabla} \cdot \vec{\pi}) - \frac{1}{m^2} (\vec{\nabla} \cdot \vec{\pi})^{+} \cdot (\vec{\nabla} \cdot \vec{\pi}) + \tag{I. 92}$$

$$+ (\vec{\nabla} \times \vec{\varphi}^{+}) \cdot (\vec{\nabla} \times \vec{\varphi}) + m^2 \vec{\varphi}^{+} \cdot \vec{\varphi}$$

So that, after a partial integration we obtain the hamiltonian :

$$H = \int d^3x \; T^{00} = \int d^3x \left\{ \vec{\pi}^{+} \cdot \vec{\pi} + \frac{1}{m^2} (\vec{\nabla} \cdot \vec{\pi}^{+})(\vec{\nabla} \cdot \vec{\pi}) + (\vec{\nabla} \times \vec{\varphi}^{+}) \cdot (\vec{\nabla} \times \vec{\varphi} \;) + \right.$$

$$\left. + m^2 \vec{\varphi}^{+} \cdot \vec{\varphi} \right\} \tag{I. 93}$$

The linear momentum is :

$$P^{k} = \int d^3x \; T^{ok}$$

with

$$T^{ok} = \mathcal{G}^{o\ell +} \partial^{k} \varphi_{\ell} + \partial^{k} \varphi_{\ell}^{+} \mathcal{G}^{o\ell}$$

that is, according to (I. 88)

$$P^{k} = \int d^3x \left\{ (\vec{\pi}^{+} \cdot \partial_k \vec{\varphi}) + (\partial_k \vec{\varphi}^{+}) \cdot \vec{\pi} \right\} \tag{I. 94}$$

The angular momentum tensor density is, according to equation (I. 58):

$$M_{\lambda\mu\eta} = L_{\lambda\mu\eta} + S_{\lambda\mu\eta}$$

where

$$L_{\lambda\mu\eta} = x_\mu T_{\lambda\eta} - x_\eta T_{\lambda\mu}$$

$$S_{\lambda\mu\eta} = \mathscr{G}^{+}_{\lambda\alpha} (\delta^\alpha{}_\mu \varphi_\eta - \delta^\alpha{}_\eta \varphi_\mu) + \text{h.c.}$$

Thus the spin vector is :

$$\vec{S} = \int d^3x \left\{ \left[\vec{\pi}^+ \times \vec{\varphi} \right] + \left[\vec{\varphi}^+ \times \vec{\pi} \right] \right\} \qquad \text{(I. 95)}$$

3) Spinor field

$$L = \frac{i}{2} \left\{ \bar{\psi}(\gamma \cdot \partial \psi) - (\partial \bar{\psi} \cdot \gamma) \psi \right\} - m \bar{\psi} \psi$$

Energy-momentum tensor :

$$T^{\alpha\beta} = \frac{i}{2} \left\{ \bar{\psi} \gamma^\alpha \partial^\beta \psi - (\partial^\beta \bar{\psi}) \gamma^\alpha \psi \right\} - \qquad \text{(I. 96)}$$

$$- \left\{ \frac{i}{2} \left[\bar{\psi} (\gamma \cdot \partial) \psi - (\partial \bar{\psi} \cdot \gamma) \psi \right] - m \bar{\psi} \psi \right\} g^{\alpha\beta}$$

The symmetrical tensor turns out to be :

$$\Theta_{\alpha\beta} = \frac{1}{2} (T^{\alpha\beta} + T^{\beta\alpha})$$

Hamiltonian :

$$H = \int d^3x \, T^{00} = \int d^3x \, \psi^+ \left\{ \vec{\alpha} \cdot (-i \vec{\nabla}) + m \beta \right\} \psi \qquad \text{(I. 97)}$$

where :

$$\vec{\alpha} = \gamma^{0} \ \vec{\gamma}$$

$$\beta \equiv \gamma^{0}$$

Linear momentum :

$$P^{k} = \int d^{3}x \ \psi^{+} \ (- i \ \partial_{k}) \ \psi \qquad (I.\ 98)$$

The angular momentum density is, according to equation (I.58)

$$M^{\lambda}{}_{\mu\eta} = \frac{i}{2} \Big\{ x_{\mu} \left[\bar{\psi} \gamma^{\lambda} \partial_{\eta} \psi - (\partial_{\eta} \bar{\psi}) \gamma^{\lambda} \psi \right] - $$
$$- x_{\eta} \left[\bar{\psi} \gamma^{\lambda} \partial_{\mu} \psi - (\partial_{\mu} \bar{\psi}) \gamma^{\lambda} \psi \right] \Big\} +$$
$$+ \frac{1}{4} (\bar{\psi} \gamma^{\lambda} \sigma_{\mu\eta} \psi + \bar{\psi} \ \sigma_{\mu\eta} \gamma^{\lambda} \psi) \qquad (I.\ 99)$$

where to the last term of equation (I. 58) we have added its hermitian conjugate.

Therefore the angular momentum tensor is :

$$J_{\mu\eta} = \int d^{3}x \ M^{0}{}_{\mu\eta} = L_{\mu\eta} + S_{\mu\eta} \qquad (I.\ 100)$$

with :

$$L_{\mu\eta} = \frac{i}{2} \int d^{3}x \ \Big\{ x_{\mu} \left[\psi^{+} \partial_{\eta} \psi - (\partial_{\eta} \psi^{+}) \ \psi \right] -$$
$$- x_{\eta} \left[\psi^{+} \partial_{\mu} \psi - (\partial_{\mu} \psi^{+}) \psi \right] \Big\} ,$$

$$S_{\mu\eta} = \frac{1}{4} \int d^{3}x \ \left\{ \psi^{+} \sigma_{\mu\eta} \psi + \psi^{+} \gamma^{0} \sigma_{\mu\eta} \gamma^{0} \psi \right\}$$

Thus the ordinary angular momentum has the form (after a partial integration in $L_{k\ell}$) :

$$J_{k\ell} = L_{k\ell} + S_{k\ell} \quad , \tag{I. 101}$$

$$L_{k\ell} = \int d^3x \left\{ x^k \psi^+(- i \partial_\ell \psi) - x^\ell \psi^+ (- i \partial_k \psi) \right\}$$

$$S_{k\ell} = \int d^3x \, \psi^+ \frac{1}{2} \sigma_{k\ell} \psi$$

The current of the spinor field will be, according to (I. 73) :

$$j^\mu(x) = \bar{\psi}(x) \gamma^\mu \psi(x) \tag{I. 102}$$

whence the charge

$$Q = \int j^0(x) \, d^3x = \int d^3x \;\; \psi^+ (x) \, \psi(x) \tag{I. 103}$$

These will refer to electric charge or the other conserved quantum numbers such as baryon number, lepton number, etc.

I. 9 - SOLITON SOLUTIONS OF CLASSICAL NON-LINEAR FIELD EQUATIONS AND TOPOLOGICAL QUANTUM NUMBERS

As stated in (I. 2), there exist solutions of classical non-linear field equations which behave like stable lumps of finite energy which propagate without diffusion. The most important examples of these solutions are the so-called topological solitons. They may exist only for fields which have an internal degree of freedom. It is convenient to study the case of fields defined on a Minkowski space with D spacial dimensions and one time dimension :

$$x^2 = (x^0)^2 - \sum_{k=1}^{D} (x^k)^2$$

A theorem called Derrick's theorem, states that there exist no time-independent non-singular solutions for a scalar field theory described by the lagrangean

$$\mathscr{L} = \frac{1}{2} (\partial_\mu \varphi)(\partial^\mu \varphi) - U(\varphi)$$

where $U(\varphi) > 0$ and $U(\varphi_0) = 0$ for the vacuum states, except for $D = 1$. The theorem applies to several such scalar fields.

Research has therefore been carried out for different types of non-linear equations and for different possible or convenient values of D. For the simplest example of the soliton solution, that of the scalar field with the lagrangean (I. 42), $D = 1$, the solution is called the kink. Other examples are the vortex solution for fields defined in a Minkowski space with $D = 2$, these fields being those in the lagrangean (I. 44) with $\mu^2 < 0$; the monopole or hedgehog solution discovered by 't Hoft and Polyakov, corresponds to the lagrangean (I. 45) ; the instanton is a solution to the equations derived from lagrangean (I. 43), for pure gauge fields, but for an imaginary time coordinate.

The internal degree of freedom of the field gives rise to an internal field space. It turns out that the solutions -the manifold of the internal field space- can define a non-trivial mapping onto the manifold of the spatial D-dimensional space ; a trivial mapping is a correspondence that maps all points of one manifold into one single point of the other. Now each mapping is characterized by an integral number which defines the so-called topological charge. The vacuum states have vanishing topological charge ; a field defined by a solution with non-vanishing topological number is stable, cannot decay into the vacuum. As we shall see in some specific examples, the topological charges are absolutely conserved.

Topological quantum numbers are therefore concepts associated to the topological structure of the lump solutions and have nothing to do with Noether's theorem. The study of the maps between the internal field space manifold and the manifold of the spacial D dimensional space of the theory, is part of a branch in mathematics called <u>theory of homotopies</u>.

Two continuous mappings f(x), g(x), each from a manifold M_x into a manifold M_y are said to be homotopic if there exists a family of continuous maps H(x, t) depending on a parameter t defined in the interval (0, 1) such that

$$H(x, o) = f(x), \quad H(x, 1) = g(x) \tag{I. 104}$$

The set of maps H(x, t) for the different values of t are so to say the possible configurations of the function f(x) [or g(x)] in the act of being continuously deformed into g(x) [or f(x)]. If f is homotopic to g by H : H : f ∿ g and if g is homotopic to k by L : L : g ∿ k

then the map $$\mathcal{G}(x,t) = \begin{cases} H\,(x,\, 2t),\ 0 \le t \le \frac{1}{2} \\ L\,(x,\, 2t - 1),\ \frac{1}{2} \le t \le 1 \end{cases}$$

will make f homotopic to k : $\mathcal{G}$: f ∿ k ; homotopy is thus an equivalance relation. Homotopically equivalent maps form a class {f}. The set of homotopy classes forms a group. An example is provided by maps from the closed line interval (0, 1) with the extremal points 0 and 1 identified, into an enclidean plane without the origin, a given point y_0 in the plane being associated to the point 0 or 1 of the interval, $f(0) = f(1) = y_0$. The maps from (0, 1) (or a circle S^1 with a point x_0 on its circumference identified with the point (0 , 1) into the above plane may be represented by closed curves starting and ending at y_0. The loops which avoid the origin may be deformed to the point y_0 ; therefore the set of all such loops forms the identity class {e}. The loops that enclose the origin once in a certain sense, for instance clock wise,form a class designated by {1} ; the class {n} is the set of loops which enclose the origin n times in correspondence to the interval (0, 1). The number n is called the winding number and is an example of a topogical number (negative numbers correspond to the loops which enclose the origin in an opposite sense).

Other examples will be found by the reader in the literature which he is invited to study.

Before giving some examples here we shall show Derrick's theorem. If $\varphi_s(x)$ is a soliton solution of the lagrangean (I. 42) its energy will be given by the hamiltonian :

$$H = \frac{1}{2}\int d^Dx\ (\nabla\ \varphi_s(x))^2 + \int d^Dx\ U(\varphi_s(x) \equiv H_1 + H_2$$

If we change x into a x, H will be expressed as

$$H(a) = a^{-(D - 2)}\ H_1 + a^{-\ D}H_2$$

As the solution energy must be stable under arbitrary variations of the field we have, in correspondence to the above scale change :

$$\left.\frac{\delta\ H}{\delta\ a}\right]_{a\ =\ 1} = 0$$

or :

$$(D - 2) H_1 + D H_2 = 0$$

which is only possible for $D = 1$ since $H_1 > 0$, $H_2 > 0$

Let us therefore consider

a) <u>The kink solution and the kink quantum number</u>.

The kink lagrangean is :

$$\mathcal{L} = \frac{1}{2} \left\{ (\partial_0 \varphi)^2 - (\partial_x \varphi)^2 \right\} + \frac{m^2}{2} \varphi^2 - \frac{\lambda}{4} \varphi^4$$

where $\varphi = \varphi(x_0, x)$ since the number of space dimensions is restricted to one by Derrick's theorem.

If we add a constant term to L, namely, $- \frac{m^4}{4 \lambda}$ we may write :

$$\mathcal{L}_1 = \frac{1}{2} (\partial_\mu \varphi)(\partial^\mu \varphi) - \frac{\lambda}{4} (\varphi^2 - \frac{m^2}{\lambda})^2$$

sum over $\mu = 0, 1$. The field equations are, from the lagragean $\mathcal{L}$:

$$(\partial_0{}^2 - \partial_x{}^2 - m^2) \varphi + \lambda \varphi^3 = 0$$

(note the sign of the term in m^2).

A time-independent solution of this equation is such that :

$$\varphi'' + m^2 \varphi - \lambda \varphi^3 = 0$$

The minimal energy solutions are :

$$\varphi_{vacuum} = \pm \frac{m}{\sqrt{\lambda}} \qquad (I.\ 105)$$

which define the two classical vacuum states . The kink solutions are :

$$\varphi_{\substack{kink \\ anti\text{-}kink}} (x) = \pm \frac{m}{\sqrt{\lambda}} \tanh \frac{mx}{\sqrt{2}} \qquad (I.\ 106)$$

the + sign corresponds to the kink, the - sign to the anti-kink. The energy of the kink is

$$H(\varphi) = \int dx \frac{1}{2} \left\{ (\partial_0 \varphi)^2 + (\partial_x \varphi)^2 + \frac{\lambda}{2} (\varphi^2 - \frac{m^2}{\lambda})^2 \right\}$$

$$H_{kink} = \frac{\sqrt{8}}{3} \frac{m^3}{\lambda}$$

The Fig. I. 1 gives the form of the solutions. The kink tends to $+ \frac{m}{\sqrt{\lambda}}$, a value of one of the vacuum states, when $x \to + \infty$, while it tends to $- \frac{m}{\sqrt{\lambda}}$, corresponding to the other vacuum state as $x \to - \infty$.

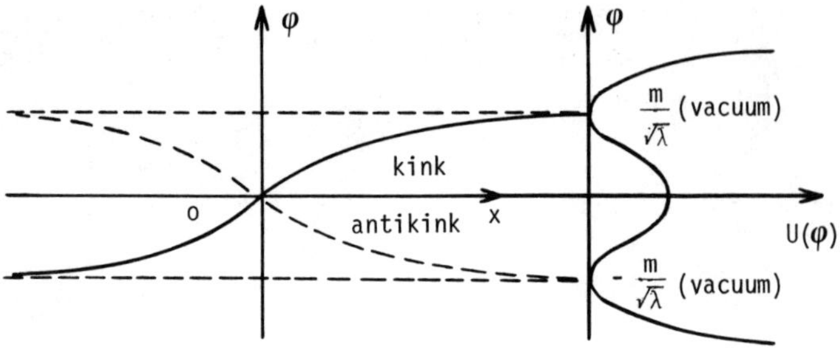

Figure I. 1

The energy density is represented in Fig. I. 2

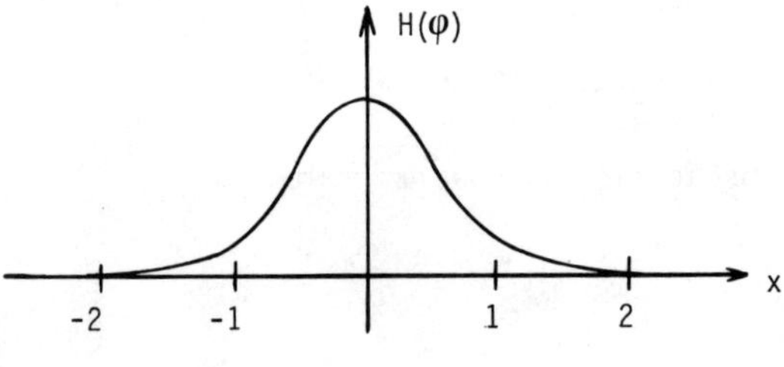

Figure I. 2

and the energy is seen to be concentrated around the origin. What is the internal degree of freedom here ? It is given by the invariance of the lagrangean under the reflection $\phi \rightarrow - \phi$.

The homotopic mapping in this case is the correspondence between the two vacuum states $\pm \frac{m}{\sqrt{\lambda}}$ and the points $x = \pm \infty$ as seen in Fig. I. Thus $\frac{m}{\sqrt{\lambda}}$ corresponds to $x = \infty$ for the kink solution and to $x = - \infty$ for the anti-kink.

There exists then a topological number which, if it is equal to 1 for the kink it will be -1 for the anti-kink and zero for the vacuum states. We may define it as :

$$k = \frac{\sqrt{\lambda}}{m} \frac{1}{2} \int_{-\infty}^{\infty} dx \frac{d\varphi}{dx} = \frac{\sqrt{\lambda}}{m} \frac{1}{2} \left[\varphi(\infty) - \varphi(-\infty) \right] \qquad (I.\ 107)$$

which is the charge

$$k = \int k_o(x)\, dx \qquad (I.\ 107a)$$

of the current

$$k_\mu(x) = \frac{\sqrt{\lambda}}{m} \frac{1}{2} \epsilon_{\mu\nu}\, \partial^\nu \varphi(x) \qquad (I.\ 107b)$$

where :

$$\epsilon_{01} = - \epsilon_{10} = 1$$

$$\epsilon_{00} = \epsilon_{11} = 0$$

For the kink :

$$k = 1,$$

for the anti-kink

$$k = - 1$$

and for the vacuum states :

$$k = 0$$

The topological current k_μ is conserved

$$\partial_\mu k^\mu = 0 \qquad \text{(I. 107c)}$$

and so the topological number k is absolutely conserved - a kink cannot decay into a vacuum.

b) <u>The vortex solutions and the winding number</u>

The second example of topological number is the winding number associated to the vortex solutions of the equations for the complex scalar field $\phi(x)$ in interaction with an electromagnetic field defined by the lagrangean (where the number of spatial dimensions is two) :

$$\mathscr{L} = - \frac{1}{4} F_{\mu\nu} F^{\mu\nu} + (D_\mu \phi)^* (D^\mu \phi) - \frac{\eta}{2} (\phi^* \phi - \frac{m^2}{\eta})^2 \qquad \text{(I. 108)}$$

with :

$$F_{\mu\nu} = \partial_\mu A_\mu - \partial_\mu A_\nu$$

$$D_\mu \phi = (\partial_\mu + i e A_\mu) \phi \qquad \text{(I. 108a)}$$

The field equations are :

$$\partial_\nu F^{\mu\nu} = e j^\mu ;$$

$$D_\mu D^\mu \phi = - \eta (\phi^* \phi - \frac{m^2}{\eta}) \phi ;$$

$$j^{\mu} = i\,(\phi^{*}\, D^{\mu}\, \phi - \phi (D^{\mu}\, \phi)^{*}) = i\,(\phi^{*} \partial^{\mu}\, \phi - \phi\, \partial^{\mu}\, \phi^{*}) - 2\, e\, A^{\mu}\, \phi^{*} \phi$$

These equations are gauge invariant under the U(1) group of phase transformations.

The name vortex solution comes from the theory of superconductors. The Meissner effect is the fact that if an external magnetic field has strenth smaller than a certain critical value H_0, then the field cannot penetrate inside the superconductor ; if $H > H_0$ the field can go through a kind of hole through a superconductor of type II and there is a magnetic flux across the superconductor. These are called vortices of magnetic flux which is quantized.

The above equations, as shown by Nielsen and Olesen, admit of vortex solutions. The magnetic flux quantum number is the winding number which characterises a mapping between the set of vacuum states and the two-dimensional geometrical plane. To see this, consider a magnetic field B in the z-direction and a circle C around the origin in the plane (x, y) over the circumference of which, C', the current is zero. From the equations for the magnetic field above we deduce the value of the vector potential $\vec{A}$ as a function of the current $\vec{j}$, that is

$$e\, j = e\, i\, (\phi\, \vec{\nabla}\, \phi^{*} - \phi^{*}\, \vec{\nabla}\, \phi) - 2\, e^{2}\, \vec{A}\, \phi^{*} \phi$$

gives :

$$\vec{A} = \frac{i}{2\, e\, \phi^{*} \phi}\, (\phi\, \vec{\nabla}\, \phi^{*} - \phi^{*}\, \vec{\nabla}\, \phi)$$

at points where $\vec{j} = 0$ (such as the circumference C'). If we set

$$\phi = f(r)\, e^{i\, \chi(\theta)}, \qquad r^{2} = x^{2} + y^{2}, \ \tan\theta = \frac{y}{x}$$

then the magnetic flux is, through the circle C (r fixed) :

$$\text{flux} = \int_C B\,dx\,dy = \oint_{C'} \vec{A} \cdot \vec{dx} = \frac{1}{e} \oint_{C'} \vec{dx} \cdot \vec{\nabla} \chi$$

that is :

$$\text{flux} = \frac{1}{e} \left[\chi(2\pi) - \chi(o) \right] = \frac{2\pi\, n}{e}, \qquad n = 0, \pm 1, \pm 2, \ldots \qquad \text{(I. 108b)}$$

The fact that the field $\phi(x)$ must be single-valued imposes the quantization of the flux. Thus the number n arises from the map of the circle in the x, y plane on the circle described by the phase χ.

The field energy (per unit length) :

$$H = \int dx\,dy\, \frac{1}{2} \left\{ \vec{B}^2 + (\vec{D}\,\phi) \cdot^{*} (\vec{D}\,\phi) + \frac{\eta}{2} \left(\phi^{*}\phi - \frac{m^2}{\eta}\right)^2 \right\}$$

will be finte if ϕ tends to its vacuum value $\phi \to e^{i\chi} \frac{m}{\sqrt{\eta}}$ at infinity and $|\vec{D}\,\phi| \to 0$ and $\vec{B} \to 0$. If $\vec{B} \to 0$ asymptotically, $\vec{A}$ will tend to a purely gauge form $\vec{A} \to \vec{\nabla}\,\alpha$:

$$\phi \to e^{i\chi(\theta)} \frac{m}{\sqrt{\eta}}$$

$$\vec{A} \to \vec{\nabla}\,\alpha$$

We see that the asymptotic solutions are characterized by a mapping between the circle at infinity on the (x, y) plane and the phase of $\chi(\theta)$ of the vacuum states of the field ϕ. When one describes a circle in the (x, y) plane the phase $\chi(\theta)$ can change from 0 to $2\pi n$. The number n is the winding number characteristic of each homotopy class in this map.

It is to be noted that the electric charge e in equations (I. 108a) and (I. 108b) is the field coupling constant, which we may designate by e_f. It is related to the particle electric charge e_p by the relation $e_f = \frac{1}{\hbar} e_p$ (see remark at the end of Chapter II). Thus Planck's constant is implied in the quantisation of the flux (I. 108b)(see Coleman[24]).

c) The 't Hooft-Polyakov monopole

Consider the lagrangean :

$$\mathscr{L} = -\frac{1}{4} F^{\mu\nu}_{a} F_{\mu\nu a} + \frac{1}{2}(D_\mu \phi_a)(D^\mu \phi_a) - \frac{\eta}{4}\left(\phi_a \phi_a - \frac{m^2}{\eta}\right)^2$$

of an isovector ϕ_a in interaction with an isovector gauge field A_a^{μ}, the internal symmetry group being the SO(3) group, where :

$$F_{\mu\nu a} = \partial_\nu A_{\mu a} - \partial_\mu A_{\nu a} + e\, \varepsilon_{abc} A_{\mu b} A_{\nu c}$$

$$D_\mu \phi_a = \partial_\mu \phi_a + e\, \varepsilon_{abc} A_{\mu c} \phi_b$$

The classical field equations are :

$$D_\nu F^{\mu\nu}_{a} = -e\, \varepsilon_{abc} \phi_b D^\mu \phi_c$$

$$D_\mu D^\mu \phi_a = -\eta\, \phi_a \left(\phi_b \phi_b - \frac{m^2}{\eta}\right) \qquad \text{(I. 109)}$$

A solution, corresponding to a magnetic monopole was found by 't Hooft and Polyakov. By introducing the following definitions :

$$B_\mu \equiv \phi_a A_{\mu a}$$

and

$$F_{\mu\nu} = \partial_\nu B_\mu - \partial_\mu B_\nu - \frac{1}{e} \varepsilon_{abc} \hat{\phi}_a \partial_\mu \hat{\phi}_b \partial_\nu \hat{\phi}_c, \qquad \hat{\phi}_a = \frac{\phi_a}{|\phi_a|}$$

this solution has the form :

$$F_{jk} = \frac{\varepsilon_{jk\ell}\, x_\ell}{e\, r^3}$$

and is the magnetic field of a magnetic charge $\frac{1}{e}$ at the origin.

The dual $\tilde{F}_{\mu\nu}$ of the field $F_{\mu\nu}$ is :

$$\tilde{F}_{\mu\nu} = \frac{1}{2}\ \varepsilon_{\mu\nu\alpha\beta}\ F^{\alpha\beta} = \frac{1}{2e}\, \varepsilon_{\mu\nu\alpha\beta}\ \ \varepsilon_{abc}\ \hat{\phi}_a\ \partial^\alpha\ \hat{\phi}_b\ \partial^\beta\ \hat{\phi}_c$$

so that the dual current is :

$$\tilde{j}^\mu = \partial^\nu\ \tilde{F}_{\mu\nu} = \frac{1}{2e}\ \varepsilon_{\mu\nu\alpha\beta}\ \varepsilon_{abc}\ \partial^\nu\ (\hat{\phi}_a\ \partial^\alpha\ \hat{\phi}_b\ \partial^\beta\ \hat{\phi}_c)$$

This is a conserved current $\partial_\mu{}^{*}j^\mu = 0$. It is the magnetic current, which does not follow from any application of Noether's theorem to the lagrangean. The magnetic charge is :

$$Q_{magn} = \int d^3x\ \tilde{j}_0 = \frac{1}{2e} \int d^3x\ \varepsilon_{ijk}\ \varepsilon_{abc}\ \partial^i\ (\hat{\phi}_a\ \partial^j\ \hat{\phi}_b\ \partial^k\ \hat{\phi}_c) \qquad \text{(I. 110)}$$

Integration gives :

$$Q_{magn} = \frac{1}{2e}\ \oint ds_i\ \varepsilon_{ijk}\ \varepsilon_{abc}\ (\hat{\phi}_a\ \partial_j\ \hat{\phi}_b\ \partial_k\ \hat{\phi}_c)$$

where the integration is over a spherical surface in the limit of infinite radius. This is 4π times an integral number so that

$$Q_{magn} = \frac{4\pi n}{e} \qquad \text{(I. 110a)}$$

The integer n is the so-called Kronecker index of the map between a two-dimensional sphere in the (x, y, z) space and a two-dimensional sphere in the (ϕ_1, ϕ_2, ϕ_3) space.

The 't Hooft-Polyakov monopople solution exists only for non-abelian gauge groups such as SO(3) and distinguishes itself from the case of the abelian group U(1) in which case the magnetic monopole has a string singularity, the Dirac magnetic monopole (see Chap. II, § II. 5).

d) Instantons

If we consider the Yang-Mills gauge fields $A_{\mu a}(x)$ where the internal index a = 1, 2, 3 is associated to the SU(2) group, the lagrangean for this field in the absence of other sources is :

$$\mathscr{L} = -\frac{1}{4} F_{\mu\nu a} F^{\mu\nu}$$

where :

$$F_{\mu\nu a} = \partial_\nu A_{\mu a} - \partial_\mu A_{\nu a} + g\, \varepsilon_{abc} A_{\mu b} A_{\nu c}$$

The equations of motion are :

$$\partial_\nu F^{\mu\nu}_a = - g\, \varepsilon_{abc} A_{\nu c} F^{\mu\nu}_b$$

or :

$$D_{\nu\, ;\, ab} F^{\mu\nu}_b = 0$$

if the covariant derivative on the isovector $F^{\mu\nu}_b$ is

$$D_{\nu;ab} = \partial_\nu \delta_{ab} + g\, \varepsilon_{abc} A_{\nu c}$$

Let us define the matrix fields :

$$A_\mu = A_{\mu a} \frac{\tau_a}{2}$$

$$F_{\mu\nu} = F_{\mu\nu a} \frac{\tau_a}{2} = \partial_\nu A_\mu - \partial_\mu A_\nu - i\, g \left[A_\mu , A_\nu \right]$$

then the field equations will take the form :

$$\partial_\nu F^{\mu\nu} + i\, g \left[A_\nu , F^{\mu\nu} \right] = 0 \qquad \text{(I. 111)}$$

The first regular solution investigated of this equation, with spherical symmetry is of the form :

$$A_\mu(x) = \frac{-\, i\, r^2}{r^2 + \beta^2}\; u^{-1}(x)\, \partial_\mu\, u(x) \qquad \text{(I. 112)}$$

where

$$r^2 = (x^1)^2 + (x^2)^2 + (x^3)^2 + (x^4)^2 \qquad \text{(I. 112a)}$$

$$u(x) = \frac{1}{r}\,(x^4 - i\, \vec{x} \cdot \vec{\tau})$$

and β is a parameter. Here we have considered an imaginary time coordinate :

$$x^4 = i\, x^0$$

For large r, $A_\mu(x)$ tends to the form $g^{-1}\, \partial_\mu\, g(x)$ which is a pure gauge since it gives $F_{\mu\nu} = 0$ in the limit $r \to \infty$:

$$r \to \infty, \quad A_\mu(x) \to -\, i\, u^{-1}(x)\, \partial_\mu\, u(x), \quad F_{\mu\nu} \to 0 \qquad \text{(I. 113)}$$

The action of this field in Minkwoski space is :

$$S = -\frac{1}{4}\int d^4x\, F_{\mu\nu a}\, F^{\mu\nu a}$$

which is equal to :

$$S = \frac{1}{2}\int (\vec{E}^2{}_a - \vec{B}^2{}_a)\, d^4x$$

where we define :

$$E^k{}_a = F^{ok}_a \quad ; \quad B^j{}_a = \frac{1}{2}\,\varepsilon^{jk\ell}\, F_a{}^{k\ell}$$

Now in Euclidean coordinates, that is, with the choice $x^4 = i\, x^0$ we have :

$$\partial^0 = i\,\partial^4,\ A^0 = -\, i\, A^4,\ F^{ok}_a = -\, i\, F^{4k}_a = -\, i\, E^k_{(4)a}$$

hence

$$-\frac{1}{2}\,(\vec{E}^2{}_a - \vec{B}^2{}_a) \rightarrow \frac{1}{2}\,(\vec{E}^2_{(4)a} + \vec{B}^2{}_a)$$

The euclidean action is thus taken to be

$$S = \frac{1}{4}\int F_{\mu\nu a}\, F^{\mu\nu}_a\, d^4x = \frac{1}{2}\,\mathrm{Tr}\int d^4x\, F_{\mu\nu}\, F^{\mu\nu} \qquad \text{(I. 114)}$$

It is positive definite and the condition that $F^a_{\mu\nu} \rightarrow 0$ as $r \rightarrow \infty$ is a condition for S to be finite ; this is satisfied, as stated above, for A_μ a pure gauge.

Now we see that asymptotically there is a correspondence between the ^3S-sphere of the euclidean (x^1, x^2, x^2, x^4) space with radius r given in

(I. 112a) and a sphere S^3 in the internal space which is equivalent to the group SU(2) which is the set of all matrices 2×2 of the form :

$$A = \begin{pmatrix} a_4 + i\, a_3 & i\, a_1 + a_2 \\ i\, a_1 - a_2 & a_4 - i\, a_3 \end{pmatrix}$$

such that

$$a_1^2 + a_2^2 + a_3^2 + a_4^2 = 1$$

The function u(x), (I. 112a), maps the sphere S^3 of the euclidean four-dimensional space on the space of the group SU(2), which is also a sphere S^3. Note the difference between this map and that defined by the trivial pure gauge $A_\mu = 0$ (both this and the asymptotic A_μ given in (I. 113) lead to a vanishing field $F_{\mu\nu}$ and thus define vacuum states). The trivial gauge $A_\mu = 0$ maps the euclidean S^3 sphere into a single point of the internal space and therefore is defined by the identity with winding number equal to zero.

The topological charge is given by :

$$Q = \frac{g^2}{16\pi^2} \int d^4x \; \mathrm{Tr}\,(F_{\mu\nu}\, \tilde{F}_{\mu\nu}) \tag{I. 115}$$

where

$$\tilde{F}_{\mu\nu} = \frac{1}{2}\, \varepsilon_{\mu\nu\alpha\beta}\, F_{\alpha\beta}$$

(note that $\mu, \nu = 1, 2, 3, 4$ and that in euclidean space there is no difference between contravariant and covariant tensors). From the inequality

$$\frac{1}{2} \int d^4x \; \mathrm{Tr}\,(F_{\mu\nu} \pm \tilde{F}_{\mu\nu})^2 \geq 0$$

it follows that S (I. 114) has a lower bound determined by the topological charge :

$$S \geq \frac{8\pi^2 Q}{g^2}$$

The lower bound is attained by S for self-dual or anti-self-dual fields :

$$F_{\mu\nu} = \pm \tilde{F}_{\mu\nu}$$

It is these solutions that are called instantons or anti-instantons.

The integrand is a total divergence :

$$\mathrm{Tr}\,(F_{\mu\nu}\,\tilde{F}_{\mu\nu}) = \partial_\mu\, J_\mu$$

$$J_\mu = 2\,\varepsilon_{\mu\nu\alpha\beta}\,\mathrm{Tr}\,(A_\nu\,\partial_\alpha\,A_\beta - \frac{2g}{3i}\,A_\nu\,A_\alpha\,A_\beta)$$

for regular A_μ's such that $\partial_\alpha\,\partial_\beta\,A_\mu = \partial_\beta\,\partial_\alpha\,A_\mu$.

Therefore :

$$Q = \frac{g^2}{16\pi^2} \oint_{S^3} J_\mu(x)\; d\,\sigma_\mu \qquad \text{(I. 116)}$$

where $d\,\sigma_\mu$ is the surface element of the sphere ^3S.

The integrals (I. 114) and (I. 115) converge for the asymptotic behaviour of A_μ in (I. 113) and in this case it is shown that Q will be of the form $n\,g^2$ where $n = 0, \pm 1, \pm 2, \ldots$ is the winding number of the homotopy class of the mapping $S^3 \rightarrow SU(2)$.

We mention that the importance of solutions such as the instantons, expressed in spaces with an imaginary time coordinate is tied to the fact that only in such euclidean space integral forms essential to define generating functionals of Green functions are well defined. In such theories the physical Green functions result from the former by analytic continuation.

We invite the interested reader to consult the literature[24-38]. We hope have arisen his interest in the question of topological quantum numbers and soliton and instanton solutions of the non-linear equations of gauge field theories. In these theories, "the topology of the internal symmetry group conspires with the topology of space-time in such a way that particularly stable structures (for these solutions) appear", in the words of Hans Joos.

PROBLEMS

I - 1. a) Establish the conditions on the matrix $D(\ell)$ which transforms a Dirac spinor $\psi(x)$, $\psi'(x') = D(\ell)\ \psi(x)$, in correspondence to a proper orthochronous Lorentz transformation so that Dirac's equation be invariant under these transformations.

b) Give the twenty - six equations which are satisfied by the thirty two real matrix elements of $D(\ell)$.

c) Show that $\gamma^0\ D^+(\ell)\ \gamma^0 = -\ D^{-1}(\ell)$ in the case of heterochronous transformations $(\ell^0{}_0 \leq -1)$.

d) Give the equation and the transformation law of the adjoint spinor $\bar{\psi}(x)$

I - 2. A- Deduce the transformation laws under proper or improper orthochronous Lorentz transformations a) of the dual tensor field $\tilde{F}_{\mu\nu}(x) = \frac{1}{2}\,\varepsilon_{\mu\nu\alpha\beta}\ F^{\alpha\beta}(x)$ of an antisymmetric tensor $F^{\alpha\beta}$; b) of the following Dirac bilinear forms in the spinors $\psi(x)$ and $\eta(x)$:

$$S(x) = \bar{\psi}(x)\ \eta(x),$$

$$V^\mu(x) = \bar{\psi}(x)\ \gamma^\mu\ \eta(x),$$

$$A^\mu(x) = \bar{\psi}(x)\ \gamma^\mu\ \gamma^5\ \eta(x)$$

$$S^{\mu\nu}(x) = \bar{\psi}(x)\ \sigma^{\mu\nu}\ \eta(x)$$

$$P(x) = i\ \bar{\psi}(x)\ \gamma^5\ \eta(x)$$

B- a) Deduce the infinitesimal and the finite forms of the matrix $D(\ell)$ in terms of the Lorentz transformation parameters $\alpha_{\mu\nu} = g_{\mu\alpha}\, g_{\nu\beta}\, \ell^{\alpha}{}_{\lambda}\, g^{\lambda\beta}$ and the γ-matrices.

b) Find the four Dirac matrices in the Majorana representation in which all their elements are 0, + i, - i, that is :

$$(\gamma^{\mu})^{*} = - \gamma^{\mu}$$

where the star means complex conjugation. Write the representation in which

$$\gamma^{0} = \begin{pmatrix} 0 & \sigma^{2} \\ \sigma^{2} & 0 \end{pmatrix}$$

c) Show that if a (Majorana) spinor is chosen as real in this representation, it will remain real in all Lorentz frames.

I - 3. Show that the following equation proposed by Rarita and Schwinger to describe free vector spinor fields $\psi^{\mu}(x)$,

$$(i\, \gamma^{\alpha}\, \partial_{\alpha} - m)\psi^{\mu}(x) - \frac{i}{3}\,(\gamma^{\mu}\, \partial^{\nu} + \gamma^{\nu}\, \partial^{\mu})\psi_{\nu} + \frac{1}{3}\,(\gamma^{\mu}\,[\, i\, \gamma^{\alpha}\, \partial_{\alpha} + m\,]\, \gamma^{\nu})\, \psi_{\nu} = 0$$

is equivalent to Dirac's equation for ψ^{μ} and the subsidiary conditions given in equation (I. 6), namely :

$$(i\, \gamma^{\alpha}\, \partial_{\alpha} - m)\, \psi^{\mu} = 0,$$

$$\gamma_{\mu}\, \psi^{\mu} = 0,$$

$$\partial_{\mu}\, \psi^{\mu} = 0$$

I - 4. a) Consider a two dimensional Minkowski space-time (one time, one space coordinates). What are the possible forms of Dirac's equation for a free spinor (with how many components) ? What are the matrices corresponding to γ^{0} and γ^{1}, say ? Is the commutation relation $\frac{1}{2}\,(\gamma^{\mu}\,\gamma^{\nu} + \gamma^{\nu}\,\gamma^{\mu}) = g^{\mu\nu}$ where $g_{00} = - g_{11} = 1$, $g_{ik} = 0$ for $i \neq k$, still satisfied ? What is the matrix corresponding to γ^{5}, $(\gamma^{5})^{+} = \gamma^{5}$, $(\gamma^{5})^{2} = I$, $\gamma^{5}\,\gamma_{\mu} = - \gamma_{\mu}\,\gamma^{5}$, $\mu = 0, 1$? What are the possible conserved currents ? What are the equations for massless spinors, for left-handed and righ-handed components ?

I - 5. A scalar field φ has a non-linear interaction with a spinor field ψ defined by the term L' in the lagrangean :

$$L = L_0 + L'$$

where

$$L_0 = \frac{1}{2}(\partial^\mu \varphi \, \partial_\mu \varphi - m^2 \varphi^2) + \frac{1}{2}\bar{\psi}(i \gamma^\alpha \partial_\alpha - M)\psi - \frac{1}{2}(i(\partial_\mu \bar{\psi})\gamma^\mu + M \bar{\psi}, \psi,$$

$$L' = g \bar{\psi} \psi \varphi \frac{1}{1 + K \varphi}$$

g and K are coupling constants.

Find a) the equations of motion for φ, ψ, $\bar{\psi}$; b) the energy-momentum tensor of the system ; c) the hamiltonian as a power series in $K \varphi$.

I - 6. Calculate the energy of the anti-kink solution of the equation :

$$(\partial^2{}_0 - \partial^2{}_x - m^2)\, \varphi (x, x_0) + \lambda \varphi^3 (x, x_0) = 0$$

I - 7. Show that the time-independent function of the coordinate x :

$$\varphi_0(x) = \frac{4}{\beta} \text{ arc tan } (x \sqrt{\alpha})$$

is a static solution of the two dimensional sine-Gordon equation :

$$(\partial^2{}_0 - \partial^2{}_x)\, \varphi (x, x_0) + \frac{\alpha}{\beta} \sin \beta \varphi (x, x_0) = 0$$

and that the corresponding energy is :

$$H_0 = \frac{8 \sqrt{\alpha}}{\beta^2}$$

I - 8. Show that the equation of Probl. I - 3. is a particular case of the family of equations :

$$\left\{ (i \gamma \cdot \partial - m) g^{\alpha\beta} - iA (\gamma^\alpha \partial^\beta + \gamma^\beta \partial^\alpha) + \gamma^\alpha \left[B (i \gamma \cdot \partial) + Cm \right] \gamma^\beta \right\} \psi_\beta = 0$$

where the constants A, B, C satisfy conditions so that equations (I. 6) are satisfied. a) What are these conditions ?

b) For what values of A, B, C is the above equation invariant for $m = 0$ under the gauge transformation $\psi'_\alpha = \psi_\alpha + \partial_\alpha \varphi$ where φ is an arbitrary spinor ?

c) Show that with the definition of γ^5 in § I - 1. g one has :

$$\varepsilon_{\alpha\beta\mu\nu} \gamma^5 \gamma^\beta = i \left\{ \gamma_\alpha \gamma_\mu \gamma_\nu - g_{\mu\nu} \gamma_\alpha - g_{\mu\alpha} \gamma_\nu + g_{\nu\alpha} \gamma_\mu \right\}$$

CHAPTER II

The Electromagnetic Gauge Field

II. 1 - FIELD INTERACTIONS

The examples given in the preceding paragraphs referred all to free fields. Physical phenomena, however, are due to interactions ; therefore we must seek how to construct the interactions among fields. These may be defined by adding to the free field lagrangeans convenient Poincaré-invariant terms which depend on both fields. Thus for an interaction of a scalar field with itself we have the lagrangean* :

$$L = \partial^\mu \varphi^+ \partial_\mu \varphi - m^2 \varphi^+ \varphi - \frac{\eta}{2}(\varphi^+ \varphi)^2 \qquad \text{(II. 1)}$$

which gives rise to a non-linear equation of motion for φ :

$$(\Box + m^2 + \eta\, \varphi^+ \varphi)\, \varphi = 0 \qquad \text{(II. 1a)}$$

An interaction between a scalar and a spinor field may be described by a term $f\, \bar{\psi}\, \psi\, \varphi$ so that the lagrangean of such a system is :

$$L = \bar{\psi}(i\, \gamma \cdot \partial - M)\, \psi + \frac{1}{2}\, (\partial_\mu \varphi\, \partial^\mu \varphi - m^2\, \varphi^2) -$$

$$- \frac{\lambda}{4!}\, \varphi^4 + f\, \bar{\psi}\, \psi\, \varphi \qquad \text{(II. 2)}$$

where f and λ are coupling constants and φ is real.

In the beta-decay of the lepton-muon :

$$\mu \to \nu_\mu + e + \bar{\nu}_e \qquad \text{(II. 3)}$$

it is known that this process is described by a coupling of two expressions, the so-called weak currents, for the μ-field and for the e-field :

$$\ell^\alpha_{(\mu)} = \bar{\nu}_\mu(x)\, \gamma^\alpha(1 - \gamma^5)\, \mu\, (x)$$

$$\ell^{\alpha +}_{(\ell)} = \bar{e}(x)\, \gamma^\alpha(1 - \gamma^5)\, \nu_e(x) \qquad \text{(II. 3a)}$$

---

* Following the convention usually adopted we reserve the term $\frac{\lambda}{4!}$ for the coupling constant of the quartic self-interaction of a real scalar field (the denominator is convenient for combinatorial reasoms in the study of Feynman diagrams for this theory). For complex fields we usually denote the coupling constant by η.

so that the corresponding lagrangean has the form :

$$L = \bar{\mu}(i\,\gamma \cdot \partial - m_\mu)\mu + \bar{e}(i\,\gamma \cdot \partial - m_e)\,e \quad +$$

$$+ \bar{\nu}_\mu\, i\,\gamma \cdot \partial\,(\nu_\mu)_L + \bar{\nu}_e\, i\,\gamma \cdot \partial\,(\nu_e)_L \quad +$$

$$+ \frac{G}{\sqrt{2}}\,(\bar{\nu}_\mu\,\gamma^\alpha(1-\gamma^5)\mu\,)(\bar{e}\,\gamma_\alpha(1-\gamma^5)\nu_e) \qquad \text{(II. 3b)}$$

where the two neutrinos ν_μ, ν_e are assumed to be massless and left-handed polarized

$$\nu_L = \frac{1}{2}\,(1-\gamma^5)\,\nu \quad ; \quad \gamma^5 = i\,\gamma^0\,\gamma^1\,\gamma^2\,\gamma^3$$

The electromagnetic interaction with an electron field is described by the lagrangean :

$$L = -\frac{1}{4}\,F^{\mu\nu}\,F_{\mu\nu} + \bar{e}\,(i\,\gamma \cdot \partial - m)\,e \;-\; e\,j_\mu\,A^\mu \qquad \text{(II. 4)}$$

where :

$$j^\mu\,(x) = \bar{e}(x)\,\gamma^\mu\,e(x)$$

$$F^{\mu\nu} = \partial^\nu\,A^\mu - \partial^\mu\,A^\nu \qquad \text{(II. 4a)}$$

and $A^\mu\,(x)$ is the electromagnetic field ; the corresponding equations of motion are :

$$\partial_\nu\,F^{\mu\nu}(x) = e\,j^\mu\,(x) \qquad \text{(II. 4b)}$$

$$(i\,\gamma^\mu\,D_\mu - m)\,e\,(x) = 0, \qquad D_\mu = \partial_\mu + ie\,A_\mu$$

We shall now study the important notion of gauge fields and show that the electromagnetic field is the simplest example of a gauge field.

II. 2 - THE ELECTROMAGNETIC FIELD AS A GAUGE FIELD

We have seen that the lagrangeans of complex fields $\psi^{\alpha}(x)$ are constructed in such a way as to be hermitian since the observables derived from the lagrangean, such as the energy-momentum tensor, the hamiltonian, etc, must have real eigenvalues. They are therefore invariant under the global gauge group U(1), that is if

$$\begin{aligned} \psi^{\alpha}(x) &\rightarrow \psi^{\alpha\,'}(x) = e^{i\omega}\psi^{\alpha}(x), \\ \psi^{\alpha+}(x) &\rightarrow \psi^{\alpha+\,'}(x) = e^{-i\omega}\psi^{\alpha+}(x) \end{aligned} \tag{II. 5}$$

ω = real constant

then :

$$\begin{aligned} & L(\psi^{\alpha\,'}(x), \psi^{\alpha+\,'}(x), \partial^{\mu}\psi'^{\alpha}(x), \partial^{\mu}\psi'^{\alpha+}(x)) = \\ = \; & L(\psi^{\alpha}(x), \psi^{\alpha+}(x), \partial^{\mu}\psi^{\alpha}(x), \partial^{\mu}\psi^{\alpha+}(x)) \end{aligned} \tag{II. 6}$$

The notion of gauge field first arose from the postulate that these lagrangeans be also invariant under the local gauge group, that is, when the parameter ω depends on x :

$$\omega = \omega(x)$$

This means that we must be free to choose a phase for $\psi^{\alpha}(x)$ in each point of space-time which is not necessarily the same when we change the point ; and that the observables do not depend on this choice.

Let us for convenience write :

$$\omega(x) = e\ \Lambda(x)$$

where e is the elementary electric charge. Under the transformation :

$$\begin{aligned} \psi^{\alpha\prime}(x) &= U(\Lambda(x))\,\psi^{\alpha}(x) \\ \psi^{\alpha+\prime}(x) &= U^{*}(\Lambda(x))\psi^{\alpha+}(x), \\ U(\Lambda(x)) &= e^{ie\Lambda(x)} \end{aligned} \tag{II. 7}$$

the derivatives of the field will transform as follows :

$$\partial^\mu \psi^{\alpha\prime}(x) = U(\Lambda(x))\partial^\mu \psi^\alpha(x) + \left[\partial^\mu U(\Lambda(x))\right]\psi^\alpha(x)$$

$$\partial^\mu \psi^{\alpha+\prime}(x) = U^*(\Lambda(x))\partial^\mu \psi^{\alpha+}(x) + \left[\partial^\mu U^*(\Lambda(x))\right]\psi^{\alpha+}(x) \qquad \text{(II. 8)}$$

and products of the form :

$$\partial^\mu \psi^{+\alpha}(x)\, \Omega_{\alpha\beta}\, \partial_\mu \psi^\beta(x)$$

are not invariant :

$$\partial^\mu \psi^{+\prime\alpha}(x)\, \Omega_{\alpha\beta}\, \partial_\mu \psi^{\prime\beta}(x) \neq \partial^\mu \psi^{+\alpha}(x)\, \Omega_{\alpha\beta}\, \partial_\mu \psi^\beta(x) \qquad \text{(II. 8a)}$$

What is the change of the lagrangean under the transformations (II. 7) ? We have, for an infinitesimal transformation :

$$\delta \psi = i\, e\, \Lambda(x)\, \psi(x),$$

$$\delta \psi^+ = -\, i\, e\, \Lambda(x)\, \psi^+(x),$$

$$\delta (\partial^\mu \psi) = i\, e\, \Lambda(x)\, \partial^\mu \psi(x) + i\, e\, \psi(x)\, \partial^\mu \Lambda(x)$$

$$\delta (\partial^\mu \psi^+) = -\, i\, e\, \Lambda(x)\, \partial^\mu \psi^+(x) - i\, e\, \psi^+(x)\, \partial^\mu \Lambda(x)$$

so that

$$\delta L = \frac{\partial L}{\partial \psi}\, \delta \psi + \frac{\partial L}{\partial(\partial^\mu \psi)}\, \delta(\partial^\mu \psi) + \text{h.c.} =$$

$$= i\, e\, \Lambda \left\{ \frac{\partial L}{\partial \psi}\, \psi - \psi^+ \frac{\partial L}{\partial \psi^+} + \right.$$

$$\left. + \frac{\partial L}{\partial(\partial^\mu \psi)}\, \partial^\mu \psi - (\partial^\mu \psi^+)\, \frac{\partial L}{\partial(\partial^\mu \psi^+)} \right\} +$$

$$+ i\, e\, (\partial^\mu \Lambda) \left\{ \frac{\partial L}{\partial(\partial^\mu \psi)}\, \psi - \psi^+ \frac{\partial L}{\partial(\partial^\mu \psi^+)} \right\}$$

But as L is invariant for Λ constant

$$\delta \; L = 0 \quad \text{for} \quad \Lambda(x) = \omega = \text{const.}$$

then

$$\frac{\partial L}{\partial \psi} \psi + \frac{\partial L}{\partial(\partial^\mu \psi)} \partial^\mu \psi -$$

$$- \psi^+ \frac{\partial L}{\partial \psi^+} - \frac{\partial L}{\partial(\partial^\mu \psi^+)} (\partial^\mu \psi^+) = 0$$

Therefore we see that under the group (II. 7) L changes as follows (see equation (I. 46)) :

$$\delta L = - e \, j_\mu(x) \, \partial^\mu \Lambda(x) \tag{II. 8b}$$

The reason for the non-invariance of the lagragean under infinitesimal or finite local phase transformations such as (II. 7) is the occurence of the term in $\partial_\mu U(x)$ in the derivative of ψ, (II. 8). A generalization of this lagrangean which remains invariant under the group of local transformations (II. 7) will be obtained if we can find a generalized derivative of ψ, $D_\mu \psi$, such that $D_\mu \psi$ transforms like ψ :

$$(D_\mu \psi^\alpha(x))' = U(\Lambda(x))(D_\mu \psi^\alpha(x)) \; ; \; (D'_\mu \psi'^\alpha)^+ = U^*(\Lambda(x))(D^*_\mu \psi^{\alpha+}(x)) \tag{II. 9}$$

As the derivative D_μ must contain the usual derivative ∂_μ as a particular case we introduce a new field $A^\mu(x)$, a real vector field such that :

$$\begin{aligned} D^\mu \psi^\alpha(x) &= (\partial^\mu + i \, e \, A^\mu(x)) \psi^\alpha(x) \\ (D^\mu \psi^\alpha(x))^+ \equiv D^{\mu *} \psi^{\alpha+}(x) &= (\partial^\mu - i \, e \, A^\mu(x)) \psi^{\alpha+}(x) \end{aligned} \tag{II. 10}$$

Equations (II. 9) and (II. 10) will give us the transformation law of the so-called vector gauge field $A_\mu(x)$ corresponding to equations (II. 7) :

$$e A'_\mu = U \, e \, A_\mu \, U^{-1} + i \, (\partial_\mu U) \, U^{-1}$$

or :

$$A'_\mu = U \left\{ A_\mu - \frac{i}{e} \partial_\mu \right\} U^{-1} \; ; \; U^+U = UU^+ = I, \; (\partial_\mu U)U^{-1} = - U(\partial_\mu U^{-1})$$

For $U(x) = e^{ie\Lambda(x)}$ we have :

$$A^{\mu\prime}(x) = A^\mu(x) - \partial^\mu \Lambda(x) \qquad \text{(II. 11)}$$

therefore the lagrangean constructed with the fields $\psi^\alpha(x)$, $\psi^{\alpha+}(x)$, the new vector gauge field $A^\mu(x)$ and the covariant derivative (II. 9) is gauge invariant since now, as a result of the equations (II. 7), (II. 9), (II. 10) and (II. 11) :

$$(D'^\mu \psi'^\alpha)^+ \; \Omega_{\alpha\beta} \; (D'_\mu \; \psi'^\beta) = (D^\mu \; \psi^\alpha) \; \Omega_{\alpha\beta} \; (D_\mu \; \psi^\beta) \qquad \text{(II. 12)}$$

since :

$$U^* \; (\Lambda) \; U \; (\Lambda) = 1$$

To this lagrangean

$$L \; (\psi^\alpha, \; D^\mu \; \psi^\alpha, \; \psi^{\alpha+}, \; (D^\mu \; \psi^\alpha)^+)$$

we must now add a term referring to the field $A^\mu(x)$ alone and this term, invariant under the transformations (II. 12), is :

$$L_A = - \frac{1}{4} \; F^{\mu\nu}(x) \; F_{\mu\nu}(x)$$

$$F^{\mu\nu}(x) = \partial^\nu \; A^\mu(x) - \partial^\mu \; A^\nu \; (x) \qquad \text{(II. 13a)}$$

Indeed the lagrangean (II. 13) contains a new field A^μ in interaction with the matter fields ψ^α. The complete lagrangean must contain a piece which describes the gauge field in the absence of the field ψ^α.

The covariant derivative components as defined by equations (II. 10) do not commute ; one has :

$$\left[D_\mu \; , \; D_\nu \right] = - \; i \; e \; (\partial_\nu \; A_\mu - \partial_\mu \; A_\nu) \qquad \text{(II. 13b)}$$

We therefore take this invariant curl as the definition of $F_{\mu\nu}$:

$$F_{\mu\nu} = \partial_\nu A_\mu - \partial_\mu A_\nu = \frac{i}{e}[D_\mu, D_\nu] \tag{II. 13c}$$

and this is the only gauge-invariant term, with first derivatives of the gauge-field $A^\mu(x)$; a term of the form $A_\mu A^\mu$ would give a mass to the field A_μ and would not be gauge invariant.

The gauge invariant lagrangean is thus :

$$\begin{aligned}\mathcal{L} \equiv \mathcal{L}(A^\mu, \partial^\nu A^\mu, \psi^\alpha, \psi^{\alpha+}, D^\mu \psi^\alpha, (D^\mu \psi^\alpha)^+) &= \\ = -\frac{1}{4} F^{\mu\nu}(x) F_{\mu\nu}(x) + L(\psi^\alpha, \psi^{\alpha+}, D^\mu \psi^\alpha, (D^\mu \psi^\alpha)^+)\end{aligned} \tag{II. 14}$$

Clearly, the covariant variational principle requires that the covariant deriva $D^\mu \psi^\alpha$, and not $\partial^\mu \psi^\alpha$, be one of the field variables to be varied.

If we vary the fields $\psi^\alpha(x)$, $D^\mu \psi^\alpha(x)$ for <u>fixed</u> $A^\mu(x)$ (that is, $\delta A^\mu = 0$) the variation principle (I. 2), (I. 2a) will now be :

$$\begin{aligned}\delta S &= \int d^4x \left\{ \frac{\partial \mathcal{L}}{\partial \psi^\alpha(x)} \delta \psi^\alpha(x) + \frac{\partial \mathcal{L}}{\partial(D^\mu \psi^\alpha(x))} \delta(D^\mu \psi^\alpha(x)) \right\} \\ &= \int d^4x \left\{ \frac{\partial \mathcal{L}}{\partial \psi^\alpha(x)} - D_\mu^* \frac{\partial \mathcal{L}}{\partial(D^\mu \psi^\alpha(x))} \right\} \delta \psi^\alpha(x) \\ &+ \int d^4x \, \partial^\mu \left(\frac{\partial \mathcal{L}}{\partial(D^\mu \psi^\alpha(x))} \delta \psi^\alpha(x) \right)\end{aligned} \tag{II. 14a}$$

One thus obtains the equations of motion

$$D^{\mu *} \frac{\partial \mathcal{L}}{\partial(D^\mu \psi^\alpha(x))} - \frac{\partial \mathcal{L}}{\partial \psi^\alpha(x)} = 0 \tag{II. 15}$$

and similarly, if one varies $\psi^{\alpha+}$, $D^{\mu *} \psi^{\alpha+}$:

$$D^\mu \frac{\partial L}{\partial(D^{\mu *} \psi^{\alpha+}(x))} - \frac{\partial \mathcal{L}}{\partial \psi^{\alpha+}(x)} = 0 \tag{II. 15}$$

Variation of the field A^{μ} for fixed ψ^{α} will lead to the A-field equations :

$$\partial_{\nu} \frac{\partial \mathscr{L}}{\partial(\partial_{\nu} A_{\mu}(x))} - \frac{\partial \mathscr{L}}{\partial A_{\mu}(x)} = 0 \qquad \text{(II. 16)}$$

which have the form :

$$\partial_{\nu} F^{\mu\nu}(x) = - \frac{\partial \mathscr{L}}{\partial A_{\mu}(x)} \qquad \text{(II. 16a)}$$

From the defintion of $F^{\mu\nu}$ it follows that* :

$$\partial^{\nu} F^{\alpha\beta} + \partial^{\beta} F^{\nu\alpha} + \partial^{\alpha} F^{\beta\nu} = 0 \qquad \text{(II. 16b)}$$

The latter, (II. 16a) and (II. 16b) are the Maxwell equations for the electromagnetic field if we define the conserved current :

$$e\, j^{\mu}(x) = - \frac{\partial \mathscr{L}}{\partial A_{\mu}(x)} \qquad \text{(II. 16c)}$$

$$\partial_{\mu} j^{\mu}(x) = 0$$

In view of equations (II. 9) and (II. 14) we have :

$$j^{\mu}(x) = i \left\{ \psi^{\alpha +}(x) \frac{\partial L}{\partial (D_{\mu} \psi^{\alpha})^{+}} - \frac{\partial L}{\partial (D_{\mu} \psi^{\alpha})} \psi^{\alpha}(x) \right\} \qquad \text{(II. 16d)}$$

Equations (II. 16) can be written

$$\partial_{\nu} \frac{\partial \mathscr{L}}{\partial F_{\mu\nu}(x)} = \frac{\partial \mathscr{L}}{\partial A_{\mu}(x)} \qquad \text{(II. 16e)}$$

---

* If $\tilde{F}_{\mu\nu}(x) = \frac{1}{2} \varepsilon_{\mu\nu\alpha\beta} F^{\alpha\beta}(x)$ is the dual of $F^{\alpha\beta}(x)$, equation (II. 16b) becomes :

$$\partial_{\nu} \tilde{F}^{\mu\nu}(x) = 0$$

The second-hand side is null since there is no conserved axial vector electromagnetic current for massive spinor fields.

which shows that the current is gauge-invariant

$$e\, j^{\mu}(x) = - \partial_{\nu} \frac{\partial \mathscr{L}}{\partial F_{\mu\nu}(x)} \qquad \text{(II. 16f)}$$

Note that the following identity holds

$$\partial^{\mu}(f^{+}(x)\, g(x)) = (D^{\mu *} f^{+}(x))\, g(x) + f^{+}(x)(D^{\mu}\, g(x)) \qquad \text{(II. 16g)}$$

which is helpful to show immediately the conservation of the current from its expression (II. 16d). Indeed the invariance of the lagrangean under the gauge transformations

$$\psi'^{\alpha}(x) = e^{ie\Lambda(x)}\, \psi^{\alpha}(x)$$

$$D^{\mu\prime}\, \psi'^{\alpha}(x) = e^{ie\Lambda(x)}\, D^{\mu}\, \psi^{\alpha}(x)$$

means that

$$\frac{\partial\, \delta \mathscr{L}}{\partial\, \Lambda(x)} = ie \left\{ \frac{\partial L}{\partial \psi^{\alpha}} \psi^{\alpha} + \frac{\partial L}{\partial(D^{\mu} \psi^{\alpha})} (D^{\mu} \psi^{\alpha}) - \right. \qquad \text{(II. 16h)}$$

$$\left. - \psi^{\alpha +} \frac{\partial L}{\partial \psi^{\alpha +}} - (D^{\mu} \psi^{\alpha})^{+} \frac{\partial L}{\partial(D^{\mu} \psi^{\alpha})^{+}} \right\} = 0$$

The equations (II. 15), (II. 15a) describe the evolution of the field ψ^{α} in interaction with the electromagnetic field $A^{\mu}(x)$.

In view of the transformations (II. 11) and of the gauge invariance, if $A^{\mu}(x)$ is a solution of these equations, then an infinite set of solutions will be obtained by formula (II. 11), depending on an arbitrary function $\Lambda(x)$.

In order to obtain a definite solution we may make a particular choice of the gauge function and thus eliminate (some of) the ambiguity in A^{μ}. This can best be done by imposing the Lorentz gauge condition :

$$\partial_{\mu}\, A^{\mu}(x) = 0 \qquad \text{(II. 17)}$$

If a given solution $\mathscr{a}^{\mu}(x)$ does not satisfy this condition, then with $\mathscr{a}^{\mu}$ we construct another solution

$$A^{\mu}(x) = \mathscr{a}^{\mu}(x) + \partial^{\mu}\, \Lambda(x)$$

and choose $\Lambda(x)$ so that $A^\mu(x)$ will satisfy equation (II. 17), that is $\Lambda(x)$ must be a solution of the equation :

$$\Box \Lambda(x) = - \partial_\mu \, a^\mu(x)$$

We may, for instance, choose a retarded solution of this equation :

$$\Lambda(x) = \int \mathcal{G}_{ret}(x - y) \, (- \partial_\mu \, a^\mu(y)) \, d^4y$$

where

$$\Box \, \mathcal{G}_{ret}(x - y) = \delta^4(x - y)$$

$$\mathcal{G}_{ret}(x - y) = 0 \quad \text{for} \quad x^0 < y^0$$

Physically a subsidiary condition is required by the fact that a free electromagnetic field as described by $A^\mu(x)$, contains four components and therefore in the quantized theory $A^\mu(x)$ would describe transverse photons (components A_1, A_2), longitudinal photons (component A_3) and time-like photons (component A_0). Only transverse photons are observed so the spurious components A_3, and A_0 must give no contribution to physical observations.

II. 3 - MAXWELL'S EQUATIONS AND THE PHOTON PROPAGATOR. GAUGE FIXING CONDITIONS

Maxwell's equations (II. 16a), (II. 16b) may be written in virtue of the definition of $F^{\mu\nu}$, equation (II. 13) :

$$\Box A^\mu(x) - \partial^\mu(\partial_\alpha A^\alpha(x)) = j^\mu(x) \qquad \text{(II. 18)}$$

Now this equation admits of no Green's function. Indeed a Green's function of equation (II. 18)

$$\left\{ \Box \, g^{\mu\nu} - \partial^\mu \partial^\nu \right\} A_\nu(x) = j^\mu(x)$$

would satisfy the equation

$$\left\{ \Box \, g^{\mu\nu} - \partial^\mu \partial^\nu \right\} \mathcal{G}_{\nu\lambda}(x - x') = \delta^\mu{}_\lambda \, \delta^4(x - x')$$

In Fourier representation

$$\mathcal{G}_{\nu\lambda}(x - x') = \int \frac{d^4k}{(2\pi)^4} \mathcal{G}_{\nu\lambda}(k)\, e^{-ik(x - x')}$$

$$\delta^4(x - x') = \int \frac{d^4k}{(2\pi)^4} e^{-ik(x - x')}$$

we have, from the latter equation :

$$\left\{ - k^2 g^{\mu\nu} + k^\mu k^\nu \right\} \mathcal{G}_{\nu\lambda}(k) = \delta^\mu{}_\lambda \tag{II. 19}$$

Now the function $\mathcal{G}_{\nu\lambda}(k)$ does not exist. For it would have to be of the form :

$$\mathcal{G}_{\nu\lambda}(k) = A(k)\, k^2\, g_{\nu\lambda} + B(k)\, k_\nu\, k_\lambda \tag{II. 19a}$$

with $A(k)$ and $B(k)$ functions of k^2. But then this would give :

$$\left\{ - k^2 g^{\mu\nu} + k^\mu k^\nu \right\} \mathcal{G}_{\nu\lambda}(k) =$$

$$= A(k)\, k^2 \left\{ - k^2 \delta^\mu{}_\lambda + k^\mu k_\lambda \right\}$$

$$\neq \delta^\mu{}_\lambda$$

no function $A(k)$ and $B(k)$ exists that satisfy equations (II. 19) and (II. 19a).

We therefore need to impose a condition on the field A_μ which amounts to fixing the gauge, for obtaining a solution of Maxwell's equations. The choice of gauge breaks the initial gauge invariance. Lorentz condition still maintains gauge invariance for those functions $\Lambda(x)$ which satisfy the wave equation :

$$\Box\, \Lambda(x) = 0 \tag{II. 19b}$$

Other gauge choices are possible such as the Coulomb gauge :

$$\vec{\nabla} \cdot \vec{A} = 0 \qquad \text{(II. 17a)}$$

If one imposes the subsidiary condition (II. 17) then equation (II. 16a), (II. 16b) are equivalent to

$$\Box \; A^{\mu}(x) = j^{\nu}(x), \qquad \partial_{\alpha} A^{\alpha}(x) = 0 \qquad \text{(II. 20)}$$

The Green's function of this equation is therefore :

$$\mathcal{G}_{\mu\nu}(k) = - \frac{g_{\mu\nu}}{k^2} \equiv g_{\mu\nu} \; D_F(k) \qquad \text{(II. 21)}$$

where the denominator means $k^2 + i\varepsilon$. This is the Feynman propagator for the photon field.

An alternative way to obtain the equations (II. 20), (II. 21) is to include a term :

$$- \frac{1}{2a} \left(\partial^{\lambda} A_{\lambda}(x)\right)^2 \qquad \text{(II. 22)}$$

in the lagrangean (II. 14) and use the method of the Lagrange multipliers :

$$\mathcal{L}' = - \frac{1}{4} F^{\mu\nu} F_{\mu\nu} + L(\psi^{\alpha}, \psi^{\alpha +}, D^{\mu} \psi^{\alpha}, (D^{\mu} \psi^{\alpha})^{+}) - \frac{1}{2a} (\partial_{\lambda} A^{\lambda})^2 \qquad \text{(II. 23)}$$

The variation of $\mathcal{L}'$ with fixed a gives the equation :

$$\left\{ \Box \, g^{\mu\nu} + \left(\frac{1}{a} - 1\right) \partial^{\mu} \partial^{\nu} \right\} A_{\nu}(x) = j^{\mu}(x) \qquad \text{(II. 23a)}$$

of which the Green's function is such that

$$- \left\{ k^2 g^{\mu\nu} + \left(\frac{1}{a} - 1\right) k^{\mu} k^{\nu} \right\} \mathcal{G}'_{\nu\lambda}(k) = \delta^{\mu}{}_{\lambda} \qquad \text{(II. 23b)}$$

and is therefore :

$$\mathcal{G}'_{\mu\nu}(k) = - \frac{1}{k^2} \left\{ g_{\mu\nu} - (1 - a) \frac{k_{\mu} k_{\nu}}{k^2} \right\} \qquad \text{(II. 23c)}$$

For a = 1 one obtains the previous result. The case lim a = 0 gives the Landau propagator ; the physical results must, however, be independent from the gauge choice, i e, from a. Note that the Landau propagator satisfies the equation

$$k^{\mu}\mathscr{G}'_{\mu\nu}(k) = 0$$

similar to the Lorentz condition for the field A_{μ}.

II. 4 - THE ENERGY-MOMENTUM TENSOR OF FIELDS IN INTERACTION WITH THE ELECTROMAGNETIC FIELD

The energy-momentum tensor of the matter fields ψ^{α} in the presence of an electromagnetic field is defined by

$$T_m^{\mu\nu} = \frac{\partial\mathscr{L}}{\partial(D_{\mu}\psi^{\alpha})} D^{\nu}\psi^{\alpha} + \text{h.c.} - \mathscr{L}g^{\mu\nu} \qquad \text{(II. 24)}$$

as a generalization of equation (I. 32).

The divergence of $T_m^{\mu\nu}$ is :

$$\partial_{\mu} T_m^{\mu\nu} = \partial_{\mu}\left(\frac{\partial\mathscr{L}}{\partial(D_{\mu}\psi^{\alpha})} D^{\nu}\psi^{\alpha}\right) + \text{h.c.} - \partial^{\nu}\mathscr{L}$$

If one takes account of the identity (II. 16g), of the field equations (II. 15), (II. 15a) and of the equation (II. 16h) one obtains :

$$\partial_{\mu} T_m^{\mu\nu} = \frac{\partial\mathscr{L}}{\partial(D^{\mu}\psi^{\alpha})}\left[D^{\mu}, D^{\nu}\right]\psi^{\alpha} + $$

$$+ \left[D^{\mu *}, D^{\nu *}\right]\psi^{\alpha +}\frac{\partial\mathscr{L}}{\partial(D^{\mu}\psi^{\alpha})^{+}}$$

Now

$$\left[D^{\mu}, D^{\nu}\right]\psi^{\alpha} = ie\, F^{\nu\mu}\psi^{\alpha} \qquad \text{(II. 25)}$$

So, in view of the equation (II. 16d) :

$$\partial_\mu T_m^{\mu\nu} = e F^{\mu\nu} j_\mu(\tilde{x}) \qquad \text{(II. 26)}$$

The second hand-side is the Lorentz force. One sees easily that, for instance :

$$\frac{d}{dx^0} \vec{P} = \rho \vec{E} + \left[\vec{j} \times \vec{B} \right]$$

where $\vec{P}$ is given by (I. 57) and

$$\rho = j^0(x) \quad E^k = F^{0k}(x), \qquad B^k = \varepsilon_{k\ell m}(\partial_\ell A^m(x) - \partial_m A^\ell(x))$$

Note that the total energy-momentum tensor is in fact :

$$T^{\mu\nu}_{\text{total}} = \frac{\partial \mathscr{L}}{\partial(D_\mu \psi^\alpha)} (D^\nu \psi^\alpha) + \text{h.c.} + \frac{\partial \mathscr{L}}{\partial F_{\alpha\mu}} F^{\beta\nu} g_{\alpha\beta} - \mathscr{L} g^{\mu\nu}$$

(II. 27)

which gives for the electromagnetic part

$$T^{\mu\nu}_\gamma = - F^{\alpha\mu} F^{\beta\nu} g_{\alpha\beta} + \frac{1}{4} (F^{\alpha\beta} F_{\alpha\beta}) g^{\mu\nu} \qquad \text{(II. 28)}$$

whence :

$$\partial_\mu T^{\mu\nu}_\gamma = - e F^{\alpha\nu} j_\alpha$$

a result which follows from the equations (II. 16a), (II. 16b) and (II. 16c).

Therefore :

$$\partial_\mu T^{\mu\nu}_{\text{total}} = 0 \qquad \text{(II. 29)}$$

II. 5 - NON-INTEGRABLE PHASE FACTOR AND THE INTEGRAL FORMULATION OF GAUGE FIELD THEORIES

In the § II. 2 the notion of electromagnetic field was presented as a vector gauge field necessary to the invariance of the equations of motion of complex fields under the local gauge group U(1). Another formulation of gauge field theory, mathematically based on the so-called fiber-bundle theory, has been developed by Wu and Yang, which sheds a deeper light on the mathematical nature of gauge field theories.

As it follows from the gauge transformation equation for the electromagnetic potential field $A_\mu(x)$, (II. 11), the electromagnetic field tensor $F_{\mu\nu}$ is gauge invariant :

$$F'_{\mu\nu} = F_{\mu\nu}$$

This fact has led to the traditional conception that the field tensor $F_{\mu\nu}$ determines all electromagnetic effects, their descripton by the vector field A_μ being a convenient auxiliary method for determining $F_{\mu\nu}$.

In 1959, Aharonov and Bohm proposed an experiment, which was carried out, and which definitely shows that if $F_{\mu\nu}$ determines all electromagnetic phenomena in classical theory, this is not true whenever the electrons (or any other charged particles) are in conditions such as to present quantum effects and are therefore described by quantum mechanics.

The principle of the Aharonov - Bohm experiment is indicated in Fig. II. 1, electrons are incident from the left towards a cylindrical region from which they are excluded.

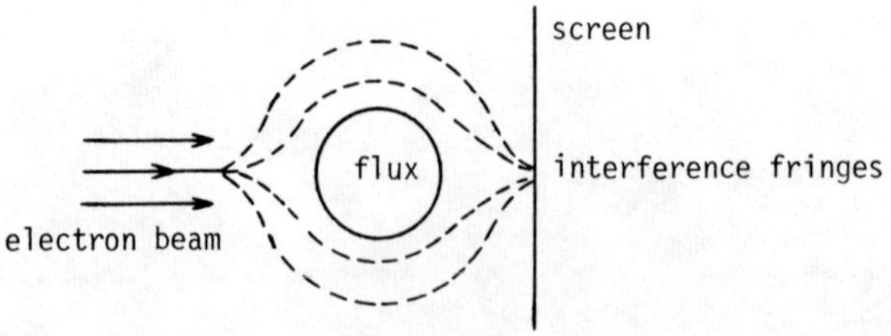

Figure II. 1

Inside the cylinder one may set up a magnetic flux which is confined to this region, the fields vanishing outside the cylinder. Experiment shows that the interference fringes produced by the electrons on a screen (due to the fact that each electon may follow a path above or below the cylinder) when there is no magnetic flux, are shifted when a magnetic flux is set up in the cylinder, the electric and magnetic field being kept zero outside the cylinder. Aharonov and Bohm predicted therefore that there may exist electromagnetic action on electrons in region where $F_{\mu\nu} = 0$

Wu and Yang have shown that if the amplitude for the experiment with no magnetic flux in the cylinder is :

$$S = f^+ + f^-$$

where f^+ (f^-) is the amplitude when the lower (upper) paths are blocked, then the amplitude when there is magnetic flux will be of the form :

$$S' = f^+ + f^- \, e^{ie\oint A_\mu dx^\mu} \qquad \text{(II. 30)}$$

where the integration is over the closed loop on the cylinder wall. Thus, the flux of the magnetic field through a surface which cuts the cylinder determines the interference effects.

The relative phase difference between f^+ and f^- in (II. 30) namely :

$$e^{ie\oint A_\mu dx^\mu} \qquad \text{(II. 31)}$$

determines the phase shift in the interference fringes. This effect is, however, gauge-invariant if the gauge phase factor is a simple-valued function. Thus, if two fields $A_\mu^{\ a}$ and $A_\mu^{\ b}$ are gauge-transformed from one another :

$$A_\mu^{\ b} = A_\mu^{\ a} - \partial_\mu \Lambda(x)$$

they give the same interference effect on the screen provided that the function :

$$e^{i e \Lambda(x)}$$

be single-valued. Indeed :

$$e^{i e \oint A_\mu{}^a \, dx^\mu} = e^{i e \oint A_\mu{}^b \, dx^\mu}$$

if

$$e^{i e \oint \partial_\mu \Lambda(x) \, dx^\mu} = 1$$

and so, if :

$$\Lambda(x)\Big]_0^{2\pi} = 2\pi n$$

(the gauge function itself need not be single valued).

The fundamental concept of a non- integrable phase factor

$$\Phi_{QP} = e^{i e \int_P^Q A_\mu \, dx^\mu} \qquad \text{(II . 32)}$$

is thus introduced, where the integration is over any path from point P to point Q, and electromagnetism, in the words of Wu and Yang, is the gauge-invariant manifestation of this phase factor, if the physical observables are independent from an arbitrary gauge transformation :

$$e^{i e \int_P^Q A_\mu \, dx^\mu} \longrightarrow e^{-i e \Lambda(Q)} \; e^{i e \int_Q^P A_\mu \, dx^\mu} \; e^{i e \Lambda(P)}$$

The gauge-transformed of the phase-factor (II. 32) contains the gauge function $\Lambda(x)$ only at the end points of the path. For a closed loop,

$$\Lambda(P) = \Lambda(Q)$$

and the phase factor is invariant.

Of course, the consideration of this phase factor raises the problem of the analytic behaviour of the field $A_\mu(x)$ in space-time. If $A_\mu(x)$ has singularities along some paths the non-integrable phase factor is not defined through these paths and conditions have to be found to define the integral in these cases.

A notable example is that of Dirac's magnetic monopole, namely a magnetic field $\vec{B}(x)$, solution of Maxwell's equation over all space excluding the origin and which has the form

$$\vec{B}(x) = \frac{g}{r^2} \frac{\vec{r}}{r} \qquad \text{(II. 33)}$$

where $r = |\vec{x}|$. This represents a monopole of strength (magnetic charge) g located at the origin of the coordinates. The region of this solution excludes the origin in three-dimensional space or the world line of the monopole in space-time.

As in the region of definition we have $\vec{\nabla}\cdot\vec{B} = 0$, a vector potential $\vec{A}(x)$ may be found such that

$$\vec{B} = \vec{\nabla} \times \vec{A}$$

One possible choice for $\vec{A}$ is, in polar coordinates :

$$A^I_\theta = A^I_r = 0, \qquad A^I_\varphi = \frac{g}{r \sin\theta} (1 - \cos\theta) \qquad \text{(II. 34)}$$

and another one is :

$$A^{II}_{\theta} = A^{II}_{r} = 0, \qquad A^{II}_{\varphi} = -\frac{g}{r \sin\theta}(1 + \cos\theta) \qquad \text{(II. 35)}$$

so that :

$$B_r = (\vec{\nabla} \times \vec{A}^{I,II})_r \equiv \frac{1}{r \sin\theta}\left(\frac{\partial}{\partial\theta}(\sin\theta\, A^{I,II}_{\varphi}) - \frac{\partial A^{I,II}_{\theta}}{\partial\varphi}\right) = \frac{g}{r^2}$$

$$B_\theta = (\vec{\nabla} \times \vec{A}^{I,II})_\theta \equiv \frac{1}{r \sin\theta}\left(\frac{\partial A^{I,II}_{r}}{\partial\varphi} - \sin\theta\, \frac{\partial(r A_\varphi)}{\partial r}\right) = 0$$

$$B_\varphi = (\vec{\nabla} \times \vec{A}^{I,II})_\varphi \equiv \frac{1}{r}\left(\frac{\partial(r A^{I,II}_{\theta})}{\partial r} - \frac{\partial A^{I,II}_{r}}{\partial\theta}\right) = 0$$

The first solution (II. 34) is singular only at $\theta = \pi$ whereas the second solution (II. 35) is singular only at $\theta = 0$. Both types of singularities may be called strings of singularities. Therefore the region of definition of the monopole solutions will be the intersection of two pieces of space, one, R_1, which excludes $\theta = \pi$ (the lower z half-axis) the other, R_2, which excludes $\theta = 0$ (the upper z half-axis) : $R_1 \cap R_2$.

If we take a sphere with center at the origin and consider a parallel over its upper surface defined by r, θ, φ the integral of $\vec{A}$ along this parallel will give the magnetic flux through the upper cap of the sphere since

$$\int_{\text{upper cap}} \vec{ds} \cdot \vec{B} = \oint_{\text{parallel}} \vec{A}^{I} \cdot \vec{dr} = \oint d\varphi\, r \sin\theta\, A^{I}_{\varphi} = 2\pi g\,(1 - \cos\theta) \qquad \text{(II. 36)}$$

This is correct since it vanishes when the parallel shrinks to a point, $\theta \to 0$.

If we had not excluded the lower half-axis from the region of definition, in the limit $\theta \to \pi$ the parallel would tend to a point, the south pole, the integral along the parallel would tend to zero but the formula (II. 36) would give $4 \pi g$ at this limit.

The flux through the lower cap cannot be calculated with $\vec{A}^I$ since it would be, for this solution :

$$\int_{\text{lower cap}} \vec{ds} \cdot \vec{B} = - \oint d \varphi \;\; r \sin \theta \;\; A^I_{\varphi} = - 2 \pi g (1 - \cos \theta)$$

and this does not tend to zero as $\theta \to \pi$.

The correct flux is obtained with the solution (II. 35) :

$$\int_{\text{lower cap}} \vec{ds} \cdot \vec{B} = - \int d \varphi \;\; r \sin \theta \;\; A^{II}_{\varphi} = 2 \pi g (1 + \cos \theta) \qquad \text{(II. 37)}$$

which has the correct behaviour at $\theta \to \pi$; the total flux through the upper and lower caps will be $4 \pi g$.

If we consider the non-integrable phase factor between two points P and Q over the sherical surface belonging to the region $R_1 \cap R_2$ then we have two equivalent descriptions corresponding to the two solutions (II. 34), (II. 35) :

$$\Phi^I_{PQ} = e^{\, i e \int_P^Q A^I_{\mu} \, dx^{\mu}}$$

$$\Phi^{II}_{PQ} = e^{\, i e \int_P^Q A^{II}_{\mu} \, dx^{\mu}}$$

If one compares these two solutions, (II. 34), (II. 35) one sees that :

$$A^{I}_{\varphi} = A^{II}_{\varphi} + \frac{2\,g}{r \sin \theta}$$

therefore

$$\int \vec{A}^{I} \cdot \vec{dr} = \int \vec{A}^{II} \cdot \vec{dr} + \int \vec{\nabla} \Lambda \cdot \vec{dr}$$

that is, for P and Q on a parallel at r, θ, φ :

$$\int_P^Q r \sin \theta \, d\varphi \, A^{I}_{\varphi} = \int_P^Q r \sin \theta \, d\varphi \, A^{II}_{\varphi} + 2\,g \int_P^Q d\varphi$$

we see that :

$$\Phi^{I}_{PQ} = e^{2\,i\,e\,g\,\varphi\,(Q)} \, \Phi^{II}_{PQ} \, e^{-\,2\,i\,g\,\varphi\,(P)} \qquad \text{(II. 38)}$$

where $S = e^{2\,i\,e\,g\,\varphi}$ is the gauge transformation on the phase factor. If we take the closed path from $\varphi(P) = 0$ to $\varphi(Q) = 2\pi$ we shall have

$$\Phi^{I}_{PQP} = \Phi^{II}_{PQP}$$

if the following condition is satisfied

$$4\,\pi\,e\,g = 2\,\pi\,n$$

with n an integer.

This is the Dirac's quantization of the magnetic strength g :

$$g = \frac{n}{2e} \qquad \text{(II. 38a)}$$

The gauge transformation is therefore single-valued and expressed by :

$$S = e^{i n \varphi}$$

What we did was thus : 1) divide the space into two overlapping regions R_1 and R_2 ; 2) define A^I_μ and A^{II}_μ, each singularity free in R_1 and R_2 respectively, and such that 3) $F_{\mu\nu} = \partial_\nu A^I_\mu - \partial_\mu A^I_\nu =$ $= \partial_\nu A^{II}_\mu - \partial_\mu A^{II}_\nu$ and 4) A^I_μ and A^{II}_μ differ by a gauge transformation. In order to define the phase factor associated to a path we have to know if the path is entirely within one of the regions, R_a, say, then A^a_μ will be inserted in the integrand of the phase factor. If the path is enterely inside the overlapping region there will be two possible phase factors related by a gauge transformation such as (II. 38). For a path that criss-crosses in and out of the overlapping region a phase factor can be defined by considering the phase factor for each path in each region and the gauge transformation which leads from the final point in the first region to the first point in the second region, and by multiplying them together.

This conception of gauge fields and transformations has been developed by Yang and co-workers, and extended to the non-abelian gauge field theories. They showed that the mathematical basis of this integral approach to gauge theories is the theory of fiber bundles. The topopogical nature of the theory is thus revealed ; the differential approach, although useful in computation, did not bring to surface this mathematical inner property of gauge field theories [102,103,37,38].

A final remark is the following. If in classical physics the action integral corresponding to the interaction between a point electron with charge e_p and an electromagnetic field is :

$$- e_p \int A_\mu (Z(s)) \frac{dZ^\mu}{ds} ds \qquad \text{(II. 39)}$$

where s is the proper-time, then the coupling constant e in the phase factor (II. 31) is related to e_p by

$$e = \frac{e_p}{\hbar} \tag{II. 40}$$

That e_p is the particle charge results from the fact that Maxwell's equations follow from an action principle containing (II. 39) and from Gauss theorem and Poisson's equation. The relation (II. 40) follows from the rule that D_μ (which contains e and not e_p) must replace ∂_μ in the quantum-particle equations like Dirac's. Planck's constant appears therefore in the quantisation equation (II. 38a) expressed in terms of the particle charge e_p.

PROBLEMS

II - 1. From the action for a system of classical particle and electromagnetic field (we re-establish the velocity of light c) :

$$S = - m_0\, c \int ds - \frac{1}{4c} \int d^4x\, F^{\mu\nu}(x)\, F_{\mu\nu}(x) - \frac{e}{c} \int j^\mu(x)\, A_\mu(x)\, d^4x$$

where the current is :

$$j^\mu(x) = \int \frac{dZ^\mu}{ds}\, \delta^4(x - Z\,(s))\, ds$$

deduce a) the classical equations of motion of the particle by variation of the path $Z^\mu(s)$; b) the classical field equations by variation of the field ; c) given the energy-momentum tensor of a classical particle :

$$T^{\mu\nu}{}_m = m_0\, c^2 \int \frac{dZ^\mu}{ds} \frac{dZ^\nu}{ds}\, \delta^4\, (x - Z(s))\, ds$$

show that

$$\partial_\nu(T^{\mu\nu}{}_m + T^{\mu\nu}{}_{field}) = 0$$

d) with the energy-momentum tensor $T^{\mu\nu}_{\ \ m}$ above calculate the total orbital angular momentum of a classical particle.

II - 2. a) Write Maxwell's equations

$$\partial_\nu F^{\mu\nu} = e\, j^\mu$$

$$\partial_\nu \tilde{F}^{\mu\nu} = 0$$

$$\tilde{F}^{\mu\nu} = \frac{1}{2}\,\varepsilon^{\mu\nu\alpha\beta}\,F_{\alpha\beta}$$

in terms of the electric field $E^k = F^{0k}$, the magnetic field $B^k = \frac{1}{2}\,\varepsilon_{k\ell n}\,F^{\ell n}$ and the charge and current densities, $\rho(x) = j^0(x)$ and $\vec{j}(x)$.

b) Which substitution between the fields $\vec{E}$ and $\vec{B}$ leaves the free field equations invariant ?

c) In the presence of matter, a magnetic monopole current would be needed for an extension of the above symmetry to the field equations. What is the transformation law of the magnetic current under improper Lorentz transformations ?

II - 3. Discuss the electromagnetic field equations in a two dimensional Minkowski space, in interaction with a spinor field.

II - 4. a) Give the field equations and the hamiltonian of an electromagnetic field in interaction with a Dirac field in the Coulomb gauge.

b) Show that the Coulomb interaction energy between the charges is contained in the field energy.

CHAPTER III

Examples of Electrodynamical Systems

III. 1 - SCALAR ELECTRODYNAMICS

This is the case of a scalar complex field $\varphi(x)$, $\varphi^+(x)$ with mass m and which may have a self-interaction term, with lagrangean given in equation (II. 1).

The electromagnetic gauge-invariant lagrangean for this field is, according to equation (II. 14) :

$$\mathscr{L} = -\frac{1}{4} F^{\mu\nu} F_{\mu\nu} + (D^\mu \varphi)^+ (D_\mu \varphi) - m^2 \varphi^+ \varphi - \frac{\eta}{2} (\varphi^+ \varphi)^2 \qquad \text{(III. 1)}$$

The interaction lagrangean between the scalar field and the electromagnetic field is :

$$\mathscr{L}_{\varphi\gamma} = -i\, e \left\{ \varphi^+ \partial^\mu \varphi - (\partial^\mu \varphi^+)\varphi \right\} A_\mu + e^2 (A^\alpha A_\alpha) \varphi^+ \varphi$$

which is added to the φ-field lagrangean

$$\mathscr{L}_\varphi = (\partial^\mu \varphi)^+ (\partial_\mu \varphi) - m^2 \varphi^+ \varphi - \frac{\eta}{2} (\varphi^+ \varphi)^2$$

and the γ-lagrangean :

$$\mathscr{L}_\gamma = -\frac{1}{4} F^{\mu\nu} F_{\mu\nu}$$

The reader will show that under the transformations (II. 7), (II. 10, (II. 11) one will have

$$\mathscr{L} = \mathscr{L}_\gamma + \mathscr{L}_\varphi + \mathscr{L}_{\varphi\gamma} = \mathscr{L}_\gamma + \mathscr{L}'_\varphi + \mathscr{L}'_{\varphi\gamma}$$

only the sum $\mathscr{L}_\varphi + \mathscr{L}_{\varphi\gamma}$ is gauge-invariant as is $\mathscr{L}_\gamma$ alone.

The current is :

$$j^{\mu}(x) = -\frac{1}{e}\frac{\partial \mathscr{L}}{\partial A_{\mu}(x)} = i\left\{ \varphi^{+}(x)\, D^{\mu}\varphi(x) - (D^{\mu}\varphi(x))^{+}\varphi(x)\right\}$$

$$= i\left\{\varphi^{+}\partial^{\mu}\varphi - (\partial^{\mu}\varphi)^{+}\varphi\right\} - 2\,e\,A^{\mu}\,\varphi^{+}\varphi \qquad \text{(III. 2)}$$

and the field equations :

$$\partial_{\nu} F^{\mu\nu}(x) = e\, j^{\mu}(x)$$
$$\left\{(D^{\mu} D_{\mu} + m^{2} + \frac{\eta}{2}(\varphi^{+}\varphi)\right\}\varphi(x) = 0 \qquad \text{(III. 3)}$$

The energy-momentum tensor of the φ-field is :

$$T_{\varphi}^{\ \mu\nu} = (D^{\mu}\varphi)^{+}(D^{\nu}\varphi) + (D^{\nu}\varphi)^{+}(D^{\mu}\varphi) -$$
$$- g^{\mu\nu}\left\{(D^{\alpha}\varphi)^{+}(D_{\alpha}\varphi) - m^{2}\varphi^{+}\varphi - \frac{\eta}{2}(\varphi^{+}\varphi)^{2}\right\} \qquad \text{(III. 4)}$$

III. 2 - PROCA VECTOR FIELD ELECTRODYNAMICS

The lagrangean for a free complex Proca vector field φ^{μ}, $\varphi^{\mu+}$ with mass m is :

$$L = \frac{1}{2}\mathscr{G}_{o}^{\mu\nu+}\mathscr{G}_{o\mu\nu} - \frac{1}{2}\mathscr{G}_{o}^{\mu\nu+}(\partial_{\nu}\varphi_{\mu} - \partial_{\mu}\varphi_{\nu})$$
$$- \frac{1}{2}(\partial^{\nu}\varphi^{\mu+} - \partial^{\mu}\varphi^{\nu+})\,\mathscr{G}_{o\mu\nu} + m^{2}\varphi^{\mu+}\varphi_{\mu} \qquad \text{(III. 5)}$$

which gives the field equations

$$\partial_{\nu}\mathscr{G}_{o}^{\mu\nu} + m^{2}\varphi^{\mu} = 0 \qquad \text{(III. 5a)}$$

with

$$\mathcal{G}_o^{\mu\nu} = \partial^\nu \varphi^\mu - \partial^\mu \varphi^\nu \tag{III. 5b}$$

We define the gauge-covariant field

$$\mathcal{G}^{\mu\nu} = D^\nu \varphi^\mu - D^\mu \varphi^\nu \tag{III. 6}$$

where D^μ is given by equation (II. 9). This means that

$$\mathcal{G}^{\mu\nu} = \mathcal{G}_o^{\mu\nu} + i\, e\, (A^\nu \varphi^\mu - A^\mu \varphi^\nu) \tag{III. 6a}$$

The gauge invariant lagrangean is therefore :

$$\mathcal{L} = -\frac{1}{4} F^{\alpha\beta} F_{\alpha\beta} + \frac{1}{2} \mathcal{G}^{\mu\nu+} \mathcal{G}_{\mu\nu} - \frac{1}{2} \mathcal{G}^{\mu\nu+} (D_\nu \varphi_\mu - D_\mu \varphi_\nu)$$

$$- \frac{1}{2} (D^{\nu *} \varphi^{\mu +} - D^{\mu *} \varphi^{\nu +}) \mathcal{G}_{\mu\nu} + m^2 \varphi^{\mu +} \varphi_\mu \tag{III. 7}$$

from which we obtain, by varying $\mathcal{G}_{\mu\nu}$, $\mathcal{G}^+_{\mu\nu}$ the equation (III. 6), and the equations of motion, by variation of φ_μ, φ^+_μ :

$$D_\nu \mathcal{G}^{\mu\nu} + m^2 \varphi^\mu = 0 \tag{III. 8}$$

The interaction lagrangean is :

$$L_{\varphi\gamma} = -\frac{ie}{2} \left\{ \mathcal{G}_o^{\mu\nu+} (A_\nu \varphi_\mu - A_\mu \varphi_\nu) - (A_\nu \varphi_\mu^{\ +} - A_\mu \varphi_\nu^{\ +}) \mathcal{G}_o^{\mu\nu} \right\}$$

$$- \frac{e^2}{2} (A^\nu \varphi^{\mu +} - A^\mu \varphi^{\nu +}) (A_\nu \varphi_\mu - A_\mu \varphi_\nu) \tag{III. 9}$$

The current is :

$$j^\mu(x) = i \left\{ \varphi_\alpha^{\ +}(x) \mathcal{G}^{\mu\alpha}(x) - \mathcal{G}^{\mu\alpha +}(x) \varphi_\alpha(x) \right\} \tag{III. 10}$$

Note that Proca's equation for a free field is equivalent to the couple of equations

$$(\Box + m^2)\, \varphi^\mu = 0$$
$$\partial_\mu \varphi^\mu = 0 \qquad \text{(III. 11)}$$

In the presence of an electromagnetic field we deduce from equation (III. 8) by application of the operator D_μ, and from the commutator (II. 13a) :

$$\left[D_\mu , D_\nu \right] = i\, e\, F_{\nu\mu} \qquad \text{(III. 12)}$$

the supplementary condition :

$$D_\mu \varphi^\mu = \frac{i\, e}{2m^2}\, F_{\alpha\beta}\, \mathscr{G}^{\alpha\beta} \qquad \text{(III. 13)}$$

which together with equation (III. 8) replaces the equations (III. 11).

III. 3 - SPINOR FIELD ELECTRODYNAMICS

The gauge-invariant lagrangean for a spinor $\psi(x)$ with mass m is :

$$\mathscr{L} = - \frac{1}{4} F^{\mu\nu} F_{\mu\nu} + \bar{\psi} \left\{ i\, \gamma \cdot D - m \right\} \psi \qquad \text{(III. 14)}$$

from which one deduces the field equations

$$(i\, \gamma^\alpha \cdot D_\alpha - m)\, \psi = 0,$$
$$i\, D^*_\alpha\, \bar{\psi}\, \gamma^\alpha + m\, \bar{\psi} = 0 \qquad \text{(III. 15)}$$
$$\partial_\nu F^{\mu\nu} = e\, j^\mu,$$
$$j^\mu = \bar{\psi}\, \gamma^\mu\, \psi$$

The energy-momentum tensor is obtained from equation (I. 35) with the substitution of the derivatives ∂^α by the covariant derivatives D^α, $D^{\alpha *}$.

III. 4 - SCALAR AND PROCA ELECTRODYNAMICS - ALTERNATIVE FORMULATIONS

Instead of the scalar field $\varphi(x)$ and its hermitian conjugate we shall introduce two equivalent real fields $\varphi_1(x)$, $\varphi_2(x)$ by the equations :

$$\varphi(x) = \frac{1}{\sqrt{2}}(\varphi_1(x) + i\,\varphi_2(x))$$
$$\varphi^+(x) = \frac{1}{\sqrt{2}}(\varphi_1(x) - i\,\varphi_2(x)) \qquad \text{(III. 16)}$$

Clearly the transformed fields by equations (II. 7) will be equivalently expressed in terms of the fields $\varphi_a(x)$ according to :

$$\varphi'_1 = \cos(e\,\Lambda)\,\varphi_1 - \sin(e\,\Lambda)\,\varphi_2$$
$$\varphi'_2 = \sin(e\,\Lambda)\,\varphi_1 + \cos(e\,\Lambda)\,\varphi_2 \qquad \text{(III. 16a)}$$

that is :

$$\begin{pmatrix} \varphi'_1 \\ \varphi'_2 \end{pmatrix} = \begin{pmatrix} \cos(e\,\Lambda) & -\sin(e\,\Lambda) \\ \sin(e\,\Lambda) & \cos(e\,\Lambda) \end{pmatrix} \begin{pmatrix} \varphi_1 \\ \varphi_2 \end{pmatrix} \qquad \text{(III. 16b)}$$

If we consider the matrix :

$$\tau_2 = \begin{pmatrix} 0 & -i \\ i & 0 \end{pmatrix} \qquad \text{(III. 17)}$$

then the above equation may be written in compact form :

$$\phi'(x) = e^{-ie\,\Lambda\,\tau_2}\,\phi(x) \qquad \text{(III. 17a)}$$

where

$$\phi(x) = \begin{pmatrix} \varphi_1(x) \\ \varphi_2(x) \end{pmatrix} \quad , \quad \phi^+(x) = (\varphi_1(x), \varphi_2(x)) \qquad \text{(III. 17b)}$$

and, by series development :

$$e^{-ie\Lambda\tau_2} = \cos(e\Lambda) - i(\sin(e\Lambda))\tau_2 = \begin{pmatrix} \cos(e\Lambda) & -\sin(e\Lambda) \\ \sin(e\Lambda) & \cos(e\Lambda) \end{pmatrix} \qquad \text{(III. 17c)}$$

The application of the covariant derivative to $\varphi(x)$, $\varphi^+(x)$ will lead to

$$D_2{}^\mu \phi(x) = (\partial^\mu - i\, e\, A^\mu \tau_2)\, \phi(x) = \begin{pmatrix} \partial^\mu \varphi_1 - e A^\mu \varphi_2 \\ \partial^\mu \varphi_2 + e A^\mu \varphi_1 \end{pmatrix} \qquad \text{(III. 18)}$$

$$(D_2{}^\mu \phi(x))^+ = \partial^\mu \phi^+ + i\, e\, A^\mu \phi^+(x)\, \tau_2 = (\partial^\mu \varphi_1 - e A^\mu \varphi_2, \partial^\mu \varphi_2 + e A^\mu \varphi_1)$$

The gauge transformation :

$$\phi'(x) = e^{-i\, e\, \Lambda(x)\, \tau_2}\, \phi(x)$$

$$A'^\mu(x) = A^\mu(x) - \partial^\mu \Lambda(x) \qquad \text{(III. 18a)}$$

will lead to

$$(D_2{}^\mu \phi(x))' = e^{-ie\, \Lambda(x)\, \tau_2}\, (D_2{}^\mu \phi(x))$$

$$(D'^\mu_2 \phi'(x))^+ = (D_2{}^\mu \phi(x))\, e^{ie\, \Lambda(x)\, \tau_2} \qquad \text{(III. 18b)}$$

The gauge-invariant lagrangean is now

$$\mathscr{L} = - \frac{1}{4} F^{\mu\nu} F_{\mu\nu} + \frac{1}{2} \left\{ (D_2^{\mu} \phi)^{+} (D_{\mu 2} \phi) - m^2 (\phi^{+} \phi) - - \frac{\eta}{2} (\phi^{+} \phi)^2 \right\} \tag{III. 19}$$

Let us now consider the case of three Proca vector fields, two of them being complex $\varphi^{\mu}(x)$, $\varphi^{\mu +}(x)$ and the third one being real, $\varphi_3^{\mu}(x)$, say. We have then to consider the following transformations :

$$\begin{aligned} \varphi^{\mu}(x) &\rightarrow e^{ie\,\Lambda(x)}\, \varphi^{\mu}(x) \\ \varphi^{\mu +}(x) &\rightarrow e^{-ie\Lambda(x)}\, \varphi^{\mu +}(x) \\ \varphi^{\mu}{}_3(x) &\rightarrow \varphi^{\mu}{}_3(x) \end{aligned} \tag{III. 20}$$

If we introduce two real fields $\varphi^{\mu}{}_1(x)$, $\varphi^{\mu}{}_2(x)$:

$$\begin{aligned} \varphi^{\mu}(x) &= \frac{1}{\sqrt{2}} (\varphi^{\mu}{}_1(x) + i\, \varphi^{\mu}{}_2(x)) \\ \varphi^{\mu +}(x) &= \frac{1}{\sqrt{2}} (\varphi^{\mu}{}_1(x) - i\, \varphi^{\mu}{}_2(x)) \end{aligned} \tag{III. 21}$$

then we may consider the triplet

$$\phi^{\mu}(x) = \begin{pmatrix} \varphi^{\mu}{}_1(x) \\ \varphi^{\mu}{}_2(x) \\ \varphi^{\mu}{}_3(x) \end{pmatrix} \tag{III. 21a}$$

which will transform in the following way :

$$\phi'^{\mu}(x) = \begin{pmatrix} \varphi'^{\mu}_1(x) \\ \varphi'^{\mu}_2(x) \\ \varphi'^{\mu}_3(x) \end{pmatrix} = \begin{pmatrix} \cos(e\,\Lambda) & -\sin(e\,\Lambda) & 0 \\ \sin(e\,\Lambda) & \cos(e\,\Lambda) & 0 \\ 0 & 0 & 1 \end{pmatrix} \begin{pmatrix} \varphi^{\mu}{}_1(x) \\ \varphi^{\mu}{}_2(x) \\ \varphi^{\mu}{}_3(x) \end{pmatrix}$$

$$= \begin{pmatrix} \cos(e\,\Lambda)\,\varphi^{\mu}{}_1 - \sin(e\,\Lambda)\,\varphi^{\mu}{}_2 \\ \sin(e\,\Lambda)\,\varphi^{\mu}{}_1 + \cos(e\,\Lambda)\,\varphi^{\mu}{}_2 \\ \varphi^{\mu}{}_3 \end{pmatrix} \qquad \text{(III. 21b)}$$

Therefore we may write :

$$\phi^{\mu\prime}(x) = e^{-\,ie\,\Lambda(x)\,L_3}\ \phi^{\mu}\,(x) \qquad \text{(III. 21c)}$$

where

$$L_3 = \begin{pmatrix} 0 & -i & 0 \\ i & 0 & 0 \\ 0 & 0 & 0 \end{pmatrix}$$

If then we define the covariant derivative :

$$\begin{aligned} D^{\mu}{}_{(3)}\ \phi^{\nu}(x) &= (\partial^{\mu} - i\ e\ A^{\mu}\ L_3)\ \phi^{\nu}(x) \\ (D^{\mu}{}_{(3)}\phi^{\nu}(x))^{+} &= \partial^{\mu}\ \phi^{\nu +} + i\ e\ A^{\mu}\ \phi^{\nu +}\ L_3 \end{aligned} \qquad \text{(III. 22)}$$

then

$$D'^{\mu}{}_{(3)}\ \phi'^{\nu} = e^{-\,ie\,\Lambda\,L_3}\ D^{\mu}{}_{(3)}\ \phi^{\nu} \qquad \text{(III. 22a)}$$

The lagrangean for this triplet of Proca fields is then :

$$\mathcal{L} = -\frac{1}{4} F^{\mu\nu} F_{\mu\nu} + \frac{1}{4} \mathcal{G}^{\mu\nu+}_{(3)} \quad {}_{(3)\mu\nu} - \frac{1}{4} \mathcal{G}^{\mu\nu+}_{(3)} (D_{(3)\nu} \phi_\mu - D_{(3)\mu} \phi_\nu) - \tag{III. 23}$$

$$- \frac{1}{4} \left[(D^\nu_{(3)} \phi^\mu)^+ - (D^\mu_{(3)} \phi^\nu)^+ \right] \mathcal{G}_{(3)\mu\nu} + \frac{m^2}{2} \phi^{\mu+} \phi_\mu$$

from which one deduces :

$$\mathcal{G}^{\mu\nu}_{(3)} = D^\nu_{(3)} \phi^\mu - D^\mu_{(3)} \phi^\nu \equiv$$

$$\equiv \begin{pmatrix} \partial^\nu \varphi^\mu{}_1 - \partial^\mu \varphi^\nu{}_1 - e(A^\nu \varphi^\mu{}_2 - A^\mu \varphi^\nu{}_2) \\ \partial^\nu \varphi^\mu{}_2 - \partial^\mu \varphi^\nu{}_2 + e(A^\nu \varphi^\mu{}_1 - A^\mu \varphi^\nu{}_1 \\ \partial^\nu \varphi^\mu{}_3 - \partial^\mu \varphi^\nu{}_3 \end{pmatrix} \tag{III. 24}$$

and the equations :

$$D^\nu_{(3)} \quad \mathcal{G}_{(3)\mu\nu} + m^2 \phi^\mu = 0 \tag{III. 24a}$$

These give the set of equations :

$$\begin{pmatrix} D_\nu (D^\nu \varphi^\mu - D^\mu \varphi^\nu) + m^2 \varphi^\mu = 0 \\ D_\nu{*}(D^\nu{*} \varphi^{\mu+} - D^\mu{*} \varphi^{\nu+}) + m^2 \varphi^{\mu+} = 0 \\ \partial_\nu (\partial^\nu \varphi^\mu{}_3 - \partial^\mu \varphi^\nu{}_3) + m^2 \varphi^\mu{}_3 = 0 \end{pmatrix} \tag{III. 24b}$$

for the fields (III. 20). The current in Maxwell's equations

$$\partial_\nu F^{\mu\nu} = e j^\mu$$

is now :

$$j^\mu(x) = \frac{i}{2} \left\{ \phi_\nu{}^+ L_3 \mathcal{G}^{\mu\nu}_{(3)} - \mathcal{G}^{\mu\nu+}_{(3)} L_3 \phi_\nu \right\} \tag{III. 25}$$

PROBLEMS

III - 1. From Proca's equation for a free massive vector field $\phi^\mu(x)$

$$\partial_\nu \mathscr{G}^{\mu\nu}{}_o(x) + m^2 \phi^\mu(x) = 0$$

$$\mathscr{G}^{\mu\nu}{}_o = \partial^\nu \phi^\mu - \partial^\mu \phi^\nu$$

it follows that :

$$\partial_\mu \phi^\mu = 0$$

so that :

$$(\Box + m^2) \phi^\mu = 0$$

What are the corresponding equations when this field interacts with an electromagnetic field ?

III - 2. From the lagrangean :

$$L = -\frac{1}{4} F^{\mu\nu} F_{\mu\nu} + \bar{\psi}_\lambda (i \gamma^\alpha D_\alpha - m) \psi^\lambda - \frac{i}{3} \psi_\lambda (\gamma^\lambda D^\nu + \gamma^\nu D^\lambda) \psi_\nu +$$

$$+ \frac{1}{3} \bar{\psi}_\lambda \gamma^\lambda (i \gamma^\alpha D_\alpha + m) \gamma^\nu \psi_\nu$$

a) deduce the equations of motion for the interacting electromagnetic and Rarita-Schwinger fields ; b) obtain the subsidiary conditions ; c) show that the current is a power series in $F^{\alpha\beta}$ (the corresponding theory is non-renormalizable).

III - 3. Given a 2 x 2 unitary, antisymmetric matrix C such that :

$${}^t\vec{\sigma} = - C^{-1} \vec{\sigma} C$$

a) show that the basis in the space of 2 x 2 matrices can be chosen as formed by C and three symmetric matrices ; b) express a symmetric two-component

Weyl spinor :

$$\psi_{\alpha\beta}(x) = \psi_{\beta\alpha}(x), \quad \alpha, \beta = 1,2$$

in terms of a three-dimensional complex vector $\vec{F}(x)$; c) assuming that $\psi_{\alpha\beta}$ satisfies Weyl's equation :

$$- i \, (\vec{\sigma} \cdot \vec{\nabla})_{\alpha\alpha'} \, \psi_{\alpha'\beta} = i \, \partial_0 \, \psi_{\alpha\beta}$$

$$- i \, (\vec{\sigma} \cdot \vec{\nabla})_{\beta\beta'} \, \psi_{\alpha\beta'} = i \, \partial_0 \, \psi_{\alpha\beta}$$

find the equation for $\vec{F}(x)$; d) show that $\vec{F}(x)$ can be expressed as a linear combination of the electric field $\vec{E}(x)$ and the magnetic field $\vec{B}(x)$ so as to give Maxwell's equations :

$$\vec{\nabla} \cdot \vec{E} = 0 \; ; \; \vec{\nabla} \cdot \vec{B} = 0 \; ;$$

$$\vec{\nabla} \times \vec{B} - \partial_0 \, \vec{E} = 0 \; ; \; \vec{\nabla} \times \vec{E} + \partial_0 \, \vec{B} = 0$$

e) how does $\vec{F}(x)$ transform under spatial reflection ?

III - 4. Develop the canonical quantization formalism for

a) a free scalar complex field

b) a free Dirac field

c) the free electromagnetic field in the Coulomb gauge.

(Consult refs. 13-23, particularly Bjorken and Drell, ref. 15)

III - 5. a) Establish the rules for Feynman diagrams in electrodynamics.

b) Indicate the calculation of cross-sections and decay probabilities.

(Consult refs. 13-23 and specially Bjorken and Drell, ref. 15, Chaps. 15-17 ; S.M. Bilenky, Introduction to Feynman diagrams, Pergamon Press, 1974, Chaps. 5-6).

CHAPTER IV

The Yang-Mills Gauge Fields

IV. 1 - THE ISOSPIN CURRENT

Let us consider a set of two complex fields which forms a two-component spinor

$$\psi(x) = \begin{pmatrix} \psi_1(x) \\ \psi_2(x) \end{pmatrix} \qquad \text{(IV. 1)}$$

When this field is submitted to transformations of the SU(2) group in correspondence to a rotation in three-dimensional x-space, this will be the Pauli spinor. As is well-known, the Pauli-Schrödinger equation has a term, the interaction between the spin magnetic moment and a magnetic field :

$$- \frac{e}{2m} \ (\vec{\sigma} . \vec{B})$$

which must leave the equation invariant under the rotations :

$$x'_k = a_{k\ell} \, x_\ell \ , \quad a_{k\ell} \ a_{m\ell} = \delta_{km}$$

$$\psi'(x') = S(a) \ \psi(x)$$

The invariance condition amounts to the equation

$$S^{-1} \ (\vec{\sigma} . \vec{B}'(x')) \ \psi'(x') \ = \ (\vec{\sigma} . \vec{B}(x)) \ \psi(x)$$

that is

$$S^{-1} \ \sigma_k \ a_{k\ell} \ B_\ell(x) \ S \ \psi(x) = \sigma_\ell \ B_\ell \ \psi(x)$$

so that S must satisfy the relation :

$$S^{-1} \ \frac{\sigma_k}{2} \ S = a_{k\ell} \ \frac{\sigma_\ell}{2}$$

For an infinitesimal rotation :

$$a_{k\ell} = \delta_{k\ell} + \eta_{k\ell}$$

this equation is satisfied by

$$S = I + i\,\alpha_k \frac{\sigma_k}{2} \qquad \text{(IV. 2)}$$

where the parameters α_k are (the rotation angles) given by :

$$\alpha_k = \frac{1}{2}\,\varepsilon_{k\ell n}\,\eta_{\ell n}$$

the antisymmetric tensor $\varepsilon_{k\ell n}$ being the structure constants of the SU(2) group :

$$\left[\frac{\sigma_k}{2}, \frac{\sigma_\ell}{2}\right]_- = i\,\varepsilon_{k\ell m}\,\frac{\sigma_m}{2}$$

$$\left[\frac{\sigma_k}{2}, \frac{\sigma_\ell}{2}\right]_+ = \frac{1}{2}\,\delta_{k\ell} \qquad \text{(IV. 2a)}$$

A two-component Pauli spinor is therefore a pair of complex functions (IV. 1) which transform according to the law :

$$\psi'(x') = e^{i\alpha_k \frac{\sigma_k}{2}}\,\psi(x)$$

in correspondence to the rotation of the space coordinates :

$$x'_k = a_{k\ell}\,x_\ell,$$

$$a_{k\ell}\,a_{kn} = \delta_{\ell n}$$

$$\alpha_k = \frac{1}{2}\,\varepsilon_{k\ell n}\,a_{\ell n}$$

We now consider spinors of the form (IV. 1) the components of which describe two particles with (exactly or almost) the same mass and which may be regarded as <u>two different states of the same particle</u>. This assumption implies the introduction of a new degree of freedom described by matrices τ_k which obey the commutation rules (IV. 2a), the representation of the isospin generators in two-dimensional space ; and that the transformation of ψ under S is given by (IV. 2) with σ replaced by τ, the coordinates x being unchanged. The components $\psi_1(x)$, $\psi_2(x)$ may be either scalars in which case we have a description of objects like the K-meson :

$$K(x) = \begin{pmatrix} K^+(x) \\ K^0(x) \end{pmatrix} \qquad \text{(IV. 3)}$$

or spinors in which case we may have objects like the two quarks u and d :

$$q(x) = \begin{pmatrix} u(x) \\ d(x) \end{pmatrix} , \qquad \text{(IV. 3a)}$$

the proton and the neutron which form the nucleon field N(x) :

$$N(x) = \begin{pmatrix} p(x) \\ n(x) \end{pmatrix} \qquad \text{(IV. 3b)}$$

the neutrino and its associated negatively charged lepton (with vanishing mass) :

$$L(x) = \begin{pmatrix} \nu_\ell(x) \\ \ell(x) \end{pmatrix} \qquad \text{(IV. 3c)}$$

This notion of isospin was introduced by Heisenberg to describe the fact that : a) proton and neutron have almost equal mass ; b) the forces between neutrons and protons, electromagnetic forces excluded, do not depend on the charge of these particles. The charge (which is the degree of freedom under consideration) of the objects (IV. 3) and (IV. 3b) is given by :

$$Q = \frac{1}{2}\,(1 + \tau_3)e \qquad \text{(IV. 3d)}$$

For the quarks (IV. 3a) we have :

$$Q_2 = \left[\frac{1}{2}\,(1 + \tau_3) - \frac{1}{3}\right] e \qquad \text{(IV. 3e)}$$

For the leptons (IV. 3c) :

$$Q_\ell = -\frac{1}{2}\,(1 - \tau_3)\,e \qquad \text{(IV. 3f)}$$

The lagrangean for such fields <u>will be invariant</u> under a transformation of the form (IV. 2)

$$\psi'(x) = (I + i\,\alpha_k\,\frac{\tau_k}{2})\,\psi(x) \qquad \text{(IV. 4)}$$

in the case of infinitesimal transformations, or

$$\psi'(x) = e^{\,i\,\vec{\alpha}\cdot\frac{\vec{\tau}}{2}}\;\psi(x) \qquad \text{(IV. 4a)}$$

for the finite case, if <u>the masses of the two fields</u> $\psi_1(x)$, $\psi_2(x)$ <u>are equal</u>. The x-coordinates are unchanged. Now the phase transformations (IV. 4), (IV. 4a) generalize those given in equations (I. 19a), (I. 19). In the previous case, the transformation group had one parameter only, λ, and one generator, the identity, the group U(1). In the present case, the group, SU(2), has three parameters,

$\vec{\alpha}$ and three generators $\frac{\vec{\tau}}{2}$. Therefore, the Noether current which corresponds to the phase change (IV. 4) is :

$$j^{\mu}{}_{k}(x) = i \left\{ \psi^{a+}(x) \frac{\tau_k}{2} \frac{\partial \mathscr{L}}{\partial(\partial_\mu \psi^{a+}(x))} - \right.$$
$$\left. - \frac{\partial \mathscr{L}}{\partial(\partial_\mu \psi^{a}(x))} \frac{\bar{\tau}_k}{2} \psi^a(x) \right\} \qquad \text{(IV. 5)}$$

This is the so-called <u>isospin current</u> of the ψ-field. For the scalar components case (sum over $a = 1,2$) :

$$\mathscr{L} = \partial^\mu \varphi_a{}^+ \partial_\mu \varphi_a - m^2 \varphi_a{}^+ \varphi_a$$

we have for its isospinor current :

$$j^{\mu}{}_{k}(x) = i \left\{ \varphi^+(x) \frac{\tau_k}{2} \partial^\mu \varphi(x) - \partial^\mu \varphi^+(x) \frac{\tau_k}{2} \varphi(x) \right\} \qquad \text{(IV. 5a)}$$

where
$$\varphi(x) = \begin{pmatrix} \varphi_1(x) \\ \varphi_2(x) \end{pmatrix}$$

For the spinor case :

$$\mathscr{L} = \bar{\psi} (i \gamma \partial - m) \psi \qquad ,$$

where ψ is given by (IV. 1), then the isospinor current is

$$j^{\mu}{}_{k}(x) = \bar{\psi}(x) \gamma^\mu \frac{\tau_k}{2} \psi(x) \qquad \text{(IV. 5b)}$$

or :

$$j^{\mu}{}_{k}(x) = \bar{\psi}_a(x) \gamma^\mu \left(\frac{\tau_k}{2}\right)_{aa'} \psi_{a'}(x)$$

and so on.

Clearly, according to the relations given in (IV. 3d) to (IV. 3f), the electromagnetic current is :

$$j^{\mu}_{\gamma}(x) = i \left\{ \varphi^{+}(x) \frac{(1+\tau_3)}{2} \partial^{\mu} \varphi(x) - \right.$$

$$\left. - \partial^{\mu} \varphi^{+}(x) \frac{(1+\tau_3)}{2} \varphi(x) \right\} = \frac{1}{2} (j^{\mu}(x) + j^{\mu}_{3}(x))$$

for the scalar isospinor such as in (IV. 3).

For the quarks (IV. 3a) the electromagnetic current is

$$j^{\mu}_{\gamma}(x) = \bar{q}(x) \gamma^{\mu} \frac{1}{2} (\frac{1}{3} + \tau_3) q(x) =$$

$$= \frac{1}{6} j^{\mu}(x) + j^{\mu}_{3}(x)$$

where

$$j^{\mu}(x) = \bar{q}(x) \gamma^{\mu} q(x)$$

$$j^{\mu}_{3}(x) = \bar{q}(x) \gamma^{\mu} \frac{\tau_3}{2} q(x)$$

We must remember that the charges of the u and d-quarks are $\frac{2}{3}$ e and $-\frac{1}{3}$ e respectively.

We note that, as a result of the invariance of the lagrangean :

$$L = L (\psi(x), \partial^{\mu} \psi (x), \psi^{+}(x), \partial^{\mu} \psi^{+}(x))$$

under the group SU(2), equation (IV. 4), we have :

$$\partial L = \frac{\partial L}{\partial \psi} \delta \psi + \frac{\partial L}{\partial(\partial^{\mu} \psi)} \delta (\partial^{\mu} \psi) + \text{h.c.}$$

hence, as

$$\delta \psi = i \vec{\alpha} \cdot \frac{\vec{\tau}}{2} \psi$$

$$\delta(\partial^\mu \psi) = i \vec{\alpha} \cdot \frac{\vec{\tau}}{2} \partial^\mu \psi$$

then, for $\vec{\alpha}$ arbitrary

$$\frac{\partial L}{\partial \psi} \frac{\vec{\tau}}{2} \psi + \frac{\partial L}{\partial(\partial^\mu \psi)} \frac{\vec{\tau}}{2} \partial^\mu \psi = \qquad \text{(IV. 6)}$$

$$= \psi^+ \frac{\vec{\tau}}{2} \frac{\partial L}{\partial \psi^+} + \partial^\mu \psi^+ \frac{\vec{\tau}}{2} \frac{\partial L}{\partial(\partial^\mu \psi^+)}$$

IV. 2 - THE YANG-MILLS ISOSPIN GAUGE-FIELD

We saw in § II.2 that the electromagnetic gauge field was introduced in order to have the postulate of invariance under the global phase transformation group (II. 5) generalized to the local phase transformation group (II. 6). And in this way the conserved electromagnetic current is coupled to the electromagnetic field.

The idea introduced by Yang and Mills was to generalize this notion to other conserved currents. Thus the isospin current being derived from the invariance of the lagrangean under the global transformations (IV. 4a) the problem arises to search for a lagrangean invariant under a local isospin transformation of the form :

$$\psi'(x) = e^{ig \vec{\Lambda}(x) \cdot \frac{\vec{\tau}}{2}} \psi(x) \qquad \text{(IV. 7)}$$

where the three parameters $\Lambda_k(x)$ now change from point to point. The constant g is explicitly introduced by analogy with the charge e introduced in transformations (II. 7).

For an infinitesimal transformation ($\vec{\Lambda}(x)$ very small everywhere) we have :

$$\psi'(x) = (I + i\, g\, \vec{\Lambda}(x) \cdot \frac{\vec{\tau}}{2})\, \psi(x)$$
$$\psi'^{+}(x) = \psi^{+}(x)(I - i\, g\, \vec{\Lambda}(x) \cdot \frac{\vec{\tau}}{2}) \qquad \text{(IV. 7a)}$$

As before, we see that the derivative of the field does not transform like the field itself, since :

$$\partial^{\mu} \psi'(x) = (I + i\, g\, \vec{\Lambda}(x) \cdot \frac{\vec{\tau}}{2})\, \partial^{\mu} \psi(x) + i\, g\, (\partial^{\mu} \vec{\Lambda}(x)) \cdot \frac{\vec{\tau}}{2} \psi(x) \qquad \text{(IV. 7b)}$$

Therefore :

$$\partial_{\mu} \psi'^{+}(x) = \partial^{\mu} \psi^{+}(x)(I - i\, g\, \vec{\Lambda}(x) \cdot \frac{\vec{\tau}}{2}) - i\, g\, \psi^{+}(x)\, \frac{\vec{\tau}}{2} \cdot \partial_{\mu} \vec{\Lambda}(x) \qquad \text{(IV. 7c)}$$

We see that terms like

$$\partial^{\mu} \varphi^{+}(x)\, \partial_{\mu} \varphi(x)$$

go over into :

$$\partial^{\mu} \varphi'^{+}(x)\, \partial_{\mu} \varphi'(x) = \partial^{\mu} \varphi^{+}(x)\, \partial_{\mu} \varphi(x) - i\, g \left\{ \varphi^{+}(x) \frac{\vec{\tau}}{2} \partial^{\mu} \varphi(x) - \partial^{\mu} \varphi^{+}(x) \frac{\vec{\tau}}{2} \varphi(x) \right\} \cdot \partial_{\mu} \vec{\Lambda}(x)$$

Also terms like

$$\bar{\psi}(x)\, \Gamma^{\mu}\, \partial_{\mu} \psi(x)$$

go over into :

$$\bar{\psi}'(x)\, \Gamma^{\mu}\, \partial_{\mu}\, \psi'(x) = \bar{\psi}(x)\, \Gamma^{\mu}\, \partial_{\mu}\, \psi(x) + i\, g\, \bar{\psi}(x)\, \Gamma^{\mu}\, \frac{\vec{\tau}}{2}\, \psi(x)\, \partial_{\mu}\, \vec{\Lambda}(x)$$

The changes in ψ and $\partial^{\mu}\psi$ are :

$$\delta\psi = i\, g\, \vec{\Lambda}(x) \cdot \frac{\vec{\tau}}{2}\, \psi\,, \qquad \delta\psi^{+} = -\, i\, g\, \psi^{+}(x)\, \frac{\vec{\tau}}{2} \cdot \vec{\Lambda}(x)$$

$$\delta(\partial^{\mu}\psi) = i\, g \left\{ \vec{\Lambda}(x) \cdot \frac{\vec{\tau}}{2}\, \partial^{\mu}\psi + \partial^{\mu}\vec{\Lambda}(x) \cdot \frac{\vec{\tau}}{2}\, \psi \right\}$$

$$\delta(\partial^{\mu}\psi^{+}) = -\, i\, g \left\{ \partial^{\mu}\psi^{+}\, \frac{\vec{\tau}}{2} \cdot \vec{\Lambda}(x) + \psi^{+}\, \frac{\vec{\tau}}{2} \cdot \partial^{\mu}\vec{\Lambda}(x) \right\}$$

Therefore the change in the lagrangean

$$L = L(\psi,\ \partial^{\mu}\psi,\ \psi^{+},\ \partial^{\mu}\psi^{+}) \qquad \text{(IV. 8)}$$

$$\delta L = \frac{\partial L}{\partial\psi}\, \delta\psi + \frac{\partial L}{(\partial(\partial^{\mu}\psi))}\, \delta(\partial^{\mu}\psi) + \text{h.c.}$$

is

$$\delta L = -\, g\, \vec{j}^{\mu}(x) \cdot \partial_{\mu}\, \vec{\Lambda}(x) \qquad \text{(IV. 9)}$$

account being taken of equations (IV. 5) and (IV. 6).

By analogy with the electrodynamical case (U(1) was there the group of transformations now the group is SU(2)) we look for a new real vector field, $A^{\mu}{}_{k}(x)$, which will change, together with the field $\psi(x)$($k = 1, 2, 3$):

$$\begin{aligned} \psi(x) &\rightarrow \psi'(x) = e^{ig\, \vec{\Lambda}(x) . \frac{\vec{\tau}}{2}}\, \psi(x) \\ A^{\mu}(x) &\rightarrow A^{\mu\prime}{}_{k}(x) \end{aligned} \qquad \text{(IV. 10)}$$

is such a way as to have the change (IV. 9) cancelled.

Equations (II. 10) suggest us to define a new, isospin covariant-derivative :

$$D^\mu \psi(x) = (\partial^\mu + i\, g\, \vec{A}^\mu(x) \cdot \frac{\vec{\tau}}{2})\, \psi(x)$$

$$(D^\mu \psi(x))^+ = \partial^\mu \psi^+(x) - i\, g\, \psi^+(x) \frac{\vec{\tau}}{2} \cdot \vec{A}^\mu(x) \qquad \text{(IV. 11)}$$

We impose it to satisfy the equation :

$$D'^\mu \psi'(x) = e^{ig\vec{\Lambda}(x). \frac{\vec{\tau}}{2}}\, D^\mu \psi(x) \qquad \text{(IV. 11a)}$$

that is, $D^\mu \psi(x)$ must transform like $\psi(x)$. We then obtain the following transformation law for the gauge field $A^\mu{}_k(x)$ for finite gauge transformations :

$$g\, A'^\mu{}_k(x) \cdot \frac{\tau_k}{2} = U(x) \left\{ g\, A^\mu{}_k \cdot \frac{\tau_k}{2} - i\, \partial^\mu \right\} U^{-1} \qquad \text{(IV. 11b)}$$

where $U(x) = e^{ig\vec{\Lambda}(x) \cdot \frac{\vec{\tau}}{2}}$ and $U\, U^+ = I$ implies, as for the case of U(1) transformations :

$$(\partial_\mu U) U^{-1} = - U\, \partial_\mu U^{-1}$$

In the case of infinitesimal transformations we have :

$$g\, A'^\mu{}_k(x) \cdot \frac{\tau_k}{2} = (I + i\, g\, \vec{\Lambda} \cdot \frac{\vec{\tau}}{2})(g\, A^\mu{}_k \cdot \frac{\tau_k}{2} - i\, \partial_\mu)(I - i\, g\, \vec{\Lambda} \cdot \frac{\vec{\tau}}{2})$$

If we therefore write

$$\vec{A}'^\mu(x) = \vec{A}^\mu(x) - \partial^\mu \vec{\Lambda}(x) + \vec{a}^\mu\, (x) \qquad \text{(IV. 12)}$$

so as to cancel the first and second terms of the second-hand side above we get :

$$\vec{a}^\mu \cdot \frac{\vec{\tau}}{2} = i\, g \left[\tilde{\Lambda}, \tilde{A}^\mu \right] - i\, g\, \tilde{\Lambda}\, (\partial_\mu \tilde{\Lambda})$$

where $\tilde{\Lambda} \equiv \vec{\Lambda} \cdot \frac{\vec{\tau}}{2}$ and $\tilde{A}^\mu = \vec{A}^\mu \cdot \frac{\vec{\tau}}{2}$.

The second term in the right-hand side is small for infinitesimal Λ and $\partial^\mu \Lambda$ everywhere so we obtain, in view of the commutation rules (IV. 2a) :

$$a^\mu{}_a (x) = g \; \varepsilon_{abc} \; A^\mu{}_b(x) \, \Lambda_c(x) \qquad \text{(IV. 12a)}$$

The vector field therefore transforms like (equations (IV. 12), (IV. 12a)) (to first order in g) :

$$A^{\mu'}{}_a(x) = A^\mu{}_a(x) - \partial^\mu \Lambda_a(x) - g \, \varepsilon_{abc} \, \Lambda_b(x) \, A^\mu{}_c(x) \qquad \text{(IV. 13)}$$

IV. 3 - THE ISOSPIN GAUGE FIELD AS A MIXTURE OF AN ABELIAN GAUGE FIELD AND AN ISOVECTOR

We may easily find the transformation law of an isovector induced by the transformation (IV. 7), or (IV. 7a) of the isospinors.

Clearly, an isovector, which has three components, may be represented by a second-order symmetric isospinor :

$$\psi_{ab}(x) = \psi_{ba}(x)$$

which will transform like :

$$\psi'_{a'b'}(x) = (I + i\, g\, \vec{\Lambda} \cdot \frac{\vec{\tau}}{2})_{a'a} \; (I + i\, g\, \vec{\Lambda} \cdot \frac{\vec{\tau}}{2})_{b'b} \; \psi_{ab}(x)$$

Now an antisymmetric matrix C exists such that :

$${}^tC = - C, \quad C^+ C = CC^+ = I$$

$${}^t\vec{\tau} = - C^{-1} \, \vec{\tau} \, C$$

the symbol t at left means transposed matrix.

For instance, in the usual Pauli representation of the matrices $\vec{\tau}$, C may be taken as τ_2.

Then, in the isospin space, we may take as a basis the matrices $\vec{\tau}\, C$ which are symmetric :

$$^{t}(\vec{\tau}\, C) = \vec{\tau}\, C$$

and the matrix C, which is antisymmetric. Therefore we can express ψ_{ab} in this basis

$$\psi_{ab}(x) = \vec{f}(x) \cdot \left(\frac{\vec{\tau}\, C}{2} \right)_{ab} \qquad \text{(IV. 14)}$$

Now for the transformed functions one has :

$$\psi'_{a'b'}(x) = \vec{f}'(x) \cdot \left(\frac{\vec{\tau}\, C}{2} \right)_{a'b'} =$$

$$= \left(\delta_{a'a} + i\, g\, \vec{\Lambda} \cdot \frac{\vec{\tau}_{a'a}}{2}\right) \left(\delta_{b'b} + i\, g\, \vec{\Lambda} \cdot \frac{\vec{\tau}_{b'b}}{2}\right) \psi_{ab}(x) =$$

$$= \vec{f}(x) \cdot \left(\frac{\vec{\tau}\, C}{2} \right)_{a'b'} + i\, g\, \vec{\Lambda} \cdot \left\{ \frac{\vec{\tau}_{a'a}}{2} \left(\vec{f}(x) \cdot \left(\frac{\vec{\tau}\, C}{2} \right)_{ab'}\right) + \right.$$

$$\left. + \frac{\vec{\tau}_{b'b}}{2} \left(\vec{f}(x) \cdot \left(\frac{\vec{\tau}\, C}{2} \right)_{a'b}\right) \right\}$$

hence

$$f'_k \left(\frac{\tau_k}{2}\, C \right)_{a'b'} = f_k \left(\frac{\tau_k C}{2} \right)_{a'b'} +$$

$$+ \; i\, g\, \Lambda_\ell \left\{ \left(\frac{\tau_\ell}{2} \; \frac{\tau_m}{2} \right)_{a'\alpha} \; C_{\alpha b'} \; f_m + \right.$$

$$\left. + \left(\frac{\tau_\ell}{2} \; \frac{\tau_m}{2} \right)_{b'\alpha} \; C_{\alpha a'} \; f_m \right\}$$

Multiply by $C^{-1}{}_{b'\beta}$ and then by $(\frac{\tau_j}{2})_{\beta a'}$ and sum over repeated indices to get :

$$f'_j = f_j + 2i\, g\, \Lambda_\ell \left\{ \mathrm{Tr}\left(\frac{\tau_\ell}{2} \frac{\tau_m}{2} \frac{\tau_j}{2} \right) f_m - \right.$$

$$\left. - \left(C^{-1} \frac{\tau_\ell}{2} \frac{\tau_m}{2} C \right)_{\beta a'} \left(\frac{\tau_j}{2} \right)_{\beta a'} f_m \right\}$$

so that from :

$$\frac{\tau_m}{2} \frac{\tau_j}{2} = \frac{1}{4} \delta_{jm} + \frac{i}{2} \varepsilon_{mjn} \frac{\tau_n}{2} ,$$

$$\mathrm{Tr}\left(\frac{\tau_\ell}{2} \frac{\tau_m}{2} \frac{\tau_j}{2} \right) = \frac{i}{4} \varepsilon_{mj\ell} ,$$

$$- \left(C^{-1} \frac{\tau_\ell}{2} \frac{\tau_m}{2} C \right)_{\beta a'} \left(\frac{\tau_j}{2} \right)_{\beta a'} = \mathrm{Tr}\left(\frac{\tau_j}{2} \frac{\tau_\ell}{2} \frac{\tau_m}{2} \right) = \frac{i}{4} \varepsilon_{\ell mj}$$

we obtain

$$f'_j = f_j - g\, \varepsilon_{j\ell m} \Lambda_\ell f_m \qquad \text{(IV. 15)}$$

The same law is deduced from the transformation of $\psi^+ \frac{\tau_k}{2} \psi$ under (IV. 10).

For finite transformations U, the law is :

$$f'_j \frac{\tau_j}{2} = U f_k \frac{\tau_k}{2} U^{-1}. \qquad \text{(IV. 15a)}$$

This is the transformation law for an isovector. We see that the isospin gauge field is not an isovector, it behaves as a sum of an abelian gauge field and an isovector field, according to equations (IV. 12), (IV. 13), (IV. 15

IV. 4 - LAGRANGEAN OF A YANG-MILLS ISOSPIN GAUGE FIELD IN INTERACTION WITH MATTER

We see that given a lagrangean formed with an isospin field $\psi(x)$:

$$L = L\, (\psi, \partial^\mu \psi, \psi^+, \partial^\mu \psi^+)$$

we obtain a gauge-invariant form for this lagrangean if we replace the derivatives by the operators defined in equations (IV. 11). What is the part corresponding to the gauge-field $A^\mu{}_a(x)$ alone ?

In the electromagnetic case we introduced the field lagrangean :

$$L_\gamma = - \frac{1}{4} F^{\mu\nu}(x) F_{\mu\nu}(x)$$

and the field $F^{\mu\nu}$ was given by the commutator of the components of D_μ . We shall extend this procedure here and define, by analogy with equation (III. 12) :

$$\left(\frac{\tau_k}{2} \right)_{ab} F^{\mu\nu}_k = \frac{i}{g} \left[D^\mu, D^\nu \right]_{ab} \qquad \text{(IV. 16)}$$

where D^μ is given in equation (IV. 11) :

$$D^\mu{}_{aa'} = \partial^\mu \delta_{aa'} + i\, g\, A^\mu{}_k(x) \left(\frac{\tau_k}{2} \right)_{aa'}$$

We obtain :

$$F^{\mu\nu}{}_k(x) = \partial^\nu A^\mu{}_k(x) - \partial^\mu A^\mu{}_k(x) + g\, \varepsilon_{k\ell m}\, A^\mu{}_\ell(x)\, A^\nu{}_m(x) \qquad \text{(IV. 16a)}$$

that is :

$$\vec{F}^{\mu\nu} = \partial^\nu \vec{A}^\mu - \partial^\mu \vec{A}^\nu + g \left[\vec{A}^\mu \times \vec{A}^\nu \right]$$

It is the fact that the transformations (IV. 4) do not in general commute (the group of transformations is then called non-abelian) because of the matrices τ_k, that gives rise to the second-term, bilinear in the field A, in the expression for $F^{\mu\nu}{}_k(x)$.

We must see what is the transformation law of $F^{\mu\nu}{}_k(x)$ under the group (IV. 10). The electromagnetic field $F^{\mu\nu}(x)$ is invariant under the electromagnetic gauge transformation (II. 11) ; in the present case, $A^\mu{}_k(x)$

transforms according to the law (IV. 13), how does $F^{\mu\nu}{}_k(x)$ transforms itself ?

Let us define :

$$\tilde{F}^{\mu\nu} = \frac{\tau_k}{2} F^{\mu\nu}{}_k = \partial^\nu \tilde{A}^\mu - \partial^\mu \tilde{A}^\nu +$$

$$+ g \frac{\tau_k}{2} \varepsilon_{k\ell m} A^\mu{}_\ell A^\nu{}_m$$

but :

$$\frac{\tau_k}{2} \varepsilon_{k\ell m} = \varepsilon_{\ell mk} \frac{\tau_k}{2} = - i \left[\frac{\tau_\ell}{2} , \frac{\tau_m}{2} \right]$$

hence

$$\tilde{F}^{\mu\nu}(x) = \partial^\nu \tilde{A}^\mu - \partial^\mu \tilde{A}^\nu - i g \left[\tilde{A}^\mu, \tilde{A}^\nu \right]$$

Now, under the transformation (IV. 13) :

$$\delta A^\mu{}_k = - \partial^\mu \Lambda_k + g \varepsilon_{k\ell m} A^\mu{}_\ell \Lambda_m$$

so :

$$\delta \tilde{A}^\mu = - \partial^\mu \tilde{\Lambda} - i g \left[\tilde{A}^\mu , \tilde{\Lambda} \right]$$

hence :

$$\delta \tilde{F}^{\mu\nu} = \partial^\nu \delta \tilde{A}^\mu - \partial^\mu \delta \tilde{A}^\nu - i g \left[\delta \tilde{A}^\mu, \tilde{A}^\nu \right] - i g \left[\tilde{A}^\mu, \delta \tilde{A}^\nu \right]$$

In view of the Jacobi identity for three operators :

$$\left[[a, b] , c \right] + \left[[c, a] , b \right] + \left[[b, c] ,a \right] = 0$$

we obtain :

$$\delta \tilde{F}^{\mu\nu} = i g \left[\tilde{\Lambda}, (\partial^\nu \tilde{A}^\mu - \partial^\mu \tilde{A}^\nu) \right] - g^2 \left[\tilde{\Lambda}, [\tilde{A}^\nu, \tilde{A}^\mu] \right]$$

whence :

$$\delta \tilde{F}^{\mu\nu} = - g \ \varepsilon_{k\ell m} \ \Lambda_k \ F^{\mu\nu}_{\ \ \ell} \ \frac{\tau_m}{2}$$

therefore

$$\delta F^{\mu\nu}_{\ \ \ k} = - g \ \varepsilon_{k\ell m} \ \Lambda_\ell \ F^{\mu\nu}_{\ \ \ m}$$

So now $F^{\mu\nu}_{\ \ \ k}(x)$ is not invariant under the isospin gauge group. It transforms like :

$$F^{\mu\nu'}_{\ \ \ k} (x) = F^{\mu\nu}_{\ \ \ k}(x) - g \ \varepsilon_{k\ell m} \ \Lambda_\ell \ F^{\mu\nu}_{\ \ \ m}$$

that is, $F^{\mu\nu}_{\ \ \ k}$ is an isovector.

What lagrangean term are we then to postulate for the gauge field alone ? Let us try a term similar to the electrodynamical one :

$$\mathscr{L}_{Yang} = - \frac{1}{4} F^{\mu\nu}_{\ \ \ k} \ F_{\mu\nu,k}$$

It will be acceptable only if it is gauge-invariant. This can be written :

$$\mathscr{L}_{Yang} = - \frac{1}{2} \ \mathrm{Tr} \ (\tilde{F}^{\mu\nu} \ \tilde{F}_{\mu\nu})$$

since :

$$\mathrm{Tr} \ (\tilde{F}^{\mu\nu} \ \tilde{F}_{\mu\nu}) = F^{\mu\nu}_{\ \ \ k} \ F_{\mu\nu,\ell} \ \mathrm{Tr} \ \left[\frac{\tau_k}{2} \ \frac{\tau_\ell}{2} \right] = \frac{1}{2} \ F^{\mu\nu}_{\ \ \ k} \ F_{\mu\nu,k}$$

Therefore

$$\delta \mathscr{L}_{\text{-}Yang} = - \frac{1}{2} \ \mathrm{Tr} \ ((\delta \ \tilde{F}^{\mu\nu}) \ \tilde{F}_{\mu\nu} + \tilde{F}^{\mu\nu}(\delta \ \tilde{F}_{\mu\nu}))$$

$$= - \frac{1}{2} \ g \ \varepsilon_{k\ell n} \ \Lambda_k \ F^{\mu\nu}_{\ \ \ \ell} \ F_{\mu\nu,n} \ \equiv 0$$

So the term $F^{\mu\nu}_{\ \ \ k} F_{\mu\nu,k}$ is gauge invariant.

We therefore take as the gauge-invariant lagrangean for an isospinor matter field ψ in interaction with the Yang-Mills field $A^{\mu}{}_{k}(x)$ the following one :

$$\mathscr{L} = -\frac{1}{4} F^{\mu\nu}{}_{k} F_{\mu\nu,k} + L(\psi, D^{\mu}\psi, \psi^{+}, (D^{\mu}\psi)^{+}) \qquad \text{(IV. 17)}$$

where

$$F^{\mu\nu}{}_{k} = \partial^{\nu} A^{\mu}{}_{k} - \partial^{\mu} A^{\nu}{}_{k} + g\, \varepsilon_{k\ell n} A^{\mu}{}_{\ell} A^{\nu}{}_{n}$$

$$D^{\mu}\psi = (\partial^{\mu} + i g A^{\mu}{}_{k} \cdot \frac{\tau_k}{2})\, \psi \qquad \text{(IV. 17a)}$$

$$(D^{\mu}\psi)^{+} = \partial^{\mu}\psi^{+} - i g \psi^{+} A^{\mu}{}_{k} \cdot \frac{\tau_k}{2}$$

with the transformation laws (for infinitesimal transformations) :

$$\psi' = (I + i g \Lambda_k \frac{\tau_k}{2})\, \psi$$

$$\psi'^{+} = \psi^{+} (I - i g \Lambda_k \frac{\tau_k}{2}) \qquad \text{(IV. 17b)}$$

$$A^{\mu'}{}_{k} = A^{\mu}{}_{k} - \partial^{\mu}\Lambda_k - g\, \varepsilon_{k\ell n} \Lambda_{\ell} A^{\mu}{}_{n}$$

$$F^{\mu\nu'}{}_{k} = F^{\mu\nu}{}_{k} - g\, \varepsilon_{k\ell n} \Lambda_{\ell} F^{\mu\nu}{}_{n}$$

For finite gauge transformations :

$$U = e^{i g \Lambda_k \frac{\tau_k}{2}}$$

one has :

$$\psi'(x) = U \psi(x) \; ;$$

$$\psi^{+'}(x) = \psi^{+}(x)\, U^{-1}$$

$$A'_{\mu k} \frac{\tau_k}{2} = U(A_{\mu k} \frac{\tau_k}{2} - \frac{i}{g} \partial_{\mu})\, U^{-1}$$

$$F'_{\mu\nu k} \frac{\tau_k}{2} = U\, F_{\mu\nu k} \frac{\tau_k}{2}\, U^{-1}$$

The field equations are now like the equations (II. 15), (II. 15a), (II. 16), (II. 16a) having in mind the new lagrangean (IV. 17) and the definitions (IV. 17a) :

$$\left(D^{\mu} \frac{\partial \mathcal{L}}{\partial (D^{\mu} \psi)^{+}} \right)^{+} - \frac{\partial \mathcal{L}}{\partial \psi} = 0$$

$$D^{\mu} \frac{\partial \mathcal{L}}{\partial (D^{\mu} \psi)^{+}} - \frac{\partial \mathcal{L}}{\partial \psi^{+}} = 0 \qquad \text{(IV. 18)}$$

$$\partial_{\nu} F^{\mu\nu}{}_{k}(x) = - \frac{\partial \mathcal{L}}{\partial A_{\mu;k}(x)}$$

The current is :

$$g\, j^{\mu}{}_{k}(x) = - \frac{\partial \mathcal{L}}{\partial A^{\mu}{}_{k}(x)}$$

Now :

$$\frac{\partial \mathcal{L}}{\partial A^{\mu}{}_{k}(x)} = \frac{\partial \mathcal{L}}{\partial (D^{\alpha} \psi)} \frac{\partial (D^{\alpha} \psi)}{\partial A^{\mu}{}_{k}} + \text{h.c.} + \frac{1}{2} \frac{\partial \mathcal{L}}{\partial F^{\alpha\beta}{}_{k'}} \frac{\partial F^{\alpha\beta}{}_{k'}}{\partial A^{\mu}{}_{k}}$$

One obtains :

$$j^{\mu}{}_{k} = i \left\{ \psi^{+} \frac{\tau_{k}}{2} \frac{\partial \mathcal{L}}{\partial (D_{\mu} \psi)^{+}} - \frac{\partial \mathcal{L}}{\partial (D_{\mu} \psi)} \frac{\tau_{k}}{2} \psi \right\} -$$

$$- \varepsilon_{k\ell n} F^{\mu\nu}_{\ell} A_{\nu,n} \qquad \text{(IV. 19)}$$

To the current due to the ψ-field is added a term coming from the gauge vector field ; this field has an isovector part and thus contributes to the isospin current. The transformed of the current is :

$$j^{\mu'}{}_{k}(x) = j^{\mu}{}_{k} - g\, \varepsilon_{k\ell n} \left\{ \Lambda_{\ell}\, j^{\mu}{}_{n} + \frac{1}{g} (\partial_{\nu} \Lambda_{\ell}) F_{n}^{\mu\nu} \right\} \qquad \text{(IV. 19a)}$$

We note that as a result of the equality :

$$\partial_\mu (f^+ \frac{\tau_k}{2} g) = (D^\mu f)^+ \frac{\tau_k}{2} g + f^+ \frac{\tau_k}{2} (D^\mu g) +$$

$$+ g\, \varepsilon_{k\ell n}\, A_{\mu,\ell}\, (f^+ \frac{\tau_n}{2} g) \qquad \text{(IV. 20)}$$

we have, in view of equations (IV. 18) :

$$\partial_\mu j^\mu{}_k = i \left\{ \psi^+ \frac{\tau_k}{2} \frac{\partial \mathscr{L}}{\partial \psi^+} + (D^\mu \psi)^+ \frac{\tau_k}{2} \frac{\partial \mathscr{L}}{\partial (D^\mu \psi)^+} + \right.$$

$$+ \text{h.c.} + i\, g\, \varepsilon_{k\ell n}\, A_{\mu,\ell} \left(\psi^+ \frac{\tau_n}{2} \frac{\partial \mathscr{L}}{\partial (D^\mu \psi)} - \frac{\partial \mathscr{L}}{\partial (D^\mu \psi)} \frac{\tau_n}{2} \psi \right) +$$

$$+ \varepsilon_{k\ell n} \left\{ g\, j^\nu{}_\ell\, A_{\nu,n} + \frac{1}{2} F^{\mu\nu}{}_\ell (\partial_\nu A_{\mu,n} - \partial_\mu A_{\nu,n}) \right\} = 0$$

The terms in $\varepsilon_{k\ell n}$ cancel and so we are left with the relationship :

$$\frac{\partial \mathscr{L}}{\partial \psi} \frac{\tau_k}{2} \psi + \frac{\partial \mathscr{L}}{\partial (D^\mu \psi)} \frac{\tau_k}{2} D^\mu \psi -$$

$$- \psi^+ \frac{\tau_k}{2} \frac{\partial \mathscr{L}}{\partial \psi^+} - (D^\mu \psi)^+ \frac{\tau_k}{2} \frac{\partial \mathscr{L}}{\partial (D^\mu \psi)^+} = 0 \qquad \text{(IV. 21)}$$

which is the analogue of equation (II. 16h), for the electromagnetic case, and of equation (IV. 6) for the global isospin transformations.

The equation (IV. 21) is also a consequence of the invariance of the lagrangean under the transformations (IV. 17) :

$$\frac{\partial}{\partial \Lambda_k} \delta \mathscr{L} = 0$$

We note that while the total current (IV. 19) does not transform like an isovector, its matter part is an isovector :

$$j^{\mu}{}_{k}(x) = j^{\mu}{}_{k}(\text{matter}) - \varepsilon_{k\ell n} \; F^{\mu\nu}{}_{\ell} \; A_{\nu,n} \quad ;$$

$$j^{\mu}{}_{k}(\text{matter}) = i \left\{ \psi^{+} \; \frac{\tau_k}{2} \; \frac{\partial \mathscr{L}}{\partial(D_{\mu}\,\psi)^{+}} - \frac{\partial \mathscr{L}}{\partial(D_{\mu}\,\psi)} \; \frac{\tau_k}{2} \; \psi \right\} \qquad \text{(IV. 19a)}$$

$$j^{\mu}{}'_{k}\,(\text{matter}) = j^{\mu}{}_{k}(\text{matter}) - \varepsilon_{kab}\, \Lambda_a \; j^{\mu}{}_{b}(\text{matter})$$

IV. 5 - FIELD EQUATIONS AND NON-LINEARITY OF THE INTERACTION

The lagrangean and the field equations are given in equations (IV. 17) and (IV. 18) with the current given in (IV. 19), (IV. 19a).

It is however convenient to express the equations for the gauge-field in terms of the covariant derivative :

$$D_{\nu;k\ell} = \partial_{\nu} \;\; \delta_{k\ell} + g \; \varepsilon_{k\ell n} \;\; A_{\nu,n} \qquad \text{(IV. 22)}$$

The gauge-field equations are then :

$$D_{\nu;k\ell} \quad F^{\mu\nu}{}_{\ell} = g \; j^{\mu}{}_{k}(\text{matter}) \qquad \text{(IV. 22a)}$$

and the homogeneous equation :

$$D^{\nu}{}_{k\ell} \quad F^{\alpha\beta}{}_{\ell} + D^{\beta}{}_{k\ell} \;\; F^{\nu\alpha}{}_{\ell} + D^{\alpha}{}_{k\ell} \; F^{\beta\nu}{}_{\ell} = 0 \qquad \text{(IV. 22b)}$$

which is the analogue of Maxwell's equation (II. 16b).

The equation (IV. 22b) can also be written in terms of the commutator (IV. 16), namely through the Bianchi identity

$$\left[D^{\nu}, [\, D^{\alpha}, \;\; D^{\beta} \,]\right] + \left[D^{\beta}, \; [\, D^{\nu}, D^{\alpha}]\right] + \left[D^{\alpha}, \; [\, D^{\beta}, D^{\nu} \;]\right] = 0 \qquad \text{(IV. 23)}$$

since

$$\frac{i}{g} \; \left[D^{\alpha}, \; [\, D^{\beta}, D^{\nu}]\right] = D^{\alpha}{}_{k\ell} \;\; F^{\beta\nu}{}_{\ell} \;\; \frac{\tau_k}{2} \qquad \text{(IV. 23a)}$$

From equation (IV. 22a) it is easily seen that the following continuity equation holds :

$$D_{\mu;k\ell} \; j^{\mu}{}_{\ell}(\text{matter}) = 0 \qquad \text{(IV. 23b)}$$

The form of the tensor $F^{\mu\nu}{}_{k}$ as given in equation (IV. 17a) shows that there is self-interaction of the field $A^{\mu}{}_{k}$. In the absence of the matter field ψ, the gauge-field equation is :

$$D_{\nu;k\ell} \; F^{\mu\nu}{}_{\ell} = 0$$

and the field acts back on itself as seen in the equation :

$$\partial_{\nu} \; F^{\mu\nu}{}_{k} = - g \; \varepsilon_{k\ell n} \; A_{\nu,n} \; F^{\mu\nu}{}_{\ell}$$

The superposition principle is thus generally lost. This would be the case in electrodynamics if photons were charged. These properties, non-covariance of the gauge field and of the total source current, alternative forms for the gauge field equations, are also found in general relativity (Chapt. V).

IV. 6 - REMARKS ON THE COVARIANT DERIVATIVE

We note that the covariant derivative (IV. 22) which acts on an isovector is deduced from that, (IV. 11), which acts on isospinors. The procedure is similar to the one used in § IV. 3. Let us consider the equation (IV. 14) and apply the following operator, similar to (IV. 11), to each one of the indices of the second-rank spinor $\psi_{ab}(x)$:

$$(\partial_{\mu} \; \delta_{aa'} \; \delta_{bb'} + i \; g \; A_{\mu,k} \; (\frac{\tau k}{2})_{aa'} \; \delta_{bb'} + i \; g \; A_{\mu,k} \; (\frac{\tau_k}{2})_{bb'} \; \delta_{aa'}) \; \psi_{a'b'} \equiv \psi_{\mu,ab}$$

If we use the decomposition (IV. 14) we have :

$$(\partial_{\mu} \; \delta_{aa'} \; \delta_{bb'} + i \; g \; A_{\mu,k} \; (\frac{\tau_k}{2})_{aa'} \; \delta_{bb'} + i \; g \; A_{\mu,k} \; (\frac{\tau_k}{2})_{bb'} \; \delta_{aa'}) \; \cdot$$

$$\cdot \; f_{\ell} \; (\frac{\tau_{\ell} C}{2})_{a'b'} = F_{\mu,n} \; (\frac{\tau_n C}{2})_{ab}$$

Multiply by $(C^{-1})_{b\alpha}$ and by $(\frac{\tau_m}{2})_{\alpha\beta}$ and take the trace to get :

$$F_{\mu,\ell} = (\partial_\mu \delta_{\ell n} + g \varepsilon_{\ell nm} A_{\mu,m}) f_n$$

In fact the derivative (IV. 11) of isospinors and the derivative (IV. 22) of isovectors are representations of the operator

$$\partial_\mu + i g A_{\mu,k} T_k$$

where T_a are the generators of the SU(2) gauge group. The representation of the latter in the two-dimensional space is :

$$(T_k)_{aa'} = (\frac{\tau_k}{2})_{aa'}$$

In three-dimensional space it is :

$$(T_k)_{\ell n} = i \varepsilon_{\ell kn} = - i \varepsilon_{\ell nk}$$

IV. 7 - ENERGY-MOMENTUM TENSOR FOR A YANG-MILLS SYSTEM

The energy-momentum tensor for a matter field ψ in interaction with a Yang-Mills field is the following :

$$T^{\mu\nu}{}_m = \frac{\partial \mathscr{L}}{\partial(D_\mu \psi)} D^\nu \psi + (D^\nu \psi)^+ \frac{\partial \mathscr{L}}{\partial(D_\mu \psi)^+} - L g^{\mu\nu} \qquad \text{(IV. 24)}$$

where the covariant derivatives are those defined in equation (IV. 11).

In view of the identity (IV. 20) the divergence of $T^{\mu\nu}$ is :

$$\partial_\mu \; T^{\mu\nu}{}_m = \left(D_\mu \frac{\partial \mathscr{L}}{\partial (D_\mu \psi)^+} \right)^+ D^\nu \psi + \frac{\partial \mathscr{L}}{\partial (D_\mu \psi)} D_\mu D^\nu \psi -$$

$$- \frac{\partial \mathscr{L}}{\partial \psi} \partial^\nu \psi - \frac{\partial \mathscr{L}}{\partial (D_\mu \psi)} \partial_\nu D_\mu \psi + \text{h. c.}$$

therefore, in view of the field equations (IV. 18) and of equations (IV. 11) and (IV. 21) :

$$\partial_\mu T^{\mu\nu}{}_m = \frac{\partial \mathscr{L}}{\partial (D^\mu \psi)} \left[D^\mu , D^\nu \right] \psi + \left[D^\mu , D^\nu \right]^+ \frac{\partial \mathscr{L}}{\partial (D^\mu \psi)^+}$$

that is, because of the definition (IV. 16) of $F^{\mu\nu}{}_k$:

$$\partial_\mu \; T^{\mu\nu}{}_m = g \; F^{\mu\nu}{}_k \;\; j_{\mu,k}(\text{matter}) \tag{IV. 24a}$$

The energy-momentum tensor of the field $A^\mu{}_k$ is

$$T^{\mu\nu}{}_Y = \frac{\partial \mathscr{L}}{\partial (\partial_\mu A_{\alpha,k})} \partial^\nu A^\beta{}_k \; g_{\alpha\beta} - \mathscr{L}_Y \; g^{\mu\nu} =$$

$$= - F^{\alpha\mu}{}_k \; F^{\beta\nu}{}_k \, g_{\alpha\beta} + \frac{1}{4} g^{\mu\nu} \left(F^{\alpha\beta}{}_k \, F_{\alpha\beta,k} \right) \tag{IV. 25}$$

(the index Y means that the quantity refers to a Yang-Mills field) from which we deduce, taking the following identity :

$$\partial_\nu(f_k\ g_k) = (D_{\nu,k\ell}\ f_\ell)\ g_k + f_k(D_{\nu,k\ell}\ g_\ell)$$

into account :

$$\partial_\mu\ T^{\mu\nu}_{Y} = -\ (D_{\mu,k\ell}\ F^{\alpha\mu}_{\ell})\ F^{\beta\nu}_{\ k}\ g_{\alpha\beta} +$$

$$+\ \frac{1}{2}\ F_{k,\alpha\beta}\ (D^{\nu}_{\ k\ell}\ F^{\alpha\beta}_{\ \ell} + D^{\beta}_{\ k\ell}\ F^{\nu\alpha}_{\ \ell} + D^{\alpha}_{\ k\ell}\ F^{\beta\nu}_{\ \ell})$$

The field equations (IV. 22 a,b) lead thus to :

$$\partial_\mu\ T^{\mu\nu}_{\ Y} = -\ g\ F^{\beta\nu}_{\ k}\ j_{\beta,k} \qquad \text{(IV. 25a)}$$

The total energy-momentum tensor is conserved :

$$\partial_\mu\ (T^{\mu\nu}_{\ m} + T^{\mu\nu}_{\ Y}) = 0$$

IV. 8 - EXAMPLES OF YANG-MILLS ISOSPIN GAUGE SYSTEMS OF FIELDS

a) Interaction of a Dirac spinor isospinor field with a Yang-Mills field :

$$L = -\ \frac{1}{4}\ F^{\mu\nu}_{\ k}\ F_{\mu\nu,k} + \bar{\psi}(x)\ \left\{ i\ \gamma^{\alpha}\ D_{\alpha} - m \right\}\ \psi(x) \qquad \text{(IV. 26)}$$

The field equations are :

$$(i\,\gamma^{\alpha}\,D_{\alpha} - m)\,\psi(x) = 0 \quad , \quad D_{\alpha} = \partial_{\alpha} + i\,g\,A_{\alpha,k}\,\frac{\tau_k}{2}$$

$$D_{\nu;k\ell}\,F^{\mu\nu}{}_{\ell} = g\,j^{\mu}{}_{k}(m) \quad , \quad D_{\nu;k\ell} = \partial_{\nu}\,\delta_{k\ell} + g\,\varepsilon_{k\ell n}\,A_{\nu,n} \qquad \text{(IV. 26a)}$$

with the matter current

$$j^{\mu}{}_{k}(m) = \bar{\psi}\,\gamma^{\mu}\,\frac{\tau_k}{2}\,\psi \qquad \text{(IV. 26b)}$$

b) Interaction of a real scalar isovector field with a Yang-Mills field :

$$L = -\frac{1}{4}\,F^{\mu\nu}{}_{k}F_{\mu\nu,k} + \frac{1}{2}\,(D^{\mu}{}_{k\ell}\,\varphi_{\ell}\,D_{\mu;kn}\,\varphi_n - m^2\,\varphi_k\,\varphi_k) - \frac{\lambda}{4!}\,(\varphi_k\,\varphi_k)^2$$

The field equations are ($D^{\mu}{}_{k\ell}$ is given in (IV. 22)) :

$$D^{\mu}{}_{k\ell}\,D_{\mu;kn}\,\varphi_n + m^2\,\varphi_k + \frac{\lambda}{12}\,(\varphi_n\,\varphi_n)\,\varphi_k = 0$$

$$D_{\nu;k\ell}\,F^{\mu\nu}{}_{\ell} = g\,j^{\mu}{}_{k}(m) \;,$$

$$j^{\mu}{}_{k}(m) = \varepsilon_{k\ell n}\,(D^{\mu}{}_{na}\,\varphi_a)\,\varphi_{\ell}$$

IV. 9 - THE GLOBAL SU(3) GROUP

The SU(2) group is the set of all unitary matrices with 2 x 2 complex elements :

$$A_2 = \begin{pmatrix} a_{11} & a_{12} \\ a_{21} & a_{22} \end{pmatrix} \qquad \text{(IV. 27)}$$

with determinant unity :

$$\det\ A_2 = 1, \qquad A_2^{+}\ A_2 = A_2\, A_2^{+} = I \tag{IV. 27a}$$

There are three independent real numbers to characterise the matrix A, so the group SU(2) is defined by three parameters. A basis to represent such matrices is formed by the identity and by the three Pauli matrices so that in general :

$$A_2 = a_4\ I + i\vec{a}\cdot\vec{\sigma} = \begin{pmatrix} a_4 + i\,a_3 & i\,a_1 + a_2 \\ i\,a_1 - a_2 & a_4 - i\,a_3 \end{pmatrix} \tag{IV. 27b}$$

with :

$$a^2_4 + \vec{a}^2 = 1 \tag{IV. 27c}$$

We consider now the SU(3) group which is the set of all unitary matrices with 3 x 3 complex elements

$$A_3 = \begin{pmatrix} a_{11} & a_{12} & a_{13} \\ a_{21} & a_{22} & a_{23} \\ a_{31} & a_{32} & a_{33} \end{pmatrix}$$

such that :

$$A_3^{+}\ A_3 = A_3\ A_3^{+} = I, \quad \det\ A_3 = 1$$

As the SU(2) matrices act on a two-component spinor, the SU(3) matrices act on a three-component complex vector

$$\psi = \begin{pmatrix} \psi_1 \\ \psi_2 \\ \psi_3 \end{pmatrix}$$

so that :

$$\psi' = A_3 \psi$$

is the tranformed vector by A_3.

As recalled in the Introduction, it is at present assumed that each variety (flavour) of quark exists in three states with the same mass which are distinguished by a new quantum number, the colour. Let us therefore consider one quark flavour q_f ; it will be represented by a three dimensional complex vector :

$$q_f = \begin{pmatrix} q_{1f} \\ q_{2f} \\ q_{3f} \end{pmatrix} , \qquad m_1 = m_2 = m_3 = m$$

and the space of these vectors transforms into itself by the colour group SU(3). Each component q_{if} is a Dirac spinor.

A quark field is then represented by a Dirac spinor with two internal indices, the flavour index f and the colour i :

$$q \equiv (q_{\alpha af}) , \qquad \begin{array}{l} f = 1, \dots n \text{ (flavour)} \\ a = 1,2,3 \text{ (colour)} \\ \alpha = 1,2,3,4 \text{ (Dirac spinor)} \end{array} \qquad \text{(IV. 28)}$$

A global SU(3) transformation is of the form :

$$q'(x) = e^{i \alpha_k \frac{\lambda_k}{2}} q(x)$$

where the λ_k's are the eight (3 x 3 matrices) generators of the group. The exponential operator acts on the color index of $q(x)$ so that, for an infinitesimal transformation we have, omitting the Dirac and flavor indices

$$q'_a(x) = (\delta_{aa'} + i \alpha_k \left(\frac{\lambda_k}{2} \right)_{aa'}) q_{a'}(x)$$

These generators have the following form :

$$\lambda_1 = \begin{pmatrix} 0 & 1 & 0 \\ 1 & 0 & 0 \\ 0 & 0 & 0 \end{pmatrix}, \quad \lambda_2 = \begin{pmatrix} 0 & -i & 0 \\ i & 0 & 0 \\ 0 & 0 & 0 \end{pmatrix}, \quad \lambda_3 = \begin{pmatrix} 1 & 0 & 0 \\ 0 & -1 & 0 \\ 0 & 0 & 0 \end{pmatrix}$$

$$\lambda_4 = \begin{pmatrix} 0 & 0 & 1 \\ 0 & 0 & 0 \\ 1 & 0 & 0 \end{pmatrix}, \quad \lambda_5 = \begin{pmatrix} 0 & 0 & -i \\ 0 & 0 & 0 \\ i & 0 & 0 \end{pmatrix}, \quad \lambda_6 = \begin{pmatrix} 0 & 0 & 0 \\ 0 & 0 & 1 \\ 0 & 1 & 0 \end{pmatrix}$$

$$\lambda_7 = \begin{pmatrix} 0 & 0 & 0 \\ 0 & 0 & -i \\ 0 & i & 0 \end{pmatrix}, \quad \lambda_8 = \frac{1}{\sqrt{3}} \begin{pmatrix} 1 & 0 & 0 \\ 0 & 1 & 0 \\ 0 & 0 & -2 \end{pmatrix} \qquad \text{(IV. 28a)}$$

There are two of these matrices which are diagonal, λ_3 and λ_8, the SU(3) is a rank-two group.

The following are the commutation rules satisfied by these matrices :

$$\left[\frac{\lambda_k}{2}, \frac{\lambda_\ell}{2} \right]_- = i \; f_{k\ell n} \frac{\lambda_n}{2} \qquad \text{(IV. 29)}$$

$$\left[\frac{\lambda_k}{2}, \frac{\lambda_\ell}{2} \right]_+ = \frac{1}{3} \delta_{k\ell} + d_{k\ell n} \frac{\lambda_n}{2}$$

They are similar to those in equations (IV. 2a) ; $f_{k\ell n}$ is totally antisymmetric in the indices, $d_{k\ell n}$ is totally symmetric. They have the following values :

$$f_{123} = 1 \; ; \; f_{147} = - f_{156} = f_{246} = f_{257} = f_{345} = - f_{367} = \frac{1}{2} \; ;$$

$$f_{458} = f_{678} = \frac{\sqrt{3}}{2} \; ; \qquad \text{(IV. 30)}$$

$$d_{118} = d_{228} = d_{338} = - d_{888} = \frac{1}{\sqrt{3}} ;$$

$$d_{448} = d_{558} = d_{668} = d_{778} = - \frac{1}{2\sqrt{3}}$$

$$d_{146} = d_{157} = - d_{247} = d_{256} = d_{344} =$$

$$= d_{355} = - d_{366} = d_{377} = \frac{1}{2} ; \ \mathrm{Tr}\,(\lambda_k \lambda_\ell) = 2\,\delta_{k\ell}$$

the other f's and d's vanish.

Similarly to equation (IV. 5), one may write down the Noether current for the global group :

$$j^\mu{}_k = i \left\{ q^+ \frac{\lambda_k}{2} \frac{\partial \mathscr{L}}{(\partial_\mu q)^+} - \frac{\partial \mathscr{L}}{\partial(\partial_\mu q)} \frac{\lambda_k}{2} q \right\}$$

IV. 10 - THE COLOUR GAUGE FIELD

We now postulate that the quark lagrangean must be invariant under the local colour SU(3) group :

$$q'(x) = e^{i g \Lambda_k(x) \frac{\lambda_k}{2}} q(x)$$

where $\Lambda_k(x)$ are eight functions which determine the transformations. In the infinitesimal case :

$$q'(x) = (I + i g \Lambda_k(x) \frac{\lambda_k}{2}) q(x) \qquad \text{(IV. 31)}$$

The derivatives have therefore to be replaced by the covariant derivatives :

$$D_\mu q(x) = (\partial_\mu + i g A_{\mu,k}(x) \frac{\lambda_k}{2}) q(x) \qquad \text{(IV. 32)}$$

where $A_{\mu,k}(x)$ are eight gauge fields.

The requirement that these derivatives transform like the chromo-spinor (IV. 31) :

$$D'_\mu \, q'(x) = e^{i \, g \, \Lambda_k(x) \frac{\lambda_k}{2}} \, D_\mu \, q(x) \qquad \text{(IV. 32a)}$$

determines the transformation law of the colour gauge fields :

$$g \, A'_{\mu,k} \frac{\lambda_k}{2} = U \left\{ g \, A_{\mu,k} \frac{\lambda_k}{2} - i \, \partial_\mu \right\} U^{-1} \qquad \text{(IV. 33)}$$

the sum is over $k = 1,2, \ldots 8$ and $U = e^{i \, g \, \Lambda_k(x) \frac{\lambda_k}{2}}$

In order to obtain the gauge field lagrangean we proceed as in § IV. 4. We define the tensor :

$$F^{\mu\nu}{}_k \left(\frac{\lambda_k}{2} \right)_{ab} = \frac{i}{g} \left[D^\mu, D^\nu \right]_{ab} \qquad \text{(IV. 34)}$$

and obtain :

$$F^{\mu\nu}{}_k = \partial^\nu A^\mu{}_k - \partial^\mu A^\nu{}_k + g \, f_{k\ell n} \, A^\mu{}_\ell \, A^\nu{}_n \, , \qquad k = 1, \ldots 8$$

With this term one then obtains the lagrangean :

$$\mathscr{L} = - \frac{1}{4} F^{\mu\nu}{}_k \, F_{\mu\nu,k} + \bar{q} \left\{ i \, \gamma^\mu \, D_\mu - m \right\} q$$

or if we include all quark flavours $f = 1, \ldots n$:

$$\mathscr{L} = - \frac{1}{4} F^{\mu\nu}{}_k \, F_{\mu\nu,k} + \sum_{f=1}^{n} \bar{q}_{af} \left\{ i \, \gamma^\mu \, (D_\mu)_{ab} - m \, \delta_{ab} \right\} q_{bf} \qquad \text{(IV. 35)}$$

the Dirac spinor indices being omitted.

The infinitesimal transformation law of $F^{\mu\nu}_k$ is

$$F'^{\mu\nu}{}_k = F^{\mu\nu}{}_k - g \, f_{k\ell n} \, \Lambda_\ell \, F^{\mu\nu}{}_n \qquad \text{(IV. 36)}$$

This is the law of transformation of a chromo-vector, like that of $\bar{q} \, \frac{\lambda_k}{2} \, q$, to first order in g.

In the quantized form of the theory, one needs to add two other terms, the gauge-fixing and the so-called Fadeev-Popov term or ghost term*.

The colour gauge field, or gluon field, is thus a set of eight massless vector fields in interaction with themselves and with the quarks through their colour degree of freedom [109-133].

The equations of this system, ignoring the Fadeev-Popov terms, are of the form of those in the § IV. 5, namely :

$$D_{\nu;k\ell} \; F^{\mu\nu}{}_{\ell} = g\, j^{\mu}{}_{k} \,,$$

$$D^{\nu}{}_{k\ell} \, F^{\alpha\beta}{}_{\ell} + D^{\beta}{}_{k\ell} \, F^{\nu\alpha}{}_{\ell} + D^{\alpha}{}_{k\ell} \; F^{\beta\nu}{}_{\ell} = 0 \,, \qquad \text{(IV. 37)}$$

$$i\,\gamma^{\alpha} \; D_{\alpha} \; q - m\, q = 0,$$

where

$$j^{\mu}{}_{k} = \sum_{f} \bar{q}_{f} \; \gamma^{\mu} \; \frac{\lambda_k}{2} \; q_{f}$$

is the current of the matter field only.

---

* The gauge fixing term is $-\frac{1}{2a}(\partial_{\mu} a^{\mu}{}_{k})^2$ similar to the term (II. 22) in electrodynamics. However, contrary to the case in electrodynamics, no simple restriction on the gauge functions $\Lambda_k(x)$ similar to equation (II. 19b), can be found in chromodynamics. Scalar and longitudinal gluons are only cancelled if a second term is added of the form $-\partial_{\mu}\varphi_k^{+}(x)\, D^{\mu}\varphi_k(x)$ where $\varphi_k(x)$ are eight scalar (ghost) fields quantized according to the Pauli principle (therefore, with negative metric in Hilbert space) and $(D^{\mu}\varphi)_k = (\partial^{\mu}\delta_{k\ell} - ig\, f_{k\ell n}\, a^{\mu}{}_{n})\,\varphi_{\ell}(x)$.

PROBLEMS

IV - 1. Let the three 2×2 hermitian matrices τ_k with complex elements, such that :

$$\left[\frac{\tau_k}{2}, \frac{\tau_\ell}{2}\right]_+ = \frac{1}{2}\,\delta_{k\ell}$$

$$\left[\frac{\tau_k}{2}, \frac{\tau_\ell}{2}\right]_- = i\,\varepsilon_{k\ell n}\,\frac{\tau_n}{2} \quad ; \quad k,\ \ell = 1,2,3.$$

Show that : a) $\mathrm{Tr}\,(\tau_k) = 0$

b) $\det\,(\tau_k) = -\,1$

c) by choosing τ_3 diagonal and of the form $\tau_3 = \begin{pmatrix} 1 & 0 \\ 0 & -1 \end{pmatrix}$ what is the most general form for τ_1 and τ_2 ?

IV - 2. Let the eight 3×3 hermitian matrices λ_k with complex elements, such that :

$$\left[\frac{\lambda_k}{2}, \frac{\lambda_\ell}{2}\right]_+ = \frac{1}{3}\,\delta_{k\ell} + d_{k\ell n}\,\frac{\lambda_n}{2},$$

$$\left[\frac{\lambda_k}{2}, \frac{\lambda_\ell}{2}\right]_- = i\,f_{k\ell n}\,\frac{\lambda_n}{2}, \quad k,\ \ell,\ n = 1,2,\ \ldots\ 8,$$

where the totally symmetric coefficients $d_{k\ell n}$ and the totally antisymmetric coefficients $f_{k\ell n}$ are given as follows :

$$d_{146} = d_{157} = -\,d_{247} = d_{256} = d_{344} = d_{355} = -\,d_{366} = -\,d_{377} = \frac{1}{2}\ ;$$

$$d_{118} = d_{228} = d_{338} = -\,d_{888} = \frac{1}{\sqrt{3}}\ ;$$

$$d_{448} = d_{558} = d_{668} = d_{778} = -\,\frac{1}{2\sqrt{3}}\ ;$$

$$f_{123} = 1\ ;$$

$$f_{147} = - f_{156} = f_{246} = f_{257} = f_{345} = - f_{367} = \frac{1}{2} \quad ;$$

$$f_{458} = f_{678} = \frac{\sqrt{3}}{2}$$

Show that : a) $Tr\ (\lambda_k) = 0, \quad k = 1,2, \dots 8$

b) $T_r\ (\lambda_k\ \lambda_\ell) = 2\ \delta_{k\ell}$ for $k, \ell = 1,2, \dots 8$

c) $\det \lambda_1 = \det \lambda_2 = \det \lambda_3 = 0$;

d) the matrices $\lambda_1, \lambda_2, \lambda_3$ can have only the numbers 0, 1, - 1 as eigen values ;

e) if λ_3 is taken as diagonal and of the form :

$$\lambda_3 = \begin{pmatrix} 1 & 0 & 0 \\ 0 & -1 & 0 \\ 0 & 0 & 0 \end{pmatrix}$$

show that :

$$\lambda_1 = \begin{pmatrix} \tau_1 & 0 \\ 0 & 0 \end{pmatrix} \quad ; \quad \lambda_2 = \begin{pmatrix} \tau_2 & 0 \\ 0 & 0 \end{pmatrix} \quad ;$$

$$\lambda_8 = \frac{1}{\sqrt{3}} \begin{pmatrix} 1 & 0 & 0 \\ 0 & 1 & 0 \\ 0 & 0 & -2 \end{pmatrix} \quad ;$$

$$\lambda_4 = \begin{pmatrix} 0 & 0 & 0^{i\alpha} \\ 0 & 0 & 0 \\ e^{-i\alpha} & 0 & 0 \end{pmatrix} \quad ; \quad \lambda_6 = \begin{pmatrix} 0 & 0 & 0 \\ 0 & 0 & e^{i\alpha} \\ 0 & e^{-i\alpha} & 0 \end{pmatrix} \quad ;$$

$$\lambda_5 = \begin{pmatrix} 0 & 0 & -ie^{i\alpha} \\ 0 & 0 & 0 \\ ie^{-i\alpha} & 0 & 0 \end{pmatrix} \quad ; \quad \lambda_7 = \begin{pmatrix} 0 & 0 & 0 \\ 0 & 0 & -ie^{i\alpha} \\ 0 & ie^{-i\alpha} & 0 \end{pmatrix}$$

where α is an arbitrary phase factor usually assumed zero.

IV - 3. Derive the transformation law for the vector gauge field $A_{\mu a}(x)$ (IV. 11b).

IV - 4. Given the 2 x 2 antisymmetric unitary matrix C such that :

$$ {}^{t}\vec{\tau} = - C^{-1} \vec{\tau} C $$

and the transformation operator

$$ U(x) = \exp (ig \vec{\Lambda}(x) \cdot \frac{\vec{\tau}}{2}) $$

deduce the transformation law for an isovector $\vec{f}(x)$:

$$ f'_k(x) = a_{k\ell} \; f_\ell(x) $$

a) Find $a_{k\ell}$ in terms of $U(x)$ and the τ matrices.

b) What is the form of $a_{k\ell}$ for infinitesimal transformations ?

IV - 5. An isospin spinor-vector can be defined as :

$$ \psi_{abc}(x) = \vec{f}_a(x) \cdot \left(\frac{\vec{\tau}\, C}{2}\right)_{bc} $$

a, b, c = 1,2, where ψ_{abc} is totally symmetric in its indices.

a) Which conditions on $\vec{f}_a$ reduce its six components to four independent components (corresponding to isospin $\frac{3}{2}$) ?

b) What are the finite and the infinitesimal transformation laws of $\vec{f}_a$?

IV - 6. Show that if the field

$$ F^{\mu\nu}{}_k = \partial^\nu A^\mu{}_k - \partial^\mu A^\nu{}_k + g\, \varepsilon_{k\ell n} A^\mu{}_\ell A^\nu{}_n $$

vanishes in all space-time, the potential $A^\mu{}_a$ can be expressed as

$$ A^\mu{}_a = \partial^\mu \Lambda_a + g\, \varepsilon_{abc} \Lambda_b A^\mu{}_c $$

and therefore can be transformed away by a gauge transformation.

IV - 7. From Dirac's equation for a spinor field $\psi(x)$ in interaction with a Yang-Mills field deduce the second order equation for ψ

CHAPTER V

The Gravitational Gauge Field

V. 1 - INTRODUCTION

The relativistic theory of gravitation was, after Maxwell's theory of the electromagnetic field, the second historical example of a gauge field theory. It was the achievement of Einstein's efforts, between the years 1905 and 1915, to generalize his special theory of relativity and answer the following question : should the independence of the physical laws of the state of motion of laboratories be restricted to uniform translations of laboratories relative to each other ? By taking into account the old empirical fact that the inertial mass of a body is equal to its gravitational mass, and by assuming that the physical laws ought to be independent of any state of motion of the laboratory, Einstein discovered the principle of equivalence and was led to postulate that the gravitational field is described by the metric tensor $g_{\mu\nu}(x)$ which determines the Riemannian geometry of space time. Thus the general theory of relativity which he looked for turned out to be the relativistic theory of gravitation. The geometry of space-time supplies us with the objects, the Riemann tensor and its contracted forms , necessary for the generalization of Poisson's equation. Einstein's gravitational field equations relates these objects to the energy-momentum tensor of all the other fields and therefore unifies the geometry of space-time and gravitational dynamics.

In this Chapter, we shall briefly review the foundations [173-177)] of this theory by following a method similar to that of the previous chapters.

V. 2 - GROUPS OF LOCAL TRANSFORMATIONS AND COVARIANT DERIVATIVES

The need of covariant derivatives in theories involving local groups of transformations follows from the fact that field functions taken at different points of space-time do not form a linear space under such transformations. Thus a given element $U(x)$ of such a group, which transforms a field $\psi(x)$ at each point of space-time

$$\psi(x) \rightarrow \psi'(x) = U(x)\, \psi(x) \qquad (V.\ 1)$$

does not transform a linear combination of fields at different points into the linear combination of the transformed fields :

$$\alpha\, \psi(x) + \beta\, \varphi(y) \rightarrow \alpha\; U(x)\; \psi(x) + \beta\, U(y)\, \varphi(y) \neq$$

$$\neq U(x) \left\{ \alpha\, \psi(x) + \beta\, \varphi(y) \right\} \qquad (V.\ 2)$$

Only locally, for fields at the same point, the linear superposition holds :

$$\alpha\, \psi(x) + \beta\, \varphi(x) \rightarrow \alpha\, U(x)\, \psi(x) + \beta\, U(x)\, \varphi(y) =$$

$$= U(x) \left\{ \alpha\, \psi(x) + \beta\, \varphi(y) \right\} \qquad (V.\ 3)$$

Therefore, as the derivative involves the comparison of values of the field at different neighbouring points, the ordinary derivative is not covariant.

The introduction of a covariant derivative results then from the notion of parallel displacement of the field. As we want to avoid the sum of fields at different points we note that we can define another kind of derivative if we substract the parallel $\bar{\psi}(x + dx)$ to the field $\psi(x)$ from $\psi(x + dx)$. Under global transformations, this is what we do for obtaining the ordinary derivative. The parallel to $\psi(x)$ at the point $x + dx$ coincides with (transforms like) $\psi(x)$ (fig. 1). To obtain the parallel, in geometric language we simply

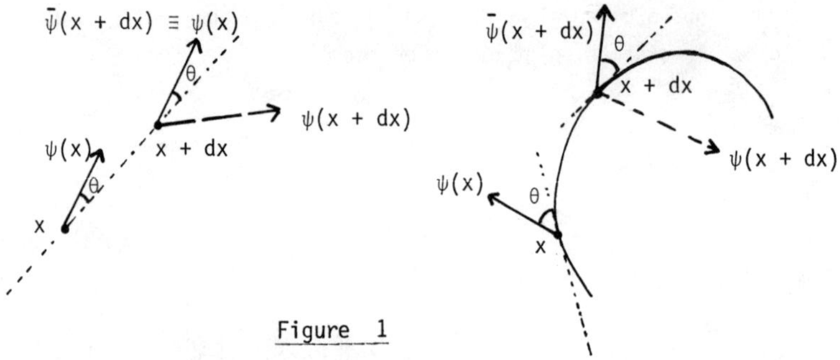

Figure 1

consider the vectors $\psi(x)$ and $\bar{\psi}(x + dx)$ which make respectively the same angle with a given line such as the one which connects the points x, $x + dx$.

Now under a local group the parallel $\bar{\psi}(x + dx)$ will be different from (transforms differently from) $\psi(x)$, so we set :

$$\bar{\psi}(x + dx) = \psi(x) - i\, g\, A_{\mu,a}(x)\, T_a\, \psi(x)\, dx^\mu \qquad (V.\ 4)$$

where the T_a are the generators of the infinitesimal transformation $U(x)$

$$U(x) = I + i\, g\, \Lambda_a(x)\, T_a \qquad \text{(sum over a)} \qquad (V.\ 5)$$

The field $A_{\mu,a}(x)$ appears as an affine connection similar to the one in Riemanian geometry.

As $\bar{\psi}(x + dx)$ transforms like $\psi(x + dx)$, the covariant derivative is then defined by subtraction between $\psi(x + dx)$ and $\bar{\psi}(x + dx)$:

$$D_\mu\, \psi(x) \cdot dx^\mu = \psi(x + dx) - \bar{\psi}(x + dx) = \left\{ \partial_\mu \psi + i\, g\, A_{\mu,a}(x)\, T_a\, \psi \right\} dx^\mu \qquad (V.\ 6)$$

In the preceding chapters, we had

$$T_a = \frac{\tau_a}{2}, \qquad a = 1,2,3$$

for the SU(2) group,

$$T_a = \frac{\lambda_a}{2}, \qquad a = 1, \ldots 8$$

for the SU(3) group and

$$T_a = I$$

for the electromagnetic U(1) group.

The covariant derivative is so chosen that it transforms like $\psi(x)$

$$D'_\mu \psi'(x) = U(x) D_\mu \psi(x) \tag{V. 7}$$

which entails for $A_{\mu,a}(x)$ the transformation law :

$$A'_{\mu a} T_a = U A_{\mu b} T_b U^{-1} + \frac{i}{g} (\partial_\mu U) U^{-1}$$

or :

$$A'_{\mu,a} T_a = A_{\mu,a} T_a - i g A_{\mu,a} \left[T_a, T_b \right] \Lambda_b + \frac{i}{g} U^{-1} \partial_\mu U \tag{V. 8}$$

for infinitesimal transformations (V. 5).

If

$$\left[T_a, T_b \right] = i C_{abc} T_c \tag{V. 8a}$$

are the commutation rules of the generators, then

$$A'_{\mu,a} = A_{\mu,a} - \partial_\mu \Lambda_a - g C_{abc} \Lambda_b A_{\mu,c}$$

The vectors under this group will transform like

$$\psi^+ T_a \psi \to \psi^+ T_a \psi - g C_{abc} \Lambda_b \psi^+ T_c \psi \tag{V. 8b}$$

under transformations (V. 5), V. 8a).

If the transformations U(x) form a group with n parameters, this group will have n generators and there will be n gauge fields.

V. 3 - COVARIANT DERIVATIVES OF TENSORS IN GENERAL RELATIVITY : THE GRAVITATIONAL GAUGE FIELD

In general relativity the principle of general covariance requires the physical equations to be invariant under general transformations of the coordinates :

$$x'^{\mu} = f^{\mu}(x) \qquad (V.\ 9)$$

where the functions f^{μ} are four independent real functions of x^{ν}, that is, their Jacobian does not vanish

$$J \equiv \left| \frac{\partial x'^{\mu}}{\partial x^{\nu}} \right| \neq 0 \qquad (V.\ 9a)$$

so that equation (V. 9) can be inverted.

The differentials dx'^{μ} and dx^{ν} are related by the equation :

$$dx'^{\mu} = \frac{\partial f^{\mu}}{\partial x^{\nu}} \quad dx^{\nu} \qquad (V.\ 10)$$

A contravariant vector is then defined as a set of four functions which transform like the differentials above :

$$F'^{\mu}(x') = \frac{\partial x'^{\mu}}{\partial x^{\nu}} \quad F^{\nu}(x) \qquad (V.\ 11)$$

and a covariant vector $\mathcal{G}_{\mu}(x)$ by the condition of invariance of the scalar product with any contravariant vector :

$$F'^{\mu}(x') \; \mathcal{G}'_{\mu}(x') = \frac{\partial x'^{\mu}}{\partial x^{\nu}} \; F^{\nu}(x) \mathcal{G}'_{\mu}(x')$$

This requires that :

$$\mathcal{G}_\nu(x) = \frac{\partial x'^\mu}{\partial x^\nu} \mathcal{G}'_\mu(x)$$

or

$$\mathcal{G}'_\mu(x') = \frac{\partial x^\nu}{\partial x'^\mu} \mathcal{G}_\nu(x) \qquad \text{(V. 12)}$$

The metric, as is well known, is now a (symmetric in its indices) function of point in space-time $g_{\mu\nu}(x)$ and the line element is :

$$ds^2 = g_{\mu\nu}(x)\, dx^\mu\, dx^\nu \qquad \text{(V. 13)}$$

A tensor of order $m + n$, m times contravariant and n times covariant transforms then like :

$$T'^{\mu_1 \cdots \mu_m}_{\nu_1 \cdots \nu_n}(x') = \frac{\partial x'^{\mu_1}}{\partial x_{\alpha_1}} \cdots \frac{\partial x'^{\mu_m}}{\partial x_{\alpha_m}} T^{\alpha_1 \cdots \alpha_m}_{\beta_1 \cdots \beta_n}(x) \frac{\partial x^{\beta_1}}{\partial x'^{\nu_1}} \cdots \frac{\partial x^{\beta_n}}{\partial x'^{\nu_n}}$$

(V. 14)

To the metric tensor is associated the co-factor of $g_{\mu\nu}(x)$, $\Delta^{\mu\nu}(x)$, such that the quantity :

$$g^{\mu\nu}(x) = \frac{\Delta^{\mu\nu}}{g}, \qquad g = \det(g_{\mu\nu}) \qquad \text{(V. 14a)}$$

is a symmetric contravariant tensor and satisfies the relation at each point x :

$$g^{\mu\nu}(x)\, g_{\nu\rho}(x) = \delta^\mu_\rho \qquad \text{(V. 15)}$$

Clearly, as in equation (V. 2), the vectors in this space form a linear space only locally :

$$a F^\alpha(x) + b\, \mathcal{G}^\alpha(x) \to \frac{\partial x'^\alpha}{\partial x^\nu}\left[a F^\nu(x) + b\, \mathcal{G}^\nu(x)\right] \qquad \text{(V. 16)}$$

this superposition does not hold for vectors at different points of space-time. One then needs to define the parallel displacement of a vector by an equation similar to (V. 4), and we shall write

$$\bar{F}^{\alpha}(x + dx) = F^{\alpha}(x) - \Gamma^{\alpha}{}_{\mu\nu}(x)\, dx^{\mu}\, F^{\nu}(x) \qquad \text{(V. 17)}$$

The covariant derivative of a vector will then be :

$$\nabla_{\mu} F^{\alpha}(x)\, dx^{\mu} = F^{\alpha}(x + dx) - \bar{F}^{\alpha}(x + dx)$$

so that :

$$\nabla_{\mu} F^{\alpha}(x) = \partial_{\mu} F^{\alpha}(x) + \Gamma^{\alpha}{}_{\mu\nu}(x)\, F^{\nu}(x) \qquad \text{(V. 18)}$$

The quantities $\Gamma^{\alpha}{}_{\mu\nu}(x)$, the affine connection, or Christoffel symbols, are the gravitational gauge fields. The requirement that (V. 18) transform like a tensor

$$\nabla'_{\mu} F'^{\alpha}(x') = \frac{\partial x^{\nu}}{\partial x'^{\mu}} \frac{\partial x'^{\alpha}}{\partial x^{\lambda}} \nabla_{\nu} F^{\lambda}(x) \qquad \text{(V. 19)}$$

determines the transformation law for $\Gamma^{\alpha}{}_{\mu\nu}(x)$:

$$\Gamma'^{\alpha}{}_{\mu\nu}(x') = \frac{\partial x'^{\alpha}}{\partial x^{\lambda}} \Gamma^{\lambda}{}_{\xi\eta} \frac{\partial x^{\xi}}{\partial x'^{\mu}} \frac{\partial x^{\eta}}{\partial x'^{\nu}} - \frac{\partial x^{\xi}}{\partial x'^{\mu}} \frac{\partial x^{\eta}}{\partial x'^{\nu}} \frac{\partial^2 x'^{\alpha}}{\partial x^{\xi}\, \partial x^{\eta}} \qquad \text{(V. 20)}$$

Like all gauge fields seen previously, this is also not covariant, it does not transform like a tensor.

To obtain the covariant derivative of tensors we remark the following identity :

$$\partial_\mu (F^\alpha(x)\, F_\alpha(x)) = (\nabla_\mu F^\alpha)\, F_\alpha + F^\alpha(\nabla_\mu F_\alpha)$$

if the covariant derivative of a covariant vector is

$$\nabla_\mu F_\alpha = \partial_\mu F_\alpha - \Gamma^\nu{}_{\mu\alpha} F_\nu \qquad \text{(V. 21)}$$

Now we may write

$$\partial_\mu (F^\alpha F^\beta) = F^\alpha \nabla_\mu F^\beta + (\nabla_\mu F^\alpha)\, F^\beta - \Gamma^\beta{}_{\mu\nu} F^\alpha F^\nu - \Gamma^\alpha{}_{\mu\nu} F^\nu F^\beta$$

Therefore the covariant derivative of a tensor of the second rank is :

$$\nabla_\mu T^{\alpha\beta} = \partial_\mu T^{\alpha\beta} + \Gamma^\alpha{}_{\mu\nu} T^{\nu\beta} + \Gamma^\beta{}_{\mu\nu} T^{\alpha\nu} \qquad \text{(V. 22)}$$

In general, we have, for a tensor of any rank :

$$\nabla_\eta T^{\alpha\beta\,\cdots}_{\mu\nu\,\cdots} = \partial_\eta T^{\alpha\beta\,\cdots}_{\mu\nu\,\cdots} + \Gamma^\alpha_{\eta\lambda} T^{\lambda\beta\,\cdots}_{\mu\nu\,\cdots} + \Gamma^\beta_{\eta\lambda} T^{\alpha\lambda\,\cdots}_{\mu\nu\,\cdots} +$$

$$+ \;\cdots\; - \Gamma^\lambda_{\eta\mu} T^{\alpha\beta\,\cdots}_{\lambda\nu\,\cdots} - \Gamma^\lambda_{\eta\nu} T^{\alpha\beta\,\cdots}_{\mu\lambda\,\cdots} - \;\cdots \qquad \text{(V. 23)}$$

The affine connection is symmetric in its lower indices :

$$\Gamma^\alpha{}_{\mu\nu} = \Gamma^\alpha{}_{\nu\mu} \qquad \text{(V. 24)}$$

In fact if one defines the quantity

$$\left\{ {\alpha \atop \mu\nu} \right\} = \frac{1}{2}\, g^{\alpha\lambda}(\partial_\mu g_{\lambda\nu} + \partial_\nu g_{\lambda\mu} - \partial_\lambda g_{\mu\nu}) \qquad \text{(V. 25)}$$

its transformation law to a new system of coordinates (V. 9) will be :

$$\begin{Bmatrix} \alpha \\ \mu\nu \end{Bmatrix}' = \frac{\partial x'^{\alpha}}{\partial x^{\lambda}} \begin{Bmatrix} \lambda \\ \xi\eta \end{Bmatrix} \frac{\partial x^{\xi}}{\partial x'^{\mu}} \frac{\partial x^{\eta}}{\partial x'^{\nu}} -$$

$$- \frac{\partial x^{\xi}}{\partial x'^{\mu}} \frac{\partial x^{\eta}}{\partial x'^{\nu}} \frac{\partial^2 x'^{\alpha}}{\partial x^{\xi} \partial x^{\eta}}$$

that is, this is the same law as (V. 20). Therefore the difference between $\Gamma^{\alpha}_{\mu\nu}$ and (V. 25) will be a tensor :

$$\Gamma'^{\alpha}_{\mu\nu} - \begin{Bmatrix} \alpha \\ \mu\nu \end{Bmatrix}' = \frac{\partial x'^{\alpha}}{\partial x^{\lambda}} \left[\Gamma^{\lambda}_{\xi\eta} - \begin{Bmatrix} \lambda \\ \xi\eta \end{Bmatrix} \right] \frac{\partial x^{\xi}}{\partial x'^{\mu}} \frac{\partial x^{\eta}}{\partial x'^{\nu}}$$

Now according to the equivalence principle there exists in every point of space-time a locally inertial system of coordinates in which the effects of gravitation are transformed away, hence $\Gamma^{\alpha}_{\mu\nu}$ vanishes in such a system, ∇_{μ} is ∂_{μ} and the first derivatives of $g_{\mu\nu}$ vanish. Therefore, as the difference above is a tensor it will vanish everywhere and so :

$$\Gamma^{\alpha}_{\mu\nu} \equiv \begin{Bmatrix} \alpha \\ \mu\nu \end{Bmatrix} = \frac{1}{2} \; g^{\alpha\lambda}(\partial_{\mu} \; g_{\lambda\nu} + \partial_{\nu} \; g_{\lambda\mu} - \partial_{\lambda} \; g_{\mu\nu}) \qquad \text{(V. 25a)}$$

and

$$\Gamma^{\alpha}_{\mu\nu} = \Gamma^{\alpha}_{\nu\mu} \qquad \text{(V. 25b)}$$

Thus $g_{\mu\nu}(x)$ is the gravitation potential in terms of which the gauge field is expressed.

V.4 - THE LAGRANGEAN OF MATTER TENSOR FIELDS IN INTERACTION WITH THE GRAVITATIONAL FIELD

In order to find this lagrangean we have first to look for the part of it which corresponds to the gravitational field alone. For this we follow the same procedure as that we used for the Yang-Mills and for the electromagnetic fields. We calculate the commutator between the covariant derivatives applied to a vector and define the tensor $\mathscr{R}^{\lambda}{}_{\alpha\mu\nu}(x)$ by the equation :

$$\mathscr{R}^{\lambda}{}_{\alpha\nu\mu} F_{\lambda} (x) = \left[\nabla_{\mu} , \nabla_{\nu} \right] F_{\alpha}(x) \qquad \text{(V. 26)}$$

We obtain

$$\mathscr{R}^{\lambda}{}_{\alpha\nu\mu} = \partial_{\nu} \Gamma^{\lambda}{}_{\alpha\mu} - \partial_{\mu} \Gamma^{\lambda}{}_{\alpha\nu} + \Gamma^{\eta}{}_{\alpha\mu} \Gamma^{\lambda}{}_{\eta\nu} - \Gamma^{\eta}{}_{\alpha\nu} \Gamma^{\lambda}{}_{\eta\mu} \qquad \text{(V. 26a)}$$

this is a fourth-rank tensor, the so-called Riemann curvature tensor which satisfies the following relations :

$$\mathscr{R}^{\lambda}{}_{\alpha\nu\mu} + \mathscr{R}^{\lambda}{}_{\mu\alpha\nu} + \mathscr{R}^{\lambda}{}_{\nu\mu\alpha} = 0$$

and if

$$\mathscr{R}_{\beta\alpha\nu\mu} = g_{\beta\lambda} \mathscr{R}^{\lambda}{}_{\alpha\nu\mu}$$

then

$$\begin{aligned} \mathscr{R}_{\beta\alpha\nu\mu} &= - \mathscr{R}_{\alpha\beta\nu\mu} \\ \mathscr{R}_{\beta\alpha\mu\nu} &= - \mathscr{R}_{\beta\alpha\nu\mu} \\ \mathscr{R}_{\beta\alpha\nu\mu} &= \mathscr{R}_{\nu\mu\beta\alpha} \end{aligned} \qquad \text{(V. 26b)}$$

These equations reduce the number of components of this tensor from 256 to only 20 independent components.

What is the invariant space-time volume element over which we must integrate the lagrangean to get the action in relativistic theory of gravitation ?

As the metric tensor transforms like :

$$g'_{\mu\nu}(x') = \frac{\partial x^{\alpha}}{\partial x'^{\mu}} \frac{\partial x^{\beta}}{\partial x'^{\nu}} g_{\alpha\beta}(x)$$

we have for the determinant of $g_{\alpha\beta}$ the following transformation law :

$$g' = \left| \frac{\partial x}{\partial x'} \right|^2 g$$

where

$$g = \det g_{\alpha\beta}$$

therefore, according to (V. 9a) :

$$g = J^2 g'$$

As

$$d^4x' = J\, d^4x$$

we see that

$$\sqrt{-g}\; d^4x = \sqrt{-g'}\; d^4x'$$

is the required invariant volume (the - sign is for the limiting case of a flat space where

$$\eta_{\alpha\beta} = \begin{pmatrix} 1 & 0 & 0 & 0 \\ 0 & -1 & 0 & 0 \\ 0 & 0 & -1 & 0 \\ 0 & 0 & 0 & -1 \end{pmatrix}, \quad \det\ \eta_{\alpha\beta} = -1$$

To obtain the action for a vector field $F^{\alpha}(x)$ in interaction with the gravitational field we first replace the ordinary derivatives of $F^{\alpha}(x)$ by the covariant derivatives, in the free field lagrangean for F^{α} :

$$L(F^{\alpha}(x),\ \partial_{\mu}\ F^{\alpha}(x)) \rightarrow L(F^{\alpha}(x),\ \nabla_{\mu}\ F^{\alpha}(x)) \qquad \text{(V. 26c)}$$

Then we form the scalar curvature $\mathscr{R}$ from the Riemann tensor, that is :

$$\mathscr{R}_{\mu\nu} = \mathscr{R}^{\lambda}{}_{\mu\alpha\nu}\ \delta^{\alpha}{}_{\lambda} \qquad \text{(V. 27)}$$

and

$$\mathscr{R} = g^{\mu\nu}\ \mathscr{R}_{\mu\nu} = \mathscr{R}^{\mu}{}_{\mu} \qquad \text{(V. 27a)}$$

The action is then

$$S = \int \sqrt{-g}\ d^4x\ (\mathscr{R} + L(F^{\alpha},\ \nabla_{\mu}\ F^{\alpha})) \qquad \text{(V. 28)}$$

V. 5 - EINSTEIN'S EQUATIONS OF THE GRAVITATIONAL FIELD

To obtain the gravitational field equations from a variation principle with the action (V. 28) it is usual to write the coupling constant

$$\kappa = \frac{8\pi}{c^4} \mathcal{G}$$

where $\mathcal{G}$ is the gravitational constant, in an explicit form in (V. 27) :

$$S = \int \sqrt{-g}\, d^4x \,(\mathcal{R} - 2\,\kappa\, L\,) = S_g + S_F \tag{V. 28a}$$

The variation principle assumes

$$\delta\, S = 0 \tag{V. 28b}$$

for arbitrary variations $\delta\, g^{\mu\nu}$, $\delta\, \partial^{\alpha}\, g^{\mu\nu}$ which vanish at the boundary of space-time.

We have :

$$\delta\, S_g = \int d^4x \left\{ \sqrt{-g}\, g^{\mu\nu}\, \delta\, \mathcal{R}_{\mu\nu} + \mathcal{R}_{\mu\nu} \sqrt{-g}\, \delta\, g^{\mu\nu} + \right.$$

$$\left. + \mathcal{R}_{\mu\nu}\, g^{\mu\nu}\, \delta\, (\sqrt{-g}) \right\} \tag{V. 29}$$

Now in a locally inertial coordinate system, the affine connections vanish so that (see (V. 26a)) :

$$\delta\, \mathcal{R}_{\mu\nu} = \delta(\partial_{\lambda}\, \Gamma^{\lambda}{}_{\mu\nu}) - \delta(\partial_{\nu}\, \Gamma^{\lambda}{}_{\lambda\mu})$$

In this system the ordinary derivatives coincide with the covariant derivatives so :

$$\delta \mathscr{R}_{\mu\nu} = \nabla_\lambda (\delta \Gamma^\lambda_{\mu\nu}) - \nabla_\nu(\delta \Gamma^\lambda_{\lambda\mu})$$

and this relationship is general since both sides of this equation are tensors.

- The first integral on the right-hand side of (V. 29) may be written :

$$\int d^4x \sqrt{-g}\, g^{\mu\nu} \delta \mathscr{R}_{\mu\nu} =$$

$$= \int d^4x \sqrt{-g} \left\{ \nabla_\lambda(g^{\mu\nu} \delta \Gamma^\lambda_{\mu\nu}) - \nabla_\nu(g^{\mu\nu} \delta \Gamma^\lambda_{\lambda\mu}) \right\} \qquad \text{(V. 30)}$$

This is so because the covariant derivative of $g^{\mu\nu}$ vanishes

$$\nabla_\lambda g^{\mu\nu} = 0$$

since it vanishes in a locally inertial coordinate system at any point. Now $\delta \Gamma^\lambda_{\mu\nu}$ is a tensor hence one can write :

$$\nabla_\lambda(g^{\mu\nu} \delta \Gamma^\lambda_{\mu\nu}) - \nabla_\nu(g^{\mu\nu} \delta \Gamma^\lambda_{\lambda\mu}) = \nabla_\lambda A^\lambda - \nabla_\nu B^\nu$$

where A^λ and B^λ are two vectors.

The divergence of a vector is :

$$\nabla_\lambda A^\lambda = \partial_\lambda A^\lambda + (\partial_\lambda \log \sqrt{-g}) A^\lambda = \frac{1}{\sqrt{-g}} \partial_\lambda (\sqrt{-g} A^\lambda)$$

since

$$\Gamma^\lambda{}_{\lambda\mu} = \partial_\mu \log \sqrt{-g}$$

Therefore (V. 30) can be transformed away :

$$\int d^4x \sqrt{-g}\, g^{\mu\nu} \delta \mathscr{R}_{\mu\nu} = \int d^4x \left\{ \partial_\lambda(\sqrt{-g} A^\lambda) - \partial_\lambda(\sqrt{-g} B^\lambda) \right\} =$$

$$= \int_{\text{boundary}} d\sigma_\nu \quad (A^\nu - B^\nu) \sqrt{-g} \;\; = 0$$

This integral vanishes because $\delta g^{\mu\nu}$ and $\delta \Gamma^\lambda{}_{\mu\nu}$ vanish at the boundary.

The third term on the right-hand side of equation (V. 29) is :

$$\int d^4x \, \mathscr{R}_{\mu\nu} g^{\mu\nu} \delta (\sqrt{-g}) = \int d^4x \sqrt{-g} \left(- \frac{1}{2} g_{\mu\nu} \mathscr{R}\right) \delta g^{\mu\nu}$$

since from (V. 14a) one deduces :

$$g = \sum_\beta g_{\alpha_0\beta} \Delta^{\alpha_0\beta}$$

(which is the development of g according to the element of a line α_0) and :

$$\frac{\partial g}{\partial g_{\alpha\beta}} = \Delta^{\alpha\beta}$$

$$g^{\mu\nu}(x) = \frac{1}{g} \frac{\partial g}{\partial g_{\mu\nu}} \qquad \text{(V. 31)}$$

so that :

$$\delta(\sqrt{-g}) = \frac{\partial(\sqrt{-g})}{\partial\, g^{\mu\nu}}\, \delta\, g^{\mu\nu} = -\frac{1}{2}\frac{1}{\sqrt{-g}}\ \frac{\partial\, g}{\partial\, g^{\mu\nu}}\ \delta\, g^{\mu\nu} =$$

$$= -\frac{1}{2}\sqrt{-g}\ g_{\mu\nu}\ \delta\, g^{\mu\nu}$$

Collecting the terms of $\delta\, S_g$ we thus have :

$$\delta\,(S_g) = \int d^4x\, \sqrt{-g}\, (\mathscr{R}_{\mu\nu} - \frac{1}{2}\, g_{\mu\nu}\, \mathscr{R})\ \delta\, g^{\mu\nu}$$

Let us calculate $\delta\, S_F$. One has :

$$\delta \int d^4x\, \sqrt{-g}\, L = \int d^4x\, (L\, \delta\, \sqrt{-g} + \sqrt{-g}\ \delta\, L)$$

Now

$$\int d^4x\, \sqrt{-g}\, \delta\, L = \int d^4x\, \sqrt{-g} \left\{ \frac{\partial\, L}{\partial\, g^{\mu\nu}}\, \delta\, g^{\mu\nu} + \frac{\partial\, L}{\partial(\partial^\alpha\, g^{\mu\nu})}\, \delta(\partial^\alpha\, g^{\mu\nu}) \right\} =$$

$$= \int d^4x\, \sqrt{-g} \left\{ \frac{\partial\, L}{\partial\, g^{\mu\nu}} - \frac{1}{\sqrt{-g}}\, \partial^\alpha\!\left(\sqrt{-g}\, \frac{\partial\, L}{\partial(\partial^\alpha\, g^{\mu\nu})}\right) \right\} \delta\, g^{\mu\nu}$$

Therefore

$$\delta \int d^4x\, \sqrt{-g}\ \ L = \int d^4x\, \sqrt{-g}\ \frac{1}{\sqrt{-g}} \left\{ \frac{\partial(\sqrt{-g}\ \ L)}{\partial\, g^{\mu\nu}} - \partial^\alpha \left(\frac{\partial(\sqrt{-g}\ L)}{\partial(\partial^\alpha\, g^{\mu\nu})} \right) \right\} \delta\, g^{\mu\nu}$$

The energy-momentum tensor of the matter field is defined by

$$T_{\mu\nu} = \frac{2}{\sqrt{-g}} \left\{ \frac{\partial(\sqrt{-g}\, L)}{\partial\, g^{\mu\nu}} - \partial^{\alpha} \left(\frac{\partial(\sqrt{-g}\, L)}{\partial(\partial^{\alpha}\, g^{\mu\nu})} \right) \right\} \qquad (V.\ 32)$$

so that

$$\delta \int d^4x\, \sqrt{-g}\, L_F = \frac{1}{2} \int d^4x\, \sqrt{-g}\, T_{\mu\nu}\, \delta\, g^{\mu\nu} \qquad (V.\ 32a)$$

According to equations (V. 28a), (V. 28b) one will then have :

$$\delta\, S = \int d^4x \sqrt{-g} \left\{ \mathcal{R}_{\mu\nu} - \frac{1}{2}\, g_{\mu\nu} \mathcal{R} - \kappa\, T_{\mu\nu} \right\} \delta\, g^{\mu\nu} = 0 \qquad (V.\ 33)$$

which entails, for arbitrary variations $\delta\, g^{\mu\nu}$ (vanishing at the boundary) :

$$\mathcal{R}_{\mu\nu} - \frac{1}{2}\, g_{\mu\nu}\, \mathcal{R} = \kappa\, T_{\mu\nu} \qquad (V.\ 34)$$

which are Einstein's equations for the gravitational field.

From the equations (V. 26),(V. 27) we see that Einstein's equations are linear in the second-order derivatives of the metric field $g_{\mu\nu}(x)$ but non-linear in $g_{\mu\nu}$ and its first derivatives. The non-linearity character of Einstein's equations (see equation (V. 26a), notably the bilinear terms in Γ) results from the self-interaction of the gravitational field : the energy-momentum tensor is the source of the gravitational field ; as the latter clearly carries energy and momentum it contributes to its own source just as a colour or an isospin gauge field having colour or isospin contributes to its source, the current.

V. 6 - THE ENERGY-MOMENTUM OF THE GRAVITATIONAL FIELD

The equations (V. 34) correspond to the equations (IV. 22a) for the Yang-Mills field : in both, the right-hand side contains only the part of the source due to matter : the matter energy-momentum tensor in (V. 34) and the matter current-vector in (IV. 22a). And in the same way that the latter is covariantly conserved, equation (IV. 23b), so is the matter energy-momentum tensor :

$$\nabla_\mu T^{\mu\nu}(x) = 0 \tag{V. 34a}$$

(Note that the covariant conservation of the matter sources contain terms coming from the field interaction).

This equation is in accord with Einstein's equation (V. 34) and its imposition was a guide for the search for a tensor :

$$G_{\mu\nu} = \mathscr{R}_{\mu\nu} - \frac{1}{2} g_{\mu\nu} \mathscr{R} \tag{V. 35}$$

with vanishing covariant divergence :

$$\nabla_\mu G^{\mu\nu} = 0 \tag{V. 35a}$$

To show that equations (V. 35), (V. 35a) are indeed satisfied, we consider the expressions (V. 26) of the curvature tensor and take its derivative :

$$\nabla_\beta \mathscr{R}^\lambda{}_{\alpha\mu\nu} = \nabla_\beta \left\{ \partial_\mu \Gamma^\lambda{}_{\alpha\nu} - \partial_\nu \Gamma^\lambda{}_{\alpha\mu} \right\} + (\nabla_\beta \Gamma^\eta{}_{\alpha\nu}) \Gamma^\lambda{}_{\eta\mu} +$$

$$+ \Gamma^\eta{}_{\alpha\nu} (\nabla_\beta \Gamma^\lambda{}_{\eta\mu}) - (\nabla_\beta \Gamma^\eta{}_{\alpha\mu}) \Gamma^\lambda{}_{\eta\nu} - \Gamma^\eta{}_{\alpha\mu} (\nabla_\beta \Gamma^\lambda{}_{\eta\nu})$$

Now in a locally inertial frame, the affine connection $\Gamma^{\lambda}_{\alpha\beta}$ vanishes and so we have

$$(\nabla_{\beta}\mathscr{R}^{\lambda}{}_{\alpha\mu\nu})_{\text{loc. inert. frame}} = \partial_{\beta}\,\partial_{\mu}\,\Gamma^{\lambda}{}_{\alpha\nu} - \partial_{\beta}\,\partial_{\nu}\,\Gamma^{\lambda}{}_{\alpha\mu}$$

In this frame therefore the following relationship holds :

$$\nabla_{\beta}\mathscr{R}^{\lambda}{}_{\alpha\mu\nu} + \nabla_{\nu}\mathscr{R}^{\lambda}{}_{\alpha\beta\mu} + \nabla_{\mu}\mathscr{R}^{\lambda}{}_{\alpha\nu\beta} = 0$$

which are the Bianchi identities (see equations (IV. 23), (IV. 23a) for the case of the Yang-Mills field) As the left-hand side is a tensor, this equation remains true in whatever system of reference.

The contraction of the indices λ and μ gives (see (V. 26a) and (V. 27)) :

$$\nabla_{\beta}\,\mathscr{R}_{\alpha\nu} - \nabla_{\nu}\,\mathscr{R}_{\alpha\beta} + \nabla_{\mu}\,\mathscr{R}^{\mu}{}_{\alpha\nu\beta} = 0$$

and a new contraction :

$$\nabla_{\beta}\,\mathscr{R} - \nabla_{\alpha}\,\mathscr{R}^{\alpha}{}_{\beta} - \nabla_{\mu}\,\mathscr{R}^{\mu}{}_{\beta} = 0 \qquad \text{(V. 36)}$$

The latter equation results from the fact that

$$g^{\alpha\nu}\,\nabla_{\beta}\,\mathscr{R}_{\alpha\nu} = \nabla_{\beta}\,(g^{\alpha\nu}\mathscr{R}_{\alpha\nu})$$

since the covariant derivative of the covariant and contravariant metric tensors is zero :

$$\nabla_\beta g_{\alpha\nu} \equiv \partial_\beta g_{\alpha\nu} - \Gamma^\lambda{}_{\beta\alpha} g_{\lambda\nu} - \Gamma^\lambda{}_{\beta\nu} g_{\alpha\lambda} = 0$$

$$\nabla_\beta g^{\alpha\nu} \equiv \partial_\beta g^{\alpha\nu} + \Gamma^\alpha{}_{\beta\lambda} g^{\lambda\nu} + \Gamma^\nu{}_{\beta\lambda} g^{\alpha\lambda} = 0 \qquad \text{(V. 37)}$$

This equation expresses the fact that in a locally inertial frame this covariant derivative vanishes and so it vanishes always since it is a tensor.

Therefore, from (V. 36) we get :

$$\nabla_\alpha(\mathscr{R}^\alpha{}_\beta - \frac{1}{2}\delta^\alpha{}_\beta \mathscr{R}) = 0 \qquad \text{(V. 37a)}$$

which proves (V. 35), (V. 35a).

Now we should like to show that Eistein's equations can be written in a form in which the energy-momentum of the gravitational field is exhibited and added to the matter energy-momentum. The equations we shall obtain correspond therefore to the equation of the Yang-Mills field (IV. 18) namely :

$$\partial_\nu F^{\mu\nu}{}_k = j^\mu{}_k \text{ (total)}$$

where the current above is the total current given by (IV. 19).

For this purpose we write for the metric tensor :

$$g_{\mu\nu}(x) = \eta_{\mu\nu} + h_{\mu\nu}(x) \qquad \text{(V. 38)}$$

where $h_{\mu\nu}(x)$ vanishes at infinity so that far away from matter the metric tends there to the flat-space metric $\eta_{\mu\nu}$.

As the curvature tensor is, according to (V. 26a), (V. 25a) :

$$\mathscr{R}_{\lambda\alpha\mu\nu} = \frac{1}{2} \left\{ \partial_\alpha \partial_\mu g_{\lambda\nu} + \partial_\nu \partial_\lambda g_{\alpha\mu} - \partial_\nu \partial_\alpha g_{\lambda\mu} - \partial_\lambda \partial_\mu g_{\alpha\nu} \right\} +$$

$$+ g_{\beta\varepsilon} \left\{ \Gamma^\beta{}_{\lambda\nu} \Gamma^\varepsilon{}_{\alpha\mu} - \Gamma^\beta{}_{\lambda\mu} \Gamma^\varepsilon{}_{\alpha\nu} \right\}$$

the substitution (V. 38) will give :

$$\mathscr{R}_{\lambda\alpha\nu\mu} = \frac{1}{2} \left\{ \partial_\alpha \partial_\mu h_{\lambda\nu} + \partial_\nu \partial_\lambda h_{\alpha\mu} - \partial_\nu \partial_\alpha h_{\lambda\mu} - \partial_\lambda \partial_\mu h_{\alpha\nu} \right\} +$$

$$+ \eta_{\beta\varepsilon} \left\{ \Gamma^\beta{}_{\lambda\nu} \Gamma^\varepsilon{}_{\alpha\mu} - \Gamma^\beta{}_{\lambda\mu} \Gamma^\varepsilon{}_{\alpha\nu} \right\} +$$

$$+ h_{\beta\varepsilon} \left\{ \Gamma^\beta{}_{\lambda\nu} \Gamma^\varepsilon{}_{\alpha\mu} - \Gamma^\beta{}_{\lambda\mu} \Gamma^\varepsilon{}_{\alpha\nu} \right\} \qquad \text{(V. 39)}$$

We separate the terms in $\mathscr{R}_{\alpha\nu}$ which are linear in $h_{\mu\nu}$:

$$\mathscr{R}^{(1)}{}_{\alpha\nu} = \frac{1}{2} \left\{ \partial_\alpha \partial_\lambda h^\lambda{}_\nu + \partial_\nu \partial_\lambda h_\alpha{}^\lambda - \partial_\nu \partial_\alpha h^\lambda{}_\lambda - \right.$$

$$\left. - \partial_\lambda \partial^\lambda h_{\alpha\nu} \right\} \qquad \text{(V. 39a)}$$

from the other terms in $\mathscr{R}_{\alpha\nu}$ and $\underline{g}_{\mu\nu} \mathscr{R}$, and we call these :

$$t_{\alpha\nu} \equiv - \frac{1}{\kappa} \left\{ \mathscr{R}_{\alpha\nu} - \frac{1}{2} g_{\alpha\nu} \mathscr{R} - \mathscr{R}^{(1)}{}_{\alpha\nu} + \frac{1}{2} \eta_{\alpha\nu} \mathscr{R}^{(1)} \right\} \qquad \text{(V. 40)}$$

(the indices in $h_{\alpha\nu}$, $\mathscr{R}^{(1)}{}_{\alpha\nu}$ and ∂_λ are raised and lowered with $\eta^{\alpha\nu}$, the indices on the full $\mathscr{R}_{\mu\alpha}$ are of course raised and lowered with $g_{\mu\alpha}$)

Einstein's equations (V. 34) assume therefore the form :

$$\mathscr{R}^{(1)}_{\alpha\nu} - \frac{1}{2}\eta_{\alpha\nu}\mathscr{R}^{(1)} = \kappa(T_{\alpha\nu} + t_{\alpha\nu}) \qquad (V.\ 41)$$

where

$$\mathscr{R}^{(1)} = \eta^{\alpha\beta}\mathscr{R}^{(1)}_{\alpha\beta}$$

These are important because they exhibit the form of an equation for a spin 2 field generated by a source which contains a part arising from the field itself (see equation (I. 8)) :

$$\begin{aligned} &\Box h_{\alpha\nu} - \partial_\alpha \partial_\lambda h^\lambda{}_\nu - \partial_\nu \partial_\lambda h^\lambda{}_\alpha + \partial_\alpha \partial_\nu h^\lambda{}_\lambda - \\ &- \eta_{\alpha\nu} \Box h^\lambda{}_\lambda + \eta_{\alpha\nu} \partial_\lambda \partial_\varepsilon h^{\lambda\varepsilon} = \qquad (V.\ 42) \\ &= -\kappa\,(T_{\alpha\nu} + t_{\alpha\nu}) \end{aligned}$$

The interpretation of this equation is this : the quantity

$$\tau_{\alpha\nu} = T_{\alpha\nu} + t_{\alpha\nu} \qquad (V.\ 43)$$

is the total energy-momentum "tensor" of matter and gravitation, $t_{\alpha\nu}$ is the part corresponding to the gravitational field. The term tensor is under quotation because $t_{\alpha\nu}$ does not transform like a tensor under general coordinate transformations. This is, however, similar to the case in the Yang-Mills field theory. In the equation (IV. 18), the right-hand side does not transform like an isovector (or a chromovector) but it is composed of the matter current which is an isovector and a field current which is not an isovector ; thus the total current "vector", as well as the Yang-Mills vector field $A^{\mu k}$ have the transformation laws given in equations (IV. 19a) and (IV. 13) and the terms in $\partial^{\mu} \Lambda_a$ are the ones which make them not to have an isovector character. Here both the gauge field $\Gamma^{\alpha}_{\mu\nu}$ and the source are also not generally covariant. From (V. 42) it follows that

$$\partial_{\nu} \tau^{\alpha\nu} = 0 \tag{V. 43a}$$

and so the total energy-momentum is conserved in the usual sense. Thus :

$$P^{\lambda} = \int d^3x \, \tau^{0\lambda} \tag{V. 43b}$$

is the conserved total energy-momentum "vector" of the system matter and gravitation. P^0, in particular, is always positive and is zero only for empty space.

As $\tau_{\alpha\nu}$ is symmetric the angular momentum tensor density

$$M^{\alpha\nu\lambda} = \tau^{\alpha\lambda} x^{\nu} - \tau^{\alpha\nu} x^{\lambda} \tag{V. 43c}$$

is conserved : $\partial_{\alpha} M^{\alpha\nu\lambda} = 0$, and

$$J^{\nu\lambda} = \int d^3x \, M^{0\nu\lambda} \tag{V. 43d}$$

is the total angular momentum of this system.

Although not covariant under the general coordinate transformation group, $\tau_{\alpha\nu}$, $t_{\alpha\nu}$ and P^{λ}, $M^{\alpha\nu\lambda}$, $J^{\nu\lambda}$ are covariant under the Lorentz group.

Thus despite such non-covariance, these quantities are conserved, Lorentz-covariant and also, as it can be shown, additive. In particular, P^{λ} plays the rôle of the usual momentum vector in collisions between systems which come from infinity and go to infinity after interaction.

V. 7 - GRAVITATIONAL INTERACTION WITH AN ELECTROMAGNETIC FIELD

This is the simplest example of the interaction of a tensor field with gravitation.

Maxwell's equations for an electromagnetic field in the presence of gravity are, according to the prescription given in (V. 26c) :

$$F_{\mu\nu}(x) = \nabla_\nu A_\mu(x) - \nabla_\mu A_\nu(x) \equiv \partial_\nu A_\mu(x) - \partial_\mu A_\nu(x)$$

$$\nabla_\nu F^{\mu\nu}(x) = j^\mu(x) \qquad (V.\ 44)$$

$$\nabla_\alpha F_{\mu\nu} + \nabla_\nu F_{\alpha\mu} + \nabla_\mu F_{\nu\alpha} = 0$$

The lagrangean L in equation (V. 28a) is, in this case

$$L = -\frac{1}{4} F_{\mu\nu}\, g^{\mu\alpha}\, g^{\nu\beta}\, F_{\alpha\beta} \qquad (V.\ 44a)$$

and the reader will then find that the energy-momentum tensor given by (V. 32) to be inserted in equation (V. 34) is :

$$T_{\mu\nu} = \frac{1}{4} F_{\lambda\alpha}\, g^{\lambda\xi}\, g^{\alpha\eta}\, F_{\xi\eta}\, g_{\mu\nu} - F_{\mu\alpha}\, g^{\alpha\beta}\, F_{\nu\beta} \qquad (V.\ 44b)$$

and is such that :

$$g^{\mu\nu}\, T_{\mu\nu} = 0$$

V. 8 - THE TETRAD FORMALISM

In order to consider spinors in general relativity it is convenient to introduce the notion of tetrad or "vierbein".

The line element at a point M is

$$ds^2 = g_{\mu\nu}(x)\, dx^\mu\, dx^\nu$$

where dx^μ is a vector between the point M and a point in its neighbourhood. According to the principle of equivalence, we may choose at every point M of space-time a locally inertial system of reference, where gravity is locally transformed away (freely falling frame) and in this system the line element has a Minkowskian structure :

$$ds^2 = \eta_{ab}\, d\xi^a\, d\xi^b$$

where $d\xi^a$ is a vector between the point M and a point in its neighbourhood in this system.

If we call :

$$v^a{}_\mu(x) = \left(\frac{\partial \xi^a(x)}{\partial x^\mu}\right)_{x = x(M)} \qquad \text{(V. 45)}$$

we then have

$$g_{\mu\nu}(x) = v^a{}_\mu(x)\, v^b{}_\nu(x)\, \eta_{ab} \qquad \text{(V. 45a)}$$

Here

$$\eta_{ab} = \begin{pmatrix} 1 & 0 & 0 & 0 \\ 0 & -1 & 0 & 0 \\ 0 & 0 & -1 & 0 \\ 0 & 0 & 0 & -1 \end{pmatrix}$$

is the flat-space metric and the latin indices $a, b, \ldots$ run from 0 to 3 but refer to coordinates in a freely falling reference frame.

In the latter frame :

$$\frac{d^2 \xi^a}{ds^2} = 0$$

so that

$$\frac{d}{ds}\left(\frac{\partial \xi^a}{\partial x^\mu}\frac{\partial x^\mu}{\partial s}\right) = 0$$

gives

$$\frac{d^2 x^\lambda}{ds^2} + \Gamma^\lambda_{\mu\nu}\frac{dx^\mu}{ds}\frac{dx^\nu}{ds} = 0$$

where

$$\Gamma^\lambda_{\mu\nu}(x) = \frac{\partial x^\lambda}{\partial \xi^a}\frac{\partial^2 \xi^a}{\partial x^\mu \partial x^\nu}$$

If $g_{\mu\nu}(x)$ and $\Gamma^\lambda_{\mu\nu}(x)$ are known at a point M in an arbitrary coordinate system x^μ, the locally inertial coordinates $\xi^a(x)$ in a neighbourhood of M can be determined. Indeed the above relation gives the equation :

$$\Gamma^\lambda_{\mu\nu}\frac{\partial \xi^a}{\partial x^\lambda} = \frac{\partial^2 \xi^a}{\partial x^\mu \partial x^\nu}$$

the solution of which is :

$$\xi^a(x) = c^a + v^a_{\ \mu} (x^\mu - x^\mu(M)) +$$

$$+ \frac{1}{2} v^a_{\ \lambda} \Gamma^\lambda_{\ \mu\nu} (x^\mu - x^\mu(M))(x^\nu - x^\nu(M)) + \dots$$

with :

$$c^a = \xi^a(M) \quad , \quad v^a_{\ \mu} = \left[\frac{\partial \xi^a}{\partial x^\mu} \right]_{x = x(M)}$$

When we change the non-inertial frame x^μ to x'^μ the coefficients $v^a_{\ \mu}(x)$ will change as follows :

$$v^a_{\ \mu}(x) \to v'^a_{\ \mu}(x') = \frac{\partial \xi^a}{\partial x'^\mu} = \frac{\partial \xi^a}{\partial x^\nu} \frac{\partial x^\nu}{\partial x'^\mu}$$

$$= v^a_{\ \nu} \frac{\partial x^\nu}{\partial x'^\mu} \tag{V. 46}$$

so these quantities may be regarded as a set of four covariant vector fields, not as a single tensor. If a change is made from the chosen locally inertial frame to another one at the same point M, $v^a_{\ \mu}(x)$ will change by a Lorentz transformation

$$v^a_{\ \mu}(x) \to \left(\frac{\partial \xi'^a (x)}{\partial x^\mu} \right)_{x = x(M)} = \ell^a_{\ b}(x)\, v^b_{\ \mu}(x) \tag{V. 46a}$$

the Lorentz coefficients $\ell^a_{\ b}$ are then a function of the point M.

The functions $v^a_{\ \mu}(x)$ constitute the tetrad or "vierbein".

Given a space-time tensor $T^{\mu_1 \mu_2 \cdots}_{\lambda_1 \lambda_2 \cdots}(x)$ we can construct with it a quantity which will be a scalar under general coordinate transformations and a tensor under the local Lorentz group at x :

$$T^{\mu_1 \mu_2 \cdots}_{\lambda_1 \lambda_2 \cdots}(x) \rightarrow T^{a_1 a_2 \cdots}_{b_1 b_2 \cdots}(x) = v^{a_1}_{\ \ \mu_1} v^{a_2}_{\ \ \mu_2} \cdots \cdot$$

$$\cdot\ T^{\mu_1 \mu_2 \cdots}_{\lambda_1 \lambda_2 \cdots} v_{b_1}^{\ \ \lambda_1} v_{b_2}^{\ \ \lambda_2} \cdots \tag{V. 47}$$

The latin indices $a, b, \ldots$ are raised by means of η^{ab} and the greek indices by $g^{\mu\nu}$:

$$v_{b_1}^{\ \ \lambda_1} = \eta_{b_1 a_1} g^{\lambda_1 \mu_1} v^{a_1}_{\ \ \mu_1} \tag{V. 47a}$$

The invariance principle is now stated as follows : 1) the action must be invariant under the group of general coordinate transformations and the fields are represented by entities which are scalar under this group ; 2) the action must also be invariant if at each point of space-time we change the locally inertial frame of reference by the Lorentz group of transformations.

Thus the physical fields will be scalars or tensors under the general group of coordinate transformations and scalars or tensors or spinors under the local Lorentz group. A Dirac spinor, in particular transforms like

$$\psi(x) \rightarrow \psi'(x') = \psi(x) \quad \text{if} \quad x^\mu \rightarrow x'^\mu = f^\mu(x) \tag{V. 48}$$

and

$$\psi(x) \rightarrow \psi'(x') = S(\ell(x))\,\psi(x)$$

if the inertial frame at x is changed by $\ell^{\alpha}{}_{\beta}(x)$.

Let us consider the action (V. 28). It can be written :

$$S = \int d^4x\; v(x) \left\{ \mathscr{R}^{\mu\nu}\, v^{a}{}_{\mu}\, v^{b}{}_{\nu}\, \eta_{ab} - 2\,\kappa\, L \right\} \qquad \text{(V. 49)}$$

where, according to (V. 45a) :

$$v(x) \equiv \det\,(v^{a}{}_{\mu}) = \sqrt{-g}$$

The variation of S corresponding to a variation $\delta\, v^{a}{}_{\mu}$ of $v^{a}{}_{\mu}(x)$, which now describes the gravitational field, will be :

$$\delta\, S = 2 \int d^4x\; v(x) \left\{ \mathscr{R}^{\mu\nu}\, v^{a}{}_{\mu}\, \eta_{ab} - \frac{1}{2}\, \mathscr{R}\, v_{b}{}^{\nu} - \kappa\, T_{b}{}^{\nu} \right\} \delta\, v^{b}{}_{\nu} \qquad \text{(V. 49a)}$$

and vanishes for arbitrary $\delta\, v^{b}{}_{\nu}$ (which are null at the outer surface). One then obtains Einstein's equations :

$$\mathscr{R}_{b}{}^{\nu} - \frac{1}{2}\, \mathscr{R}\, v_{b}{}^{\nu} = \kappa\, T_{b}{}^{\nu} \qquad \text{(V. 50)}$$

where

$$\mathscr{R}_b{}^{\nu} = \mathscr{R}^{\mu\nu} \; v^a{}_\mu \; \eta_{ab}$$

and

$$T_a{}^{\mu} = \frac{1}{v} \left\{ \frac{\partial (v\, L)}{\partial\, v^a{}_\mu} - \partial^\alpha \left(\frac{\partial (\, v\, L\,)}{\partial (\partial^\alpha \, v^a{}_\mu)} \right) \right\} \qquad \text{(V. 51)}$$

In the case of the electromagnetic field one finds

$$T^a{}_\mu = \frac{1}{4} F_{\lambda\nu}\, g^{\lambda\alpha}\, g^{\nu\beta}\, F_{\alpha\beta}\, v^a{}_\mu - F_{\mu\nu}\, v_b{}^\alpha\, \eta^{ab}\, g^{\nu\beta}\, F_{\alpha\beta}$$

so that :

$$T_{\mu\nu} = v^a{}_\mu\, \eta_{ab}\, T^b{}_\nu =$$

$$= - F_{\mu\alpha}\, g^{\alpha\beta}\, F_{\nu\beta} + \frac{1}{4} F_{\varepsilon\lambda}\, g^{\varepsilon\alpha}\, g^{\lambda\beta}\, F_{\alpha\beta}\, g_{\mu\nu}$$

$T^a{}_\mu$ therefore gives rise to the electromagnatic field tensor (V. 44b).

V. 9 - DIRAC'S EQUATION AND CURRENT IN GENERAL RELATIVITY

Let us now consider a spinor $\psi(x)$, that is, a field the transformation laws of which are those in (V. 48).

Its covariant derivative will be defined by the introduction of four 4×4 matrices $\Gamma_\mu(x)$, functions of point, the spinor affine connection :

$$\nabla_\mu \psi(x) \equiv (\partial_\mu + \Gamma_\mu(x)) \psi(x) \qquad \text{(V. 52)}$$

The Lorentz-vector and coordinate-scalar derivative operator will be :

$$\nabla_a \psi(x) = v_a{}^\mu(x) \nabla_\mu \psi(x) \quad ; \quad a = 0,1,2,3 \qquad \text{(V. 52a)}$$

If at each point x, $\psi(x)$ transforms according to the spinor representation of the Lorentz group, in correspondence to changes in the locally inertial frame :

$$\psi'(x) = S(\ell(x)) \psi(x)$$

then the condition of covariant transformation of the derivative gives :

$$v'_a{}^\mu(\partial'_\mu + \Gamma'_\mu(x)) \psi' = \ell_a{}^b v_b{}^\mu S(\ell) (\partial_\mu + \Gamma_\mu(x)) \psi$$

hence :

$$\Gamma'_\mu(x) = S \Gamma_\mu(x) S^{-1} - (\partial_\mu S) S^{-1} \qquad \text{(V. 53)}$$

in correspondence to the above change of locally freely falling system.

With the derivative (V. 52), (V. 52a), we are tempted then to postulate the following Dirac's equation for a massive spin 1/2 particle in interaction with the gravitational field :

$$(i \gamma^a \nabla_a - m) \psi (x) = 0 \qquad \text{(V. 54)}$$

This equation can also be written in the following way, where no mention is made of the tetrads :

$$(i \gamma^\mu(x) \nabla_\mu - m) \psi (x) = 0 \qquad \text{(V. 54a)}$$

where

$$\gamma^\mu(x) = v_a{}^\mu(x) \gamma^a \qquad \text{(V. 55)}$$

are four point-dependent matrices which obey the commutation rule :

$$\left[\gamma^\mu(x), \gamma^\nu(x)\right]_+ = 2 g^{\mu\nu}(x) \qquad \text{(V. 55a)}$$

if the γ^a's are the usual, flat-space, Dirac matrices :

$$\left[\gamma^a, \gamma^b\right]_+ = 2 \eta^{ab}$$

The gravitational field, which acts on the ψ field is contained in the matrix $\Gamma_\mu(x)$, which must vanish in flat space. In order to distinguish the Dirac matrices in equation (V. 54) from those which depend on x, we shall always make explicit this dependence. Thus, at each point x, we define the adjoint :

$$\bar{\psi}(x) = \psi^{+}(x)\ \gamma^{0}$$

where γ^{0} is the usual Dirac matrix and similary we define :

$$\overline{(\nabla_{\mu}\ \psi)} = \partial_{\mu}\ \overline{\psi} + \overline{\psi}\ \overline{\Gamma}_{\mu}\ (x) \qquad (V.\ 56)$$

where

$$\partial_{\mu}\ \overline{\psi}(x) = (\partial_{\mu}\ \psi^{+}(x))\ \gamma^{0}$$

$$\overline{\Gamma}_{\mu}(x) = \gamma^{0}\ \Gamma^{+}{}_{\mu}(x)\ \gamma^{0} \qquad (V.\ 56a)$$

As $\bar{\psi}(x)\ \psi(x)$ is a Lorentz-scalar and a coordinate-scalar we have in a locally inertial system, the identity :

$$\partial_{\mu}\ (\overline{\psi}\ \psi) = (\partial_{\mu}\ \overline{\psi})\ \psi + \overline{\psi}\ (\partial_{\mu}\ \psi)$$

In an arbitrary system we must have

$$\nabla_{\mu}\ (\overline{\psi}\ \psi)\ \equiv\ \partial_{\mu}\ (\overline{\psi}\ \psi) = \overline{(\nabla_{\mu}\ \psi)}\ \psi + \overline{\psi}\ (\nabla_{\mu}\ \psi)$$

and this requires that :

$$\overline{\Gamma}_{\mu}(x) = -\ \Gamma_{\mu}(x)$$

Thus if Dirac's equation is

$$(i\ \gamma^{\mu}(x)\ (\partial_{\mu} + \Gamma_{\mu}(x)) - m)\ \psi(x) = 0 \qquad \text{(V. 57)}$$

its adjoint will be :

$$i\ \overline{(\nabla_{\mu}\ \psi)}\ \gamma^{\mu}(x) + m\ \overline{\psi}(x) = 0$$

or

$$i\ (\partial_{\mu}\ \overline{\psi} - \overline{\psi}\ \Gamma_{\mu}(x))\ \gamma^{\mu}(x) + m\ \overline{\psi}(x) = 0 \qquad \text{(V. 57a)}$$

What is the current and which conservation law does it obey ?

From the two equations (V. 57), (V. 57a) we deduce the relationship :

$$(\partial_{\mu}\ \overline{\psi})\ \gamma^{\mu}(x)\ \psi(x) + \overline{\psi}(x)\ \gamma^{\mu}(x)\ \partial_{\mu}\ \psi(x)\ +$$
$$+\ \overline{\psi}(x)\left[\gamma^{\mu}(x),\ \Gamma_{\mu}(x)\right]_{-}\ \psi(x) = 0$$

Now the expression :

$$\overline{\psi}(x)\ \gamma^{\mu}(x)\ \psi(x)\ \equiv\ \overline{\psi}(x)\ v_{a}{}^{\mu}(x)\ \gamma^{a}\ \psi(x)$$

is a four-vector under general coordinate transformation (see (V. 48) and (V. 46) therefore its covariant derivative is :

$$\nabla_\lambda\ (\overline{\psi}\ (x)\ \gamma^\mu(x)\ \psi(x)) =$$

$$= \partial_\lambda\ (\overline{\psi}\ (x)\ \gamma^\mu(x)\ \psi\ (x)) + \Gamma^\mu{}_{\lambda\nu}\ (\overline{\psi}\ (x)\ \gamma^\nu(x)\ \psi\ (x))$$

Thus if the matrix $\Gamma_\mu(x)$ satisfies the following equation :

$$\partial_\lambda\ \gamma^\mu(x) + \Gamma^\mu{}_{\lambda\nu}(x)\ \gamma^\nu(x) + \left[\Gamma_\lambda(x),\ \gamma^\mu(x)\right]_- = 0 \qquad \text{(V. 58)}$$

then the current four-vector will be conserved according to the equation :

$$\nabla_\mu\ (\overline{\psi}\ (x)\ \gamma^\mu(x)\ \psi\ (x)) = 0$$

The equation (V. 58) follows from the equation for the derivative of the metric field (V. 37).

Indeed if we replace $g_{\mu\nu}(x)$ by the anticommutator (V. 55a) in equation (V. 37) we obtain :

$$\partial_\lambda\left[\gamma_\mu(x),\ \gamma_\nu(x)\right]_+ - \Gamma^\alpha{}_{\lambda\mu}\left[\gamma_\alpha(x),\ \gamma_\nu(x)\right]_+ - \Gamma^\alpha{}_{\lambda\nu}\left[\gamma_\mu(x),\gamma_\alpha(x)\right]_+ = 0$$

and this equation will be satisfied if there exists a Γ_μ matrix which satisfies equation (V. 58) and :

$$\nabla_\lambda\ \gamma_\mu \equiv \partial_\lambda\ \gamma_\mu - \Gamma^\alpha{}_{\lambda\mu}\ \gamma_\alpha + \left[\Gamma_\lambda,\ \gamma_\mu\right]_- = 0 \qquad \text{(V. 58a)}$$

This shows that the transition from Dirac's equation in flat space to the form (V. 54) is correct since $\nabla_\lambda [\gamma^\alpha \psi] = \gamma^\alpha \nabla_\lambda \psi$.

V. 10 - THE DIRAC FIELD ENERGY-MOMENTUM TENSOR

The lagrangean which generates the equations (V. 57) and which enters the action (V. 28a) is :

$$L = \overline{\psi} (x) \left\{ i \gamma^\mu(x) \nabla_\mu - m \right\} \psi (x) \tag{V. 59}$$

For the calculation of the variation of the lagrangean with respect to a variation $\delta g_{\mu\nu}$ of the gravity field :

$$\delta L = \overline{\psi} i (\delta \gamma^\mu(x)) \nabla_\mu \psi + \overline{\psi} i \gamma^\mu(x) \delta \Gamma_\mu(x) \psi \tag{V. 60}$$

we need to know $\delta \gamma^\mu(x)$ and $\delta \Gamma_\mu(x)$ as a function of $g_{\mu\nu}(x)$ and its derivatives and variations.

From the anticommutator (V. 55a) we get :

$$\left[\delta \gamma^\mu(x), \gamma^\nu(x)\right]_+ + \left[\gamma^\mu(x), \delta \gamma^\nu(x)\right]_+ = 2 \delta g^{\mu\nu}(x) \tag{V. 61}$$

The solution of this equation is :

$$\delta \gamma^\mu = \frac{1}{2} \gamma_\lambda \delta g^{\lambda\mu} \tag{V. 61a}$$

From the equation (V. 58) we obtain :

$$\partial_\lambda \, \delta \, \gamma^\mu + (\delta \, \Gamma^\mu{}_{\lambda\nu}) \, \gamma^\nu + \Gamma^\mu{}_{\lambda\nu} \, \delta \, \gamma^\nu + \left[\Gamma_\lambda, \delta \, \gamma^\mu\right]_- + \left[\delta \, \Gamma_\lambda, \gamma^\mu\right]_- = 0$$

which, together with (V. 61a) and (V. 58), gives :

$$\frac{1}{2} \left\{ (\delta \, \Gamma^\mu{}_{\alpha\beta}) \, \gamma^\beta - (\delta \, \Gamma^\nu{}_{\alpha\lambda}) \, \gamma_\nu \, g^{\mu\lambda} + \left[\delta \, \Gamma_\alpha, \gamma^\mu\right]_- \right\} = 0$$

the solution of which is :

$$\delta \, \Gamma_\alpha(x) = \frac{i}{8} \, (g_{\nu\beta} \, \delta \, \Gamma^\beta{}_{\mu\alpha} - g_{\mu\beta} \, \delta \, \Gamma^\beta{}_{\nu\alpha}) \, \sigma^{\mu\nu}(x) \qquad \text{(V. 62)}$$

where :

$$\sigma^{\mu\nu}(x) = \frac{i}{2} \left[\gamma^\mu(x), \gamma^\nu(x)\right]_-$$

and :

$$\left[\frac{\sigma^{\mu\nu}}{2}(x), \gamma^\lambda(x)\right]_- = i \, (g^{\lambda\nu}(x) \, \gamma^\mu(x) - g^{\lambda\mu}(x) \, \gamma^\nu(x))$$

We are now in possession of $\delta \, \gamma^\mu$, equation (V. 61a), of $\delta \, \Gamma_\alpha$, equation (V. 62), and of $\delta \, \Gamma^\beta{}_{\mu\alpha}$:

$$\delta \Gamma^{\beta}_{\mu\alpha} = \frac{1}{2} (\delta g^{\beta\lambda}) \left\{ \partial_{\mu} g_{\lambda\alpha} + \partial_{\alpha} g_{\lambda\mu} - \partial_{\lambda} g_{\mu\alpha} \right\} +$$

$$+ \frac{1}{2} g^{\beta\lambda} \left\{ \partial_{\mu} \delta g_{\lambda\alpha} + \partial_{\alpha} \delta g_{\lambda\mu} - \partial_{\lambda} \delta g_{\mu\alpha} \right\}$$

After substitution of these expressions in (V. 60), partial integration of the terms with $(\partial_{\mu} \delta g_{\lambda\alpha})$ and similar ones, one obtains, by comparison with (V. 32a) :

$$\delta \int d^4x \sqrt{-g} \; L = \frac{1}{2} \int d^4x \sqrt{-g} \; T_{\mu\nu} \; \delta g^{\mu\nu}$$

the following expression for the Dirac energy-momentum tensor :

$$T^{\mu\nu}(x) = \frac{i}{4} \left\{ \bar{\psi} \gamma_{\mu} \nabla_{\nu} \psi - \overline{(\nabla_{\nu} \psi)} \gamma_{\mu} \psi + \right.$$

$$\left. + \bar{\psi} \gamma_{\nu} \nabla_{\mu} \psi - \overline{(\nabla_{\mu} \psi)} \gamma_{\nu} \psi \right\}$$

for fields which are solutions of equations (V. 57) and (V. 57a).

V. 11 - GAUGE FIXING CONDITIONS

As a result of energy-momentum covariant conservation, equations (V. 34a), there are only six independent equations out of Einstein' s equations (V. 34). Four conditions have therefore to be imposed in order to complete to ten the number of equations needed to determine the ten components of $g_{\mu\nu}(x)$. As gauge invariance is here associated to arbitrariness in the coordinates the gauge fixing condition amounts to making a particular choice of a coordinate system.

A popular choice are the harmonic coordinate conditions :

$$g^{\mu\nu}(x)\, \Gamma^{\alpha}{}_{\mu\nu}(x) = 0$$

PROBLEMS

V - 1. a) Show that the contraction of the affine connection is :

$$\Gamma^{\alpha}{}_{\lambda\alpha}(x) = \partial_{\lambda}\, (\log \sqrt{-\, g(x)})$$

where $\quad g(x) = \det\, (g_{\mu\nu}(x))$

b) Calculate the gravitational convariant divergence of a vector $A^{\alpha}(x)$

c) Calculate the covariant dalembertian of a scalar field :

$$\Box\, \varphi(x) \equiv \nabla_{\alpha}\, (g^{\alpha\lambda}(x)\, \partial_{\lambda}\, \varphi(x))$$

d) show that

$$\nabla_{\beta}\, T^{\alpha\beta} = \frac{1}{\sqrt{-g}}\, \partial_{\lambda}\, (\sqrt{-g}\;\; T^{\alpha\lambda}) + \Gamma^{\alpha}{}_{\mu\lambda}\, T^{\mu\lambda}$$

what is the formula for an antisymetric tensor ?

V - 2. a) Deduce another form of Einstein's equations :

$$R_{\mu\nu} - \frac{1}{2}\, g_{\mu\nu}\, R = K\, T_{\mu\nu}$$

in terms of $T_{\mu\nu}$ and the trace $T \equiv T^{\mu}{}_{\mu}$.

b) Apply it to the case of the gravitational field generated by the energy-momentum of a Yang-Mills field ; what does it have in common with the equation of the gravitational field generated by an electromagnetic field ?

V - 3. Deduce the equation of the geodesic from a variational principle

$$\delta S = 0,$$

$$S = \int \left[g_{\mu\nu}(z) \frac{dz^{\mu}}{ds} \frac{dz^{\nu}}{ds} \right]^{1/2} ds$$

for arbitrary variations $\delta z^{\alpha}(s)$ which vanish at the boundaries of integration. Interpret this equation as an equation of motion of a classical particle. What is the force acting on the particle ? What relationship between gravitational and inertial mass does it imply ?

V - 4. Show that a matter field tensor $T_{\mu\nu}(x)$ cannot be the source of gravitation in a two dimensional space-time.

CHAPTER VI

Weak Interactions and Intermediate Vector Bosons

VI. 1 - INTRODUCTION

In the preceding chapters, we have established the basic equations for the Yang-Mills and the gravitational gauge fields. All these fields are massless . The first theory to be elaborated was that of the electromagnetic field and later on the Einstein theory of the gravitational field. We have thus seen the foundations for the study of electromagnetic and of gravitational interactions.

What is the usefulness of the other, non-abelian, gauge fields ? The developments in theoretical particle physics in the last ten years have led us to a successful description of the weak interactions by means of Yang-Mills fields Morever, as we shall see, it turns out that this is at the same time a unified description of both the electromagnetic and weak interactions in the frame work of gauge theory. More recent research work suggests that the strong interactions are most likely described by the SU(3) - colour gauge field. The corresponding theory is the so-called quantum chromodynamics. And unification of strong, electromagnetic and weak interactions may be achieved in the SU(5) model. Leptons and hadrons are also unified in the SU(4) x SU(4) model of Pati and Salam.

In this Chapter we shall briefly review the form of charged weak currents, the current-current Fermi theory and the intermediate vector boson version of the weak interaction theory[39-78]. The Fermi theory was first proposed in 1934 ; its final form followed the discovery of parity violation in weak reactions by Yang and Lee and the form of weak currents by Feynman and Gell-Mann and Marshak and Sudarshan. The Lorentz nature of these currents, namely a combination of a vector and an axial vector, suggested that the weak interactions might be du to an exchange of vector bosons between hadrons and between leptons.

VI. 2 - CHARGED WEAK CURRENTS

Weak interactions are successfully described at low-energies by an effective lagrangean which has the form

$$L = \frac{\mathcal{G}_F}{\sqrt{2}} \; j^{\mu +}(x) \; j_\mu \; (x) \qquad \text{(VI. 1)}$$

where $j^\mu(x)$ is the so-called charged weak current. This was shown by Feynman and Gell-Mann to have the form V-A (a vector minus an axial vector) and is the sum of two parts, one the leptonic weak current $\ell^\mu(x)$, the other, the hadronic weak current $h^\mu(x)$:

$$j^\mu (x) = \ell^\mu (x) + h^\mu (x) \qquad \text{(VI. 1a)}$$

In terms of the leptonic fields

$$\nu_e(x),\ e(x)\ ;\ \nu_\mu(x),\ \mu(x)\ ;\ \nu_\tau(x),\ \tau(x)\ ;\ \ldots.$$

the leptonic weak current has the form :

$$\ell^\lambda(x) = (\overline{\nu}_e \gamma^\lambda (1 - \gamma^5)e) + (\overline{\nu}_\mu \gamma^\lambda (1 - \gamma^5)\mu) + + (\overline{\nu}_\tau \gamma^\lambda (1 - \gamma^5)\tau) + \ldots \qquad \text{(VI. 1b)}$$

the points indicating contributions from other possible, not yet known, leptons (the theory is not yet able to predict the number of possible leptons).

The hadronic weak current cannot be expressed in terms of observed-hadron fields in a simple way in virtue of the strong interaction between these fields. One studies symmetry and algebraic properties of $h^\lambda(x)$ and its matrix elements between known initial and final hadronic states can be expressed in terms of kinematic variables and dynamical form factors in a Lorentz covariant form.

The low-energy hadronic current $h'^\lambda(x)$ for ordinary and strange hadronic matter has the form :

$$h'^\lambda(x) = \cos\theta_c(v'^\lambda_{(0)}(x) - a'^\lambda_{(0)}(x)) + \sin\theta_c\,(v'^\lambda_{(1)}(x) - - a'^\lambda_{(1)}(x)) \qquad \text{(VI. 2)}$$

where the subscripts (0) and (1) refer to strangeness changes $\Delta S = 0$ and $\Delta S = 1$ respectively : θ_c is the Cabibbo angle : $\sin\theta_c \cong 0.21$.

These currents satisfy the chiral $SU(3) \otimes SU(3)$ commutation relations.

If F_a, a = 1,... 8, are the generators of the SU(3) group which satisfy the algebra :

$$\left[F_a, F_b\right] = i\, f_{abc}\; F_c \tag{VI. 3}$$

then there exists an octet of vector currents $V'^{\lambda}_a(x)$ under SU(3) such that :

$$F_a = \int V'^{o}_a(x)\; d^3x \tag{VI. 3a}$$

There is also an octet of axial currents $A^{\lambda}_a\,(x)$ such that :

$$F_a^5 = \int A'^{o}_a(x)\; d^3x \tag{VI. 3b}$$

and the generators F_a^5 together with F_a form a closed algebra defined by

$$\begin{aligned} \left[F_a(t),\; F_b(t)\right] &= i\, f_{abc}\, F_c(t) \\ \left[F_a^5(t),\, F_b^5(t)\right] &= i\, f_{abc}\, F_c(t) \\ \left[F_a^5(t),\, F_b(t)\right] &= i\, f_{abc}\, F_c^5(t) \end{aligned} \tag{VI. 3c}$$

This is the SU(3) ⊗ SU(3) algebra which is also expressed by the left and right generators

$$\begin{aligned} F_a^L(t) &= \frac{1}{2}\left\{ F_a(t) - F_a^5(t) \right\} \\ F_a^R(t) &= \frac{1}{2}\left\{ F_a(t) + F_a^5(t) \right\} \end{aligned} \tag{VI. 3d}$$

which obey the commutation rules :

$$\left[F_a^{\,L}(t),\; F_b^{\,L}(t)\right] = i\, f_{abc}\, F_c^{\,L}(t)$$

$$\left[F_a^{\,R}(t),\; F_b^{\,R}(t)\right] = i\, f_{abc}\, F_c^{\,R}(t) \qquad \text{(VI. 3e)}$$

$$\left[F_a^{\,R}(t),\; F_b^{\,L}(t)\right] = 0$$

One has then

$$\left[F_a(t),\; V'^{\mu}{}_b(x)\right] = i\, f_{abc}\, V'^{\mu}{}_c(x)$$

$$\left[F_a(t),\; A'^{\mu}{}_b(x)\right] = i\, f_{abc}\, A'^{\mu}{}_c(x)$$

The vector currents $v'^{\lambda}{}_{(0)}(x)$ and $v'^{\lambda}{}_{(1)}(x)$ in equation (VI. 2) are then

$$v'^{\lambda}{}_{(0)}(x) = V'^{\lambda}{}_1(x) + i\, V'^{\lambda}{}_2(x)$$

$$v'^{\lambda}{}_{(0)}(x) = V'^{\lambda}{}_4(x) + i\, V'^{\lambda}{}_5(x)$$

The hadronic currents are expressed in terms of the fields which describe the quark constituents of hadrons.

In the preceding case, in which only strange matter was present in addition to ordinary hadronic matter, the current in terms of the quarks u,d,s is

$$h'^{\lambda}(x) = \bar{u}(x)\, \gamma^{\lambda}(1 - \gamma^5) \left\{ d(x) \cos\theta_c + s(x) \sin\theta_c \right\} \qquad \text{(VI. 4)}$$

In the case in which we consider the ordinary, strange and charmed hadronic matter, formed by the quarks u,d,s,c, the charged weak current of hadrons is :

$$h^{\lambda}(x) = \bar{u}(x)\,\gamma^{\lambda}(1-\gamma^{5})\left\{ d(x)\cos\theta_{c} + s(x)\sin\theta_{c} \right\} + \bar{c}(x)\,\gamma^{\lambda}(1-\gamma^{5})\left\{ -d(x)\sin\theta_{c} + s(x)\cos\theta_{c} \right\} \qquad \text{(VI. 5)}$$

The following are the selection rules which these currents give rise to :

Transitions with :

$\Delta c = 0,\ \Delta s = 0,\ \Delta Q = 1,\ \Delta I = 0, 1 \quad : \quad h^{\lambda} \sim \cos\theta_{c}$

$\Delta c = 0,\ \Delta s = \Delta Q = 1,\ \Delta I = 1/2 \quad : \quad h^{\lambda} \sim \sin\theta_{c}$

$\Delta c = 1,\ \Delta s = 0,\ \Delta Q = 1,\ \Delta I = 1/2 \quad : \quad h^{\lambda} \sim -\sin\theta_{c}$

$\Delta c = 1,\ \Delta s = \Delta Q = 1,\ \Delta I = 0, 1 \quad : \quad h^{\lambda} \sim \cos\theta_{c}$

that is,

$$h^{\lambda}(x) = \left\{ (v^{\lambda} - a^{\lambda})_{\substack{\Delta s = 0 \\ \Delta c = 0}} + (v^{\lambda} - a^{\lambda})_{\substack{\Delta s = 1 \\ \Delta c = 1}} \right\} \cos\theta_{c} + \left\{ (v^{\lambda} - a^{\lambda})_{\substack{\Delta s = 1 \\ \Delta c = 0}} + (v^{\lambda} - a^{\lambda})_{\substack{\Delta s = 0 \\ \Delta c = 1}} \right\} \sin\theta_{c} \qquad \text{(VI. 5a)}$$

The generalized charge operators :

$$\mathcal{F}_a = \int v^0_a(x)\, d^3x \quad , \quad a = 1, \ldots\ldots 15$$

are the generators of the flavour SU(4) group.

Together with

$$\mathcal{F}^5_a = \int a^o_a(x)\, d^3x$$

they generate the chiral algebra SU(4) ⊗ SU(4).

Calling q(x) the quark field for the quartet u,d,s,c one has for these currents :

$$v^\lambda_a(x) = \bar{q}(x)\, \gamma^\lambda\, \frac{\eta_a}{2}\, q(x)$$

$$a^\lambda_a(x) = \bar{q}(x)\, \gamma^\lambda\, \gamma^5\, \frac{\eta_a}{2}\, q(x) \qquad \text{(VI. 6)}$$

where $\frac{\eta_a}{2}$ are the fifteen generators of SU(4) in the 4 x 4 matrix representation. Sum over the colours is understood.

The weak currents are therefore (VI. 1b) for leptons and (VI. 6) for quarks which enter the expression (VI. 5a). For more than four quarks, u,d,c,s,t,b the terms with the Cabibbo linear combinations of d and s in (VI. 5) are replaced by $\bar{u}\, \gamma^\lambda(1 - \gamma^5)\, d'$, $\bar{c}\, \gamma^\lambda(1 - \gamma^5)\, s'$, $\bar{t}\, \gamma^\lambda(1 - \gamma^5)\, b'$ where d',s',b' are the transformed of d,s,b by a unitary matrix[93].

VI. 3 - THE INTERMEDIATE VECTOR BOSON FIELD

Let us consider the β-decay of the muon

$$\mu \to \nu_\mu + e + \bar{\nu}_e \qquad \text{(VI. 7)}$$

Its amplitude will be given by the following number, if we consider the lagrangean (VI. 1) and the currents (VI. 1b) :

$$S = - \frac{i \mathcal{G}_F}{\sqrt{2}} \int d^4x \left\{ \bar{\nu}_\mu(x)\gamma^\lambda(1 - \gamma^5)\mu(x) \right\} \left\{ \bar{e}(x)\gamma_\lambda(1 - \gamma^5)\nu_e(x) \right\} \qquad \text{(VI. 8)}$$

The form V-A of this interaction suggests that <u>this amplitude might result from an interaction between the currents and a vector field $W^\lambda(x)$</u> so that instead of the lagrangean (VI. 1) we would have the following one :

$$\mathcal{L} = g\, j^\lambda\,(x)\, W_\lambda\,(x) + \text{h.c.} \qquad \text{(VI. 9)}$$

The charged field $W^\lambda(x)$ would thus play a rôle similar to that of the photon field in electrodynamics. With this interaction (VI. 9) the amplitude of the reaction (VI. 7) will be :

$$S' = - ig^2 \iint d^4x\, d^4y (\bar{\nu}_\mu(x)\gamma^\lambda(1 - \gamma^5)\mu(x))(\Delta_F(x - y))_{\lambda\eta} (\bar{e}(y)\gamma^\eta(1 - \gamma^5)\nu_e(y))$$

where $(\Delta_F(x - y))_{\lambda\eta}$ is the Feynman propagator for a massive vector field :

$$(\Delta_F(x - y))_{\lambda\eta} = (g_{\lambda\eta} + \frac{1}{m_W^2} \partial_\lambda\, \partial_\eta)\, \Delta_F(x - y) \qquad \text{(VI. 10)}$$

m_W is the mass of this field and

$$\Delta_F(x - y) = - \int \frac{d^4k}{(2\pi)^4} \frac{e^{-ik(x-y)}}{k^2 - m_W^2 + i\varepsilon}$$

In momentum space we have for S and S' :

$$S = (2\pi)^4 \delta^4 (p_{\nu_\mu} + p_e + p_{\bar{\nu}_e} - p_\mu) N.M$$

$$S' = (2\pi)^4 \delta^4 (p_{\nu_\mu} + p_e + p_{\bar{\nu}_e} - p_\mu) N.M'$$

where N is a normalization factor and

$$M = - i \frac{\mathcal{G}_F}{\sqrt{2}} \left\{ \bar{\nu}(p_\mu)\gamma^\alpha(1-\gamma^5)\mu(p_\mu) \right\} \left\{ \bar{e}(p_e)\gamma_\alpha(1-\gamma^5)\nu(-p_{\nu_e}) \right\} \qquad \text{(VI. 11)}$$

$$M' = ig^2 \left\{ \bar{\nu}(p_\mu)\gamma^\alpha(1-\gamma^5)\mu(p_\mu) \right\} \frac{g_{\alpha\beta} - \frac{k_\alpha k_\beta}{m_W^2}}{k^2 - m_W^2 + i\varepsilon} \left\{ \bar{e}(p_e)\gamma^\beta(1-\gamma^5)\nu(-p_{\nu_e}) \right\} \qquad \text{(VI. 12)}$$

We see that in the low momentum-transfer approximation the two matrix elements M and M' coincide if :

$$k^2 \ll m_W^2 \quad , \quad k = p_{\nu\mu} - p_\mu$$

and

$$\frac{g^2}{m_W^2} = \frac{\mathcal{G}_F}{\sqrt{2}} \qquad \text{(VI. 13)}$$

In this way, the Fermi approach being obtained in the domain where it is valid, it would be more satisfactory to describe weak interactions by means of such massive vector field $W_\mu(x)$ -the so-called intermediate vector boson field.

The lagrangean (VI. 9) thus replaces the Fermi current-current interaction lagrangean. The total lagrangean of the naive intermediate vector boson theory of weak interactions is :

$$L_W = -\frac{1}{2} W^{\mu\nu+} W_{\mu\nu} + m_W^2 W^{\mu+} W_\mu + \sum_\ell \bar{\ell} (i \gamma^\alpha \partial_\alpha - m_\ell) \ell +$$

$$+ \sum_{\nu_\ell} \bar{\nu}_\ell i \gamma^\alpha \partial_\alpha \nu_\ell + \sum_c \bar{q} (i \gamma^\alpha \partial_\alpha - M) q + g (W^\lambda j_\lambda + W^{\lambda+} j_\lambda^{+}) \qquad \text{(VI. 14)}$$

which replaces the lagrangean of the Fermi theory :

$$L_F = \sum_\ell \bar{\ell}(i \gamma^\alpha \partial_\alpha - m_\ell \ell + \sum_{\nu\ell} \bar{\nu}_\ell i \gamma^\alpha \partial_\alpha \nu_\ell + \sum_c \bar{q} (i \gamma^\alpha \partial_\alpha - M) q + \frac{G_F}{\sqrt{2}} j^{\lambda+} j_\lambda \qquad \text{(VI. 15)}$$

The lagrangean which results from the Salam-Weinberg theory contains, among other, terms of self-interaction between the vector bosons (§ VIII. 6).

VI. 4 - HIGH-ENERGY DIVERGENCES IN THE FERMI AND VECTOR BOSON THEORIES

The Fermi lagrangean (VI. 15) leads to cross sections for the neutrino-lepton scattering which grow with the energy and violate the unitarity bound. This difficulty is overcome by the intermediate vector boson lagrangean but the latter also leads to difficulties for processes like the production of W-bosons in the neutrino-antineutrino annihilation.

The matrix element for the elastic neutrino-electron scattering due to the lagrangean (VI. 1) is similar to the expression (VI. 11) :

$$M = \frac{-i \mathcal{G}_F}{\sqrt{2}} (\bar{u}_{\nu_e} \gamma^\alpha (1 - \gamma^5) u_e)(\bar{u}_e \gamma_\alpha (1 - \gamma^5) u_{\nu_e}) \qquad \text{(VI. 16)}$$

for the reaction :

$$\nu_e + e \rightarrow \nu_e + e \qquad \text{(VI. 16a)}$$

This gives the following differential cross section :

$$\frac{d\,\sigma(\nu_e \rightarrow \nu_e)}{d\,\Omega} = \frac{G_F^2}{4\pi^2}\,\frac{(s - m_e^2)^2}{s} \quad , \qquad s = (p_{\nu e} + p_e)^2$$

and this becomes, for high energies :

$$\frac{d\,\sigma(\nu_e \rightarrow \nu_e)}{d\,\Omega} \cong \frac{G_F^2}{4\pi^2}\, s \qquad \text{for} \quad s >> m_e^2 \qquad \text{(VI. 17)}$$

The total cross-section is :

$$\sigma\,(\nu_e \rightarrow \nu_e) \cong \frac{G_F^2}{\pi}\, s \quad , \quad s >> m_e^2 \qquad \text{(VI. 17a)}$$

For the antineutrino-electron scattering

$$\bar{\nu}_e + e \rightarrow \bar{\nu}_e + e \qquad \text{(VI. 18)}$$

the differential cross-section is :

$$\frac{d\,\sigma(\bar{\nu}_e \rightarrow \bar{\nu}_e)}{d\,\Omega} = \frac{G_F^2}{4\pi^2}\,\frac{(s - m_e^2)^2}{s}\,\frac{1}{4s^2}\left\{(s - m_e^2)\cos\theta + (s - m_e^2)\right\}^2$$

which for high energies goes over to :

$$\frac{d\,\sigma(\bar{\nu}_e \to \bar{\nu}_e)}{d\,\Omega} \cong \frac{G_F^2}{4\pi^2}\ \frac{s(1 + \cos\theta)^2}{4} \qquad s \gg m_e^2 \qquad \text{(VI. 19)}$$

where θ is the scattering angle in the center of mass system. The total cross section is, for high energies :

$$\sigma\,(\bar{\nu}_e \to \bar{\nu}_e) \cong \frac{G_F^2}{3\pi}\ s \qquad \text{(VI. 20)}$$

The fact that the two cross-sections, one for $\nu_e \to \nu_e$, the other for $\bar{\nu}_e \to \bar{\nu}_e$, differ, can be understood by the following argument : for very high energies, the electron mass will be negligible and this particle will be left-handed. As ν_e is left-handed, it follows that the component of the total angular momentum on the momentum direction in the center of mass system vanishes :

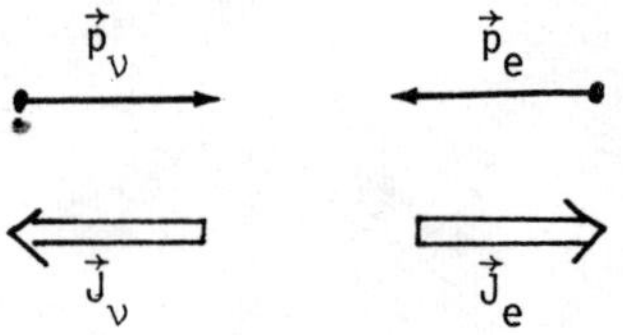

hence only S-waves contribute to σ in (VI. 17).

For the reaction (VI. 18), on the other hand, the antineutrino is right-handed, hence the total angular momentum over the momentum direction will be one :

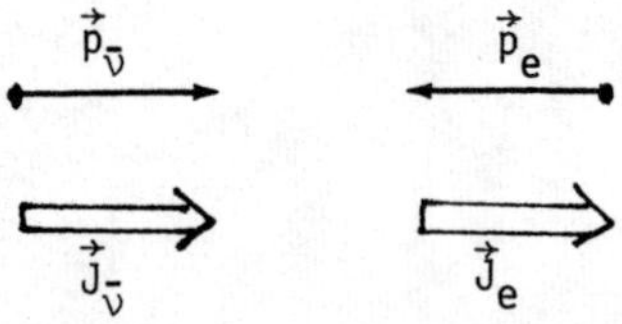

Therefore the p waves contribute to σ in (VI. 19).

Now the unitarity of the S matrix implies an upper bound for cross-sections and in the case of S-wave scattering this is

$$\frac{d\sigma}{d\Omega} \leq \frac{1}{p^2}$$

in the center of mass system where $s \approx 4p^2$ (mass neglected). Therefore this bound is attained by the cross-section (VI. 17) when :

$$\frac{G_F^{\,2}}{4\pi^2}\, s = \frac{1}{p^2} \quad , \quad s \approx 4p^2$$

that is, for the center of mass system momentum :

$$p^2 = \frac{\pi}{G_F}$$

that is, for $p \cong 500$ GeV.

This difficulty is eliminated in the intermediate vector boson theory. In this case the reaction (VI. 16) has the diagram

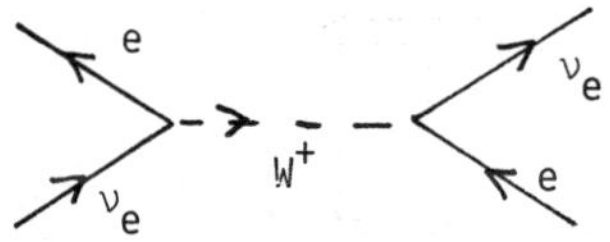

Figure VI. 1

while the reaction (VI. 18) is represented by :

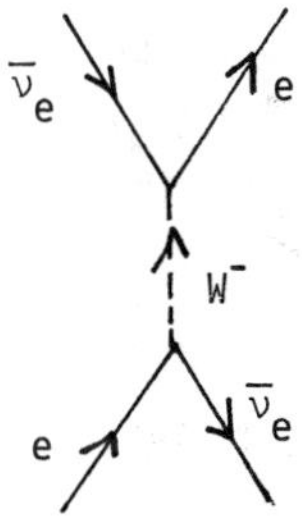

Figure VI. 2

The amplitude of the latter is :

$$M = i\, g^2(\bar{v}_{\nu_e}\ \gamma^\alpha(1-\gamma^5)u_e)\ \frac{g_{\alpha\beta} - \frac{k_\alpha k_\beta}{m_W^2}}{s - m_W^2}\ (\bar{u}_e\gamma^\beta(1-\gamma^5)v_{\nu_e}) \tag{VI. 21}$$

where :

$$s = p^2 = (p_e + p_{\nu_e})^2$$

The term in $k_\alpha\, k_\beta$ in the numerator of the propagator will give a contribution proportional to the electron mass in view of Dirac's equation for the incoming and outgoing leptons. Thus :

$$\bar{v}_{\nu_e}\ k_\alpha\, \gamma^\alpha(1 - \gamma^5)u_e = \bar{v}_{\nu_e}\ (p_e + p_{\nu_e})_\alpha\ \gamma^\alpha(1 - \gamma^5)u_e =$$

$$= \bar{v}_{\nu_e}\ (1 + \gamma^5)(p_e)_\alpha\ \gamma^\alpha\ u_e = m_e\ \bar{v}_{\nu_e}(1 + \gamma^5)u_e$$

As this term is thus proportional to $\frac{m_e^2}{m_W^2}$ we have for high energies :

$$\sigma\ (\bar{\nu}_e \to \bar{\nu}_e) \cong \frac{g^4}{s} \tag{VI. 22}$$

This s dependence is due to the term in s in the denominator of the propagator.

We see that instead of growing with s as in (VI. 20), the cross-section for antineutrino-electron scattering decreases with s as $\frac{1}{s}$.

Some comments are needed now on the origin of the energy dependence of the cross-sections (VI. 17a) and (VI. 20). The reason is that the Fermi coupling constant $\mathcal{G}_F$ has a dimension of $(\text{energy})^{-2}$

$$\mathcal{G}_F \, m_p^{\,2} \sim 1.05 \times 10^{-5}$$

Therefore as $\mathcal{G}_F$ appears as $\mathcal{G}_F^{\,2}$ in first order in the cross-section, $\mathcal{G}_F^{\,2}$ will have to be multiplied by a factor $(\text{energy})^2$ in order that the dimension ℓ^2 of the cross-section appear :

$$\sigma \sim \mathcal{G}_F^{\,2} \, E^2 \sim \frac{1}{(\text{energy})^4} \, (\text{energy})^2 \sim \frac{\ell^4}{\ell^2} \sim \ell^2$$

As at high energies the masses are neglected the available factor is really the center of mass energy.

In higher orders then the energy growth of the cross-section will be stronger. When we consider the lagrangean (VI. 14) the dimensionless coupling constant g appears in (VI. 21) and the propagator will give the needed energy dependence for the cross-section (VI. 22).

As to the unitarity violation resulting from (VI. 9), it can be seen immediately in the problem of the neutrino-antineutrino annihilation with the production of W bosons :

$$\nu + \bar{\nu} \rightarrow W^+ + W^-$$

the diagram of which is (Fig. VI. 3) :

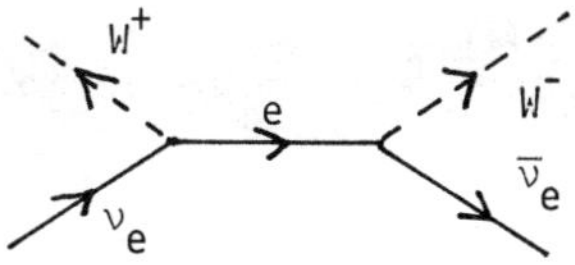

Figure VI. 3

The amplitude of this diagram is :

$$M = i\, g^2 (\bar{v}_{\nu_e} \gamma^\mu (1 - \gamma^5)\, \varepsilon^+_\mu \, \frac{\gamma^\alpha (p_\alpha - k^+_\alpha) + m_e}{(p - k^+)^2 - m_e^2} \, \varepsilon^-_\lambda \gamma^\lambda (1 - \gamma^5) u_{\nu_e})$$

where ε^+_μ, ε^-_λ are the polarization vectors of W^+ and W^- respectively.

For high energies the cross-sections for production of transverse and longitudinal vector mesons are, respectively :

$$\sigma\,(\nu\bar{\nu} \rightarrow W^+_T\, W^-_T) \sim \frac{g^4}{m_W^2}$$

$$\sigma\,(\nu\bar{\nu} \rightarrow W^+_L\, W^-_L) \sim \frac{g^4}{m_W^2}\, \frac{s}{m_W^2}$$

The latter cross-section violates unitarity.

We may, finally, compare quantum electrodynamics with the intermediate vector boson theory.

Quantum electrodynamics is renormalizable. The divergent integrals which occur in higher order perturbation terms can be eliminated by the introduction of a finite number of counter-terms which is the same whatever the order of the calculation and which have the same form as terms in the initial, bare lagrangean. This means that it is possible to redefine the parameters of the theory -in quantum electrodynamics the mass and the charge- and the scaling of the fields in such a way as to be left with finite physical quantities to all orders.

The Fermi current-current lagrangean and the lagrangean of the intermediate vector boson theory of weak interactions are not renormalizable.

It is a necessary condition for the renormalizability of an interaction lagrangean that the coupling constant be either dimensionless or of dimensions with positive powers of energy. In the case of the vector boson theory, the coupling constant is actually dimensionless, however, the propagator for the massive vector field is

$$- \frac{g_{\mu\nu} - \frac{1}{m_W^2} k_\mu k_\nu}{k^2 - m_W^2 + i \varepsilon}$$

In the case of low momentum transfer, as already remarked,

$$k^2 << m_W^2$$

the effective coupling constant will be (VI. 13) and so it is the same as in Fermi theory.

For high energies, the asymptotic behaviour of the propagator is

$$\frac{1}{m_W^2} \frac{k_\mu k_\nu}{k^2}$$

and again the effective coupling constant is (VI. 13).

This term can be transformed away in the case of neutral massive vector bosons in interaction with conserved currents but is present in the case of interactions of charged vector bosons with charge-changing currents.

Physicists were thus confronted with the following situation : on the one hand there exist the vector boson gauge theories which are renormalizable but the gauge bosons are massless ; on the other hand, we need massive vector bosons for the description of weak interactions by field theory but this theory is then not renormalizable.

The solution to this problem was found after the discovery of the rôle of spontaneous symmetry breakdown in field theory and the Higgs mechanism : it was discovered that the renormalizability character of a gauge theory is not lost when the symmetry is spontaneously broken but this mechanism leads to the introduction of massive vector fields.

The work of Brout, Englert, Guralnik, Hagen, Higgs and Kibble led to the discovery of the Higgs mechanism ; and the work of Salam and Weinberg, and of 't Hoft, opened up a whole new domain for the theoretical description of the forces of nature as different manifestations of basic gauge fields.

PROBLEMS

VI - 1. In quantum field theory, the fields are operators defined in a Hilbert space of state-vectors $|\psi>$. Corresponding to a Poincaré transformation of the geometrical coordinates, either the operators do not change but the state vectors change, and this is the so-called Poincaré-Schrödinger representation

$$x' = a + \ell x \rightarrow |\psi'>_S = U(a, \ell)\, |\psi>_S \, , \quad \theta'_S(x) = \theta_S(x)$$

where $\theta(x)$ designates a point-dependent operator ; or else the state vector does not change, it is the operators that change ; this is the so-called Poincaré - Heisenberg representation :

$$|\psi'>_H = |\psi>_H \, , \quad \theta'_H(x) \neq \theta_H(x)$$

The equivalence of the two representations is expressed by

$$|<\psi'|\, \theta'(x)|\varphi'>_S|^2 = |<\psi'|\, \theta'(x)|\varphi'>_H|^2$$

whence for unitary transformations $U(a, \ell)$

$$\theta'(x) = U^{-1}(a, \ell)\, \theta(x)\, U(a, \ell)$$

From this equation and the expression of U :

$$U(a, \ell) = \exp\left\{ i\, a_\mu P^\mu - \frac{i}{2} \varepsilon_{\mu\nu} J^{\mu\nu} \right\}$$

where P_μ and $J_{\mu\nu}$ are the momentum and angular momentum operators deduce the following equations for a Dirac spinor :

$$\left[\psi(x), P^\mu\right] = i\, \partial^\mu \psi(x)$$

$$\left[\psi(x), J^{\mu\nu}\right] = \left[\frac{\sigma^{\mu\nu}}{2} + i\,(x^\mu \partial^\nu - x^\nu \partial^\mu)\right] \psi(x)$$

VI - 2. The global phase transformation of a complex field

$$\varphi'(x) = e^{ie\alpha}\, \varphi(x)$$

induces a transformation on the field $\varphi(x)$ regarded as an operator in Hilbert space :

$$\varphi'(x) = U^{-1}(x)\, \varphi(x)\, U(x)$$

the generator of which is the charge operator Q. Deduce the commutation rule between the field operator φ and the charge Q.

VI - 3. The effective lagrangean for leptons which interact with an electromagnetic field and among themselves according to the weak current-current coupling is :

$$\mathscr{L} = \mathscr{L}_0 + \mathscr{L}_{(\gamma)} + \mathscr{L}_{(w)}$$

where

$$\mathscr{L}_0 = \frac{1}{2} \sum_\ell \left\{ \bar{\ell}\,(i\,\gamma^\alpha \partial_\alpha - m_\ell)\,\ell + \bar{\nu}_\ell\, i\,\gamma^\alpha \partial_\alpha \frac{1}{2}(1 - \gamma^5)\,\nu_\ell \right\}$$

$+ \text{h.c.} \quad ; \quad \ell = e, \mu, \tau$

$$\mathscr{L}_{(\gamma)} = e\, j^{\lambda}{}_{(\gamma)}\, A_{\lambda}\,(x) \; ;$$

$$j^{\lambda}{}_{(\gamma)} = \sum_{\ell} \bar{\ell}\, \gamma^{\lambda}\, \ell \; ;$$

is the electro magnetic current of leptons ;

$$\mathscr{L}_{(w)} = \frac{\mathscr{G}_F}{\sqrt{2}}\; j^{\lambda +}{}_{(w)}(x)\; j_{\lambda (w)}(x)$$

$$j^{\lambda}{}_{(w)} = \sum \bar{\nu}_{\ell}\, \gamma^{\lambda}\,(1 - \gamma^5)\ell$$

is the leptonique charged weak current :

a) Write $\mathscr{L}$ in terms of the left-handed and right-handed polarized components for all fields ℓ and ν_{ℓ} :

$$\psi_L = \frac{1}{2}\,(1 - \gamma^5)\,\psi \; ; \quad \psi_R = \frac{1}{2}\,(1 + \gamma^5)\,\psi$$

b) show that $\mathscr{L}$ is invariant under the chiral transformation :

$$\nu_{\ell} \rightarrow -\gamma^5\; \nu_{\ell}$$

c) show that $\mathscr{L}$ is invariant, in the limit $m_{\ell} \rightarrow 0$, under the unitary leptonic group $U_1 \otimes U_2$ where

$$\begin{pmatrix} e_L \\ \mu_L \\ \tau_L \end{pmatrix} \rightarrow U_1 \begin{pmatrix} e_L \\ \mu_L \\ \tau_L \end{pmatrix}$$

$$\begin{pmatrix} (\nu_e)_L \\ (\nu_\mu)_L \\ (\nu_\tau)_L \end{pmatrix} \rightarrow U_1 \begin{pmatrix} (\nu_e)_L \\ (\nu_\mu)_L \\ (\nu_\tau)_L \end{pmatrix}$$

$$\begin{pmatrix} e_R \\ \mu_R \\ \tau_R \end{pmatrix} \rightarrow U_2 \begin{pmatrix} e_R \\ \mu_R \\ \tau_R \end{pmatrix}$$

where U_1 and U_2 are arbitrary 3 x 3 unitary matrices.

d) Under which special choice of U_1 and U_2 does one have the (μ, e) universality, $\mu \rightleftarrows e$, in the limit $m_\mu = m_e$?

VI - 4. Consider the lagrangean of the preceding problem.

a) For which special choice of the matrices U_1, U_2 does one obtain separate conservation of lepton-number currents for each kind of lepton (ℓ, ν_ℓ) ?

b) Give the corresponding lepton numbers.

c) Is the photon decay of muons $\mu \rightarrow e + \gamma$ possible under the separate conservation of lepton numbers ?

d) Introduce the left-handed isospinor for the lepton ℓ :

$$L_\ell = \frac{1}{2}(1 - \gamma^5)\,\psi_\ell = \frac{1}{2}(1 - \gamma^5) \begin{pmatrix} \nu_\ell(x) \\ \ell(x) \end{pmatrix} \quad , \; \ell = e, \mu, \tau$$

and the right-handed isoscalar :

$$R_\ell = \frac{1}{2}(1 + \gamma^5)\,\ell(x) \; .$$

What is the form of the preceding lagrangean in terms of these fields in the limit $m \rightarrow 0$? What is the form of the electromagnetic and weak currents ?

e) What are the global phase transformations on L_ℓ and R_ℓ which give rise to these currents by Noether's theorem ?

f) Give the Nother current corresponding to the SU(2) global phase transformation :

$$L_\ell \rightarrow \exp\left(i\,\vec{\alpha}\cdot\frac{\vec{\tau}}{2}\right) L_\ell$$

$$R_\ell \rightarrow R_\ell$$

and its relation with the weak current.

VI - 5. The weak leptonic charge is defined from the weak current

$$j^\lambda{}_{(w)} = \sum_\ell \bar{\nu}_\ell \gamma^\lambda (1 - \gamma^5)\, \ell$$

by the equation :

$$Q_{(w)}(t) = \int d^3x\, j^0{}_{(w)}(x)$$

$$= \int d^3x \sum_\ell \left\{ \nu^+{}_\ell(x)\, (1 - \gamma^5)\, \ell\,(x) \right\}$$

The fields are quantized by the anticommutators :

$$[e_\alpha(\vec{x}, t), e^+{}_\beta(\vec{x}', t)]_+ = \delta^3(x - x') = [\nu_\alpha(\vec{x}, t), \nu^+{}_\beta(\vec{x}', t)]_+$$

a) Find the commutation relation :

$$\left[Q_{(w)}(t), Q^+{}_{(w)}(t) \right]$$

by using a relation between $[AD, BC]$ and $[A, B]_+$, $[A, C]_+$, $[D, B]_+$ and $[D, C]_+$.

b) Call $2\,Q_{(w)3}(t)$ the above commutator. Find the commutation relations which are satisfied by the hermitian operators $K^L{}_a$, a = 1,2,3 where

$$Q_{(w)}(t) = 2\,(K^L{}_1 + i\,K^L{}_2)$$

$$Q_{(w)3}(t) = 4\,K^L{}_3$$

c) Separate the vector and axial vector parts in K_a, call them K_a and K^5_a and define :

$$K^R_a = \frac{1}{2}(K_a + K^5_a) \quad ; \quad K^L_a = \frac{1}{2}(K_a - K^5_a)$$

Find the commutation relations between K^R_a, K^L_a, which define a $SU(2) \otimes SU(2)$ algebra.

d) What is the relation ship between the electric charge and the total lepton number ? Give the commutation relations between the lepton charge $Q_{(\gamma)}$ and the operators K.

VI - 6. a) Given a massive Proca vector field $\phi^\mu(x)$ in interaction with a vector current $j^\mu(x)$, write down the equations which generalize Maxwell's equations in terms of the fields $\mathscr{E}^k_{(v)} = \mathscr{G}^{ok}_{(v)}$, $\mathscr{B}^k_{(v)} = \frac{1}{2}\,\varepsilon_{k\ell n}\,\mathscr{G}^{\ell n}$ where $\mathscr{G}^{(v)}_{\mu\nu} = \partial_\nu\,\phi_\mu - \partial_\mu\,\phi_\nu$.

b) Obtain similar equations for an axial vector field $a^\mu(x)$ in interaction with an axial vector current $\rho^\mu(x)$ in terms of the fields $\mathscr{B}^k_{(a)} = \mathscr{G}^{ok}_{(a)}$, $\mathscr{E}^k_{(a)} = \frac{1}{2}\,\varepsilon_{k\ell n}\,\mathscr{G}^{\ell n}_{(a)}$ with $\mathscr{G}^{\mu\nu}_{(a)} = \partial^\nu\,Q^\mu - \partial^\mu\,Q^\nu$.

c) Assuming the fields in a) and b) have the same mass, obtain the equations which give a simple model for the intermediate boson field W^μ which interacts with V - A currents.

CHAPTER VII

The Higgs Mechanism

VII. 1 - THE NOTION OF SPONTANEOUS SYMMETRY BREAK-DOWN

In order to introduce the Higgs mechanism into the frame of gauge theory, let us first examine the notion of spontaneous symmetry break-down.

There are examples in nature,of symmetries which are not exact symmetries of the lagragean of a physical system. Thus the SU(3) group is an exact symmetry only if the components of the flavour multiplet of quarks have the same mass. To a lagrangean which is invariant under a group we may add a "small" term which breaks this invariance and this gives rise to the idea of broken symmetries or of almost exact symmetries.

Thus the lagrangean for three quark flavors, u,d,s assumed to have the same mass m :

$$L = \sum_{f=1}^{3} \bar{q}_f \, (i\gamma.\partial - m) \, q_f$$

is invariant under the flavour SU(3) group :

$$q_f \longrightarrow e^{i\alpha_k . \frac{\lambda_k}{2}} \, q_f$$

As the quark flavours have different masses the effective lagrangean will be of the form :

$$L = \sum_{f=1}^{3} \bar{q}_f \, (i\gamma.\partial - m) \, q_f + \bar{d} \, (m - m_d) \, d + \bar{s} \, (m - m_s) \, s$$

by assuming that $m = m_u$. The additional terms clearly break the above invariance and the symmetry is the more broken the larger are the mass differences .

There is, however, another very important example of broken symmetry : the lagrangean is invariant under a group of transformations but the ground state of the system -the vacuum state in field theory- is not invariant under this group. The well-known example is the Heinsenberg ferromagnet : an infinite crystalline system composed of spin - $\frac{1}{2}$ magnetic dipoles with spin-spin interactions among neighbouring dipoles. This interaction, although invariant under the rotation group, tends to align the dipoles in a given direction ; the ground state is thus not rotationally invariant and is one of an infinite set of possible ground states, corresponding to the continuum of directions in space.

Similar is the case of, say, the deuteron ground state. Its hamiltonian has a spin term $(\vec{\sigma}_p \cdot \vec{\sigma}_n)\ f(r)$ and a tensor force term which are invariant under rotation but the deuteron ground state has spin one and is thus three-times degenerate.

Whenever this phenomenon happens -the lagrangean admits of a symmetry group but the ground state is not invariant under this group- one speaks of a spontaneously broken symmetry.

Let us now consider the traditional example of a real scalar field with quartic self-interaction :

$$L = \frac{1}{2}\left\{ (\partial^\mu \varphi)(\partial_\mu \varphi) - \mu^2 \varphi^2 \right\} - \frac{\lambda}{4!}\,\varphi^4 \qquad \text{(VII. 1)}$$

In the classical treatment of this field, the ground state, the state of lowest energy, corresponds to the vacuum state in quantum theory.

The field equations are :

$$(\square + \mu^2)\,\varphi + \frac{\lambda}{3!}\,\varphi^3 = 0$$

A constant field φ_0, which does not depend on x, will be a solution of these equations if the following relationship holds :

$$\varphi_0 \; (\mu^2 + \frac{\lambda}{3!} \varphi_0^{\,2}) = 0 \qquad \text{(VII. 2)}$$

As the hamiltonian corresponding to the above L is :

$$H = \frac{1}{2} \left\{ \pi^2 + (\vec{\nabla} \varphi)^2 + \mu^2 \varphi^2 \right\} + \frac{\lambda}{4!} \varphi^4$$

where

$$\pi(x) = \partial^0 \varphi(x)$$

we see that the solution φ_0 of equation (VII. 2) is the one which makes the potential energy $U(\varphi)$ in (we call potential energy that part of H which survives for a constant field) :

$$H = \frac{1}{2} \left\{ \pi^2 + (\vec{\nabla} \varphi)^2 \right\} + U\,(\varphi)$$

$$U\,(\varphi) = \frac{1}{2} \mu^2 \varphi^2 + \frac{\lambda}{4!} \varphi^4$$

a minimum ; as $\pi_o = 0$, $\vec{\nabla} \varphi_o = 0$, φ_o gives the minimum of the energy H.

Now as H must be bounded from below, the constant λ is positive :

$$\lambda > 0$$

Therefore the position of the minimum depends of the sign of μ^2.

If $\mu^2 > 0$, then the only solution of equation (VII. 2) is :

$$\varphi_0 = 0$$

and the potential energy $U(\varphi)$ has the form given in the Fig. VII. 1. The ground state corresponds to this solution $\varphi_o = 0$.

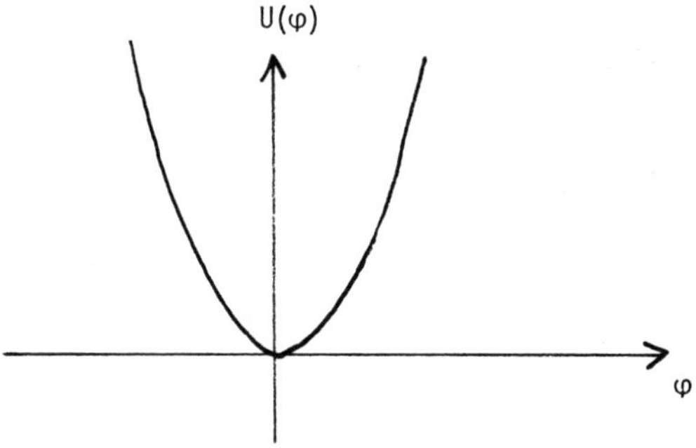

Figure VII. 1

The constant μ plays the rôle of the mass of the field (of the scalar mesons described by φ) and determines the first term in the development of $U(\varphi)$ around the minimum

$$U(\varphi) = U(\varphi_0) + (\varphi - \varphi_0)\, U'(\varphi_0) + \frac{(\varphi - \varphi_0)^2}{2!}\, U''(\varphi_0) + \ldots$$

Let us now assume that the parameter μ^2 is negative

$$\mu^2 < 0$$

In this case the equation (VII. 2) will be satisfied by $\varphi_0 = 0$ but also by :

$$\varphi_{0(+)} = + \sqrt{- \frac{6\mu^2}{\lambda}} \equiv a \qquad \text{(VII. 3)}$$

or by :

$$\varphi_{0(-)} = - \sqrt{- \frac{6\mu^2}{\lambda}} \qquad \text{(VII. 3a)}$$

so that :

$$U\ (\varphi_0) = 0$$

$$U\ (\varphi_{0(+)}) = U\ (\varphi_{0(-)}) = -\frac{3}{2}\ \frac{\mu^4}{\lambda}$$

The form of the curve will now be given by Fig. VII. 2

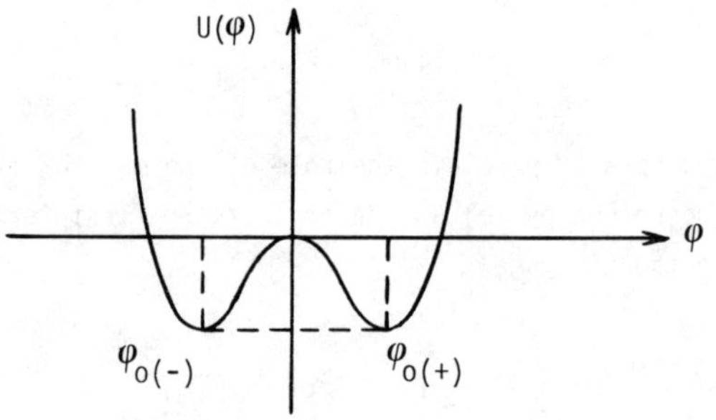

Figure VII. 2

There are two values of φ, $\varphi_{0(+)}$ and $\varphi_{0(-)}$, which give the minimum of $U(\varphi)$.

As the lagrangean is invariant under the transformation

$$\varphi \rightarrow -\varphi \qquad \text{(VII. 4)}$$

the ground state is now degenerate and each one of the two possible ground states transforms into the other one under this symmetry. In virtue of this symmetry, it is irrelevant which value of φ_0 we choose, $\varphi_{0(+)}$ or $\varphi_{0(-)}$, for studying the development of $U(\varphi)$, $H(\varphi)$ and $L(\varphi)$ around the chosen ground state. But once we make this choice the symmetry is spontaneously broken.

In order to study the behaviour of the lagrangean around the chosen ground state we define a new field :

$$\varphi' = \varphi - a \qquad \text{(VII. 5)}$$

where we call a the vacuum state which we choose :

$$a = \varphi_{0(+)}$$

Then we can express the lagrangean (VII. 1) in terms of φ' and get (we drop constant terms) :

$$\begin{aligned} L &= \frac{1}{2}\left\{(\partial^\mu \varphi')(\partial_\mu \varphi') - \mu^2 (\varphi' + a)^2\right\} - \frac{\lambda}{4!} (\varphi' + a)^4 \\ &= \frac{1}{2}\left\{(\partial^\mu \varphi')(\partial_\mu \varphi') - m^2 \varphi'^2\right\} - \frac{\lambda a}{3!} \varphi'^3 - \frac{\lambda}{4!} \varphi'^4 \end{aligned} \qquad \text{(VII. 6)}$$

where

$$m^2 = 2 |\mu^2| \qquad \text{(VII. 6a)}$$

The transformed field φ' has the properties that we require of a physical field : its vacuum state (or minimum energy value of the field) corresponds to $\varphi'_0 = 0$, its mass is the positive number $+\sqrt{2|\mu^2|}$ and it displays a cubic interaction $\sim \varphi'^3$ besides the quartic coupling. In view of the cubic interaction, the lagrangean above is not invariant under the reflection

$$\varphi' \to -\varphi'$$

and it would be difficult to guess from this lagrangean that it came from another one which possessed the symmetry (VII. 4).

Note that if we want the ground states to have zero potential energy we add a constant, $\frac{3}{2}\frac{\mu^4}{\lambda}$, to $U(\varphi)$ and obtain a function

$$U_1(\varphi) = \frac{\lambda}{4!} (\varphi^2 - a^2)^2, \quad a^2 = -\frac{6\mu^2}{\lambda} > 0$$

This example shows us that it is possible to change from the field with a mass term corresponding to an imaginary mass to another one with a physical mass.

VII. 2 - GOLDSTONE BOSONS

This type of considerations leads us to examine the case of spontaneous break-down of continuous groups of symmetry. This study gives way to the notion of Goldstone Bosons.

Let us now examine two scalar fields $\varphi_1(x)$, $\varphi_2(x)$, the lagrangean of which is :

$$L = \frac{1}{2} \left\{ (\partial^\mu \varphi_1)(\partial_\mu \varphi_1) + (\partial^\mu \varphi_2)(\partial_\mu \varphi_2) - \mu^2 (\varphi_1{}^2 + \varphi_2{}^2) \right\} - \frac{\lambda}{4!}(\varphi_1{}^2 + \varphi_2{}^2)^2 \qquad \text{(VII. 7)}$$

This lagrangean was chosen invariant under the group SO(2) of rotations in the plane φ_1, φ_2

$$\begin{aligned} \varphi'_1 &= \varphi_1 \cos\alpha - \varphi_2 \sin\alpha \\ \varphi'_2 &= \varphi_1 \sin\alpha + \varphi_2 \cos\alpha \end{aligned} \qquad \text{(VII. 8)}$$

or

$$\begin{pmatrix} \varphi'_1 \\ \varphi'_2 \end{pmatrix} = \begin{pmatrix} \cos\alpha & -\sin\alpha \\ \sin\alpha & \cos\alpha \end{pmatrix} \begin{pmatrix} \varphi_1 \\ \varphi_2 \end{pmatrix} \qquad \text{(VII. 8a)}$$

The potential energy is here :

$$U(\varphi_1, \varphi_2) = \frac{1}{2}\mu^2(\varphi_1^2 + \varphi_2^2) + \frac{\lambda}{4!}(\varphi_1^2 + \varphi_2^2)^2$$

and its minimum occurs for :

$$\frac{\partial U}{\partial \varphi_1} = \varphi_1 \left\{ \mu^2 + \frac{\lambda}{3!}(\varphi_1^2 + \varphi_2^2) \right\} = 0$$

$$\frac{\partial U}{\partial \varphi_2} = \varphi_2 \left\{ \mu^2 + \frac{\lambda}{3!}(\varphi_1^2 + \varphi_2^2) \right\} = 0$$

For $\mu^2 < 0$, $\lambda > 0$ the minima occur for φ_1 and φ_2 on the circle

$$\varphi_1^2 + \varphi_2^2 = a^2$$

where, as before :

$$a^2 = - \frac{6\mu^2}{\lambda}$$

The ground states are the points φ_1, φ_2 which lie on this circle and they transform into each other under the group SO(2). We can always choose the axis in the φ_1, φ_2 plane in such a way that

$$\varphi_{1(+)} = a \quad , \quad \varphi_{2(+)} = 0$$

is the ground state, which, as before, implies the spontaneous break-down of the symmetry.

The transformation to new fields around this vacuum state is the following

$$\varphi'_1 = \varphi_1 - a \quad ; \quad \varphi'_2 = \varphi_2$$

which leads to the lagrangean

$$L = \frac{1}{2}\left\{ (\partial^{\mu} \varphi'_1)(\partial_{\mu} \varphi'_1) - m^2 \varphi'^2_1 \right\} + \frac{1}{2} (\partial^{\mu} \varphi'_2)(\partial_{\mu} \varphi'_2) -$$

$$- \frac{\lambda}{3!} a (\varphi'^2_1 + \varphi'^2_2)\varphi'_1 - \frac{\lambda}{4!}(\varphi'^2_1 + \varphi'^2_2)^2$$

We see that one of the fields, φ'_1, acquires a positive mass :

$$m = \sqrt{2|\mu^2|} = \sqrt{\lambda \frac{a^2}{3}}$$

whereas the field φ'_2 is massless. <u>This massless field</u> is called a <u>Goldstone Boson</u>. And this result is an illustration of Goldstone's theorem : there will be N-M massless bosons in a theory in which a sub-group of dimension N-M from a symmetry group G of dimension N, is spontaneously broken.

Consider a n-dimensional real vector field φ the components of which are scalar fields

$$\varphi = \begin{pmatrix} \varphi_1 \\ \vdots \\ \varphi_n \end{pmatrix}$$

with lagrangean :

$$L = \frac{1}{2} (\partial^{\mu} \varphi)(\partial_{\mu} \varphi) - U(\varphi)$$

Let G be the continuous group which leaves the potential energy $U(\varphi)$ invariant Let T_{α} be the N generators of this group and ω_{α} the infinitesimal parameters so that

$$\varphi \longrightarrow \varphi' = (I + i\omega_{\alpha} T_{\alpha}) \varphi$$

T_α here is the $n \times n$ matrix representation of the generators. By definition, under this group :

$$\delta U = \frac{\partial U}{\partial \varphi_k} \delta \varphi_k = i \frac{\partial U}{\partial \varphi_k} \omega_\alpha (T_\alpha)_{k\ell} \varphi_\ell = 0$$

As the parameters ω_α are arbitrary, these are N equations :

$$\frac{\partial U}{\partial \varphi_k} (T_\alpha)_{k\ell} \varphi_\ell = 0 \qquad \alpha = 1, \ldots, N$$

If we differentiate this equation once we obtain :

$$\frac{\partial^2 U}{\partial \varphi_p \, \partial \varphi_k} (T_\alpha)_{k\ell} \varphi_\ell + \frac{\partial U}{\partial \varphi_k} (T_\alpha)_{kp} = 0$$

Let us take this equation at the value

$$\varphi = a$$

which minimizes $U(\varphi)$; as :

$$\left(\frac{\partial U}{\partial \varphi_k} \right)_{\varphi = a} = 0$$

we get

$$\left(\frac{\partial^2 U}{\partial \varphi_p \, \partial \varphi_k} \right)_{\varphi = a} (T_\alpha)_{k\ell} a_\ell = 0$$

Now as we expand $U(\varphi)$ around a we have :

$$U(\varphi) = U(a) + \frac{1}{2} (M^2)_{k\ell} (\varphi - a)_k (\varphi - a)_\ell + .. = U(a) + \frac{1}{2}\left(\frac{\partial^2 U}{\partial\varphi_k \partial\varphi_\ell}\right)_{\varphi = a} (\varphi - a)_k (\varphi - a)_\ell + \dots$$

so the second derivative above is just the squared-mass matrix, which gives us :

$$(M^2)_{pk} (T_\alpha)_{k\ell} a_\ell = 0 \quad ; \quad \alpha = 1, \dots, N \qquad \text{(VII. 9)}$$

If, after the choice of the ground state, a sub-group g, of dimension n, remains a symmetry of this state, then for any generator of this sub-group :

$$(T_\alpha)_{k\ell} a_\ell = 0 \quad ; \quad \alpha = 1, \dots, n \leq N$$

whereas for the generators of the group $(G - g)$, of $N-n$ dimensions, which breaks the symmetry, we have :

$$(T_\alpha)_{k\ell} a_\ell \neq 0 \quad ; \quad \alpha = n + 1, \dots, N$$

Therefore, the equation (VII. 9) for $(M^2)_{pk}$ says that there are $N-n$ zero eigenvalues for the squared-mass matrix, the massless bosons.

This result prevented for some time the consideration of spontaneous break-down of symmetries into particle theory since it would imply Goldstone Bosons and no evidence was found about massless, spinless bosons.

In the years 1964-1966, a series of papers [79-83] appeared which showed that the introduction of spontaneous break-down of symmetry for a scalar field in gauge theory leads to the disappearance of the Goldstone Bosons, which are gauged away, and the appearance of massive vector fields resulting from transformations of the gauge fields.

VII. 3 - THE HIGGS MECHANISM

Consider the lagrangean (VII. 7) formulated in terms of the complex scalar field :

$$\varphi(x) = \frac{1}{\sqrt{2}} (\varphi_1(x) + i\,\varphi_2(x))$$

$$\varphi^*(x) = \frac{1}{\sqrt{2}} (\varphi_1(x) - i\,\varphi_2(x))$$

that is :

$$L_\varphi = (\partial^\mu\varphi)^*(\partial_\mu\varphi) - \mu^2\,\varphi^*\,\varphi - \frac{\lambda}{3!}\,(\varphi^*\,\varphi)^2 \qquad \text{(VII. 7a)}$$

(we keep this coefficient for the coupling constant instead of $\frac{\eta}{2}$ because of the expression of this interaction in terms of real fields (see (VII. 7))).

This lagragean is invariant under the transformation

$$\varphi(x) \longrightarrow \tilde{\varphi}(x) = e^{i\alpha}\,\varphi(x)$$

corresponding to equations (VII. 8).

Let us now consider a local SO(2) rotation :

$$\tilde{\varphi}(x) = e^{ig\theta(x)}\,\varphi(x) \qquad \text{(VII. 10)}$$

In this case, as we have learned, we have to introduce a gauge vector field $a_\mu(x)$ (see the Chapter II) and the invariant lagrangean under the group

(VII. 10) is :

$$\mathscr{L} = -\frac{1}{4} F^{\mu\nu} F_{\mu\nu} + (D^{\mu} \varphi)^* (D_{\mu} \varphi) - \mu^2 \varphi^* \varphi - \frac{\lambda}{3!} (\varphi^* \varphi)^2 \qquad \text{(VII. 11)}$$

where

$$F^{\mu\nu} = \partial^{\nu} \mathcal{a}^{\mu}(x) - \partial^{\mu} \mathcal{a}^{\nu}(x)$$

$$D_{\mu} = \partial_{\mu} + i\, g\, \mathcal{a}_{\mu}(x)$$

and the transformations (VII. 10) are accompanied by the following transformations of the gauge field :

$$\tilde{\mathcal{a}}^{\mu}(x) = \mathcal{a}^{\mu}(x) - \partial^{\mu} \theta(x) \qquad \text{(VII. 10a)}$$

The new point now, as compared to the precedent study of the electromagnetic field (which stops at the last equations, with $g = e$) is that we assume

$$\lambda > 0 \quad , \quad \mu^2 < 0$$

so that the minima of the scalar potential energy occur for :

$$\varphi^* \varphi = \frac{1}{2} a^2 \quad , \quad a^2 = -\frac{6\mu^2}{\lambda} \quad , \quad \mu^2 < 0 \quad , \quad \lambda > 0$$

The break-down of the symmetry above, equations (VII. 10) and (VII. 10a) occurs when we choose one of the possible minima as our ground state, say :

$$\varphi_0 = \frac{1}{\sqrt{2}} a$$

or :

$$\varphi_{1(o)} = a \quad , \quad \varphi_{2(o)} = 0$$

The transformation to new fields referred to this ground state is then :

$$\varphi'(x) = \varphi(x) - \frac{a}{\sqrt{2}} \quad , \quad \underline{a} \text{ real}$$

and the corresponding lagrangean will be :

$$\mathscr{L} = -\frac{1}{4} F^{\mu\nu} F_{\mu\nu} + \frac{1}{2} g^2 a^2 \mathcal{a}^\mu \mathcal{a}_\mu + \frac{1}{2} \left\{ (\partial^\mu \varphi'_1)(\partial_\mu \varphi'_1) - m^2 \varphi'^2_1 \right\} +$$

$$+ \frac{1}{2} (\partial^\mu \varphi'_2)(\partial_\mu \varphi'_2) - \frac{\lambda}{3!} a(\varphi'^2_1 + \varphi'^2_2)\varphi'_1 - \frac{\lambda}{4!} (\varphi'^2_1 + \varphi'^2_2)^2 +$$

$$+ \text{ coupling terms between } \varphi \text{ and } \mathcal{a}^\mu \qquad \text{(VII. 12)}$$

where

$$m^2 = -2\mu^2, \quad \mu^2 < 0.$$

The new thing is the second term on the right-hand side, namely

$$\frac{1}{2} g^2 a^2 \mathcal{a}^\mu \mathcal{a}_\mu$$

which corresponds to a mass for the field $\mathcal{a}^\mu(x)$ equal to :

$$m_V = g a$$

The gauge transformations for the new fields are now :

$$\tilde{\varphi}' + \frac{a}{\sqrt{2}} = e^{ig\theta(x)} \left\{ \varphi'(x) + \frac{a}{\sqrt{2}} \right\}$$

so that

$$\tilde{\varphi}_1 = \cos g\,\theta\,(x) \left\{ \varphi'_1 + \frac{a}{\sqrt{2}} \right\} - \sin g\,\theta\,(x)\,\varphi'_2 - \frac{a}{\sqrt{2}}$$

$$\tilde{\varphi}_2 = \cos g\,\theta\,(x) \quad \varphi'_2 + \sin g\,\theta\,(x) \left\{ \varphi'_1 + \frac{a}{\sqrt{2}} \right\}$$

$$\tilde{a}^{\mu}(x) = a^{\mu}(x) - \partial^{\mu}\,\theta(x)$$

We note, however, that the lagrangean (VII. 12) has a spurious field : indeed the number of independent field components in the first lagrangean (VII. 11) is four : two components for the massless vector field, plus φ_1 and φ_2 whereas (VII. 12) has 5 components : φ'_1, φ'_2 and the three components of the massive vector field. Indeed we can transform the Goldstone boson field φ'_2 away. For this we introduce new variables, namely :

$$\varphi(x) = \frac{1}{\sqrt{2}} \left\{ \rho(x) + a \right\} e^{igw(x)/a}$$

$$a_{\mu}(x) = C_{\mu}(x) - \frac{1}{a}\,\partial_{\mu}\,w(x) \qquad \text{(VII. 13)}$$

The new fields are $\rho(x)$, $w(x)$ and C_{μ} ; and the value $\underline{a}$ of φ which renders the potential energy minimum is essential for this transformation $a \neq 0$.

We obtain :

$$D_{\mu}\,\varphi \equiv \partial_{\mu}\,\varphi + i\,g\,a_{\mu}\,\varphi =$$

$$= \frac{1}{\sqrt{2}} \left\{ \partial_{\mu}\,\rho + i\,g\,C_{\mu}\,(\rho + a) \right\} e^{igw/a}$$

and then the lagrangean (VII. 11) goes over into :

$$L = -\frac{1}{4} C^{\mu\nu} C_{\mu\nu} + \frac{1}{2} g^2 a^2 C^{\mu} C_{\mu} + \frac{1}{2} (\partial^{\mu} \rho)(\partial_{\mu} \rho) - \frac{1}{2} (m_{\rho}^2) \rho^2 -$$

$$- \frac{\lambda}{4!} \rho^4 - \frac{\lambda a}{3!} \rho^3 + \frac{g^2}{2} C^{\mu} C_{\mu} (\rho^2 + 2 \rho a) \qquad \text{(VII. 14)}$$

In this lagrangean there are the massive vector field C_{μ} with mass :

$$m_V = g a$$

and the massive scalar field ρ with mass

$$m_{\rho} = \sqrt{2|\mu^2|}$$

The Goldstone Boson, which in the new variables would be described by the field $w(x)$, disappeared. Instead, its corresponding degree of freedom contributed to the additional degree of freedom acquired by the massive vector field C_{μ}.

This is the Higgs mechanism[79-83].

PROBLEMS

VII - 1. The lagrangean :

$$L = \frac{1}{2} m (\dot{x})^2 - \frac{1}{2} m \omega^2 x^2 - \frac{\lambda}{4} x^4$$

describes a linear harmonic oscillator submitted to an x^3 attractive force if $\lambda > 0$ and if ω is a real number, the mass m being positive.

a) What are the values of x which give a minimum to the potential energy if $m > 0$, $\lambda > 0$, $\omega^2 < 0$?

b) Break the symmetry by choosing one of these values to define a new coordinate x' which vanishes at the minimum potential energy ; and show that in the new variable the lagrangean defines an oscillator with a real frequency and forces proportional to $x'2$ and to x'^3.

VII - 2. a) Deduce the complete lagrangean (VII. 12) ; b) with the variables defined by equations (VII. 13) obtain the lagrangean (VII. 14).

VII - 3. Read Coleman, ref. 24 ; Rajaraman, ref. 25 ; O'Raifeartaigh, ref. 109.

CHAPTER VIII

The Salam-Weinberg Model

VIII. 1 - UNIFICATION OF THE ELECTROMAGNETIC AND WEAK INTERACTION THEORIES : THE SALAM-WEINBERG MODEL

The aim of physics is the description of large classes of phenomena, based on a few simple ideas and postulates. Physical theories give rise to intuitive representations and images which contribute to an understanding of the corresponding class of phenomena.

A higher level of understanding is reached in physics whenever two different theories are unified, two apparently unrelated classes of facts are discovered to be intimately connected, forming two subclasses of a larger set of phenomena described by a larger, unitary theory.

In the history of physics, some notable unification efforts were achieved. They are indicated in the table VIII. 1

TABLE VIII. 1 - Unification of physical theories

Newton, 1686 : he stated that the gravity force which the Earth exerts upon all terrestrial bodies is identical to the gravitational force between any two bodies anywhere in the universe The physics of gravity on Earth is a chapter of the Newtonian theory of universal gravitation.

Maxwell, 1855 : the laws of electricity and magnetism as expressed by Maxwell's equations imply that the electric and magnetic fields have the same physical nature, transform among themselves and are components of a single physical entity, the electromagnetic field. Light waves are electromagnetic waves ; optics is a chapter of electrodynamics.

Einstein, Lorentz, Poincaré, 1905 : special relativity, as discovered by these authors and developed mainly by Einstein, is based upon the fact that time and ordinary three-dimensional coordinate space are sub-spaces of a four-dimensional Minkowski space. Physical variables have a definite transformation law under the proper orthochronous Poincaré group and these laws acomplish the unification of quantities such as momentum and energy, electric field and magnetic field, charge density and current density and so on. Matter and energy are equivalent.

Einstein, 1915 : in his theory of general relativity, the gravitational field is described by the fundamental metric tensor of a four-dimensional Riemann space. The geometry of this space is determined by an equation which involves the energy-momentum tensor of the universe. Gravitational dynamics is unified with geometry.

De Broglie, 1924 : the duality wave-particle holds not only for light but also for all particles. Mechanics became wave mechanics which evolved into the quantum mechanics of Schrödinger ; Born, Heisenberg and Jordan ; and Dirac.

After the discovery of the vectorial form of the weak currents and of the possible description of weak interactions by means of a field of intermediate vector bosons, the idea arose among physicists of a possible unified description of electromagnetic and weak forces. This idea was based on certain basic common features of these forces namely, a) representation of both types of forces by vector fields ; b) a universal coupling constant, e for the electromagnetic interactions and g for the weak interactions, related among themselves by the Fermi constant and equation (VI. 13). From the latter equation, however, it follows that if the coupling constant e of charged currents with the electromagnetic field A_μ is assumed to be the same[84)] as the coupling constant g of the weak currents with the weak vector boson field W_μ, then the mass of the vector bosons must be much larger than the proton mass, of the order of 40 GeV. Thus if the electromagnetic vector field A_μ and the weak vector boson field W_μ were to belong to some multiplet, having each the same coupling constant with the corresponding current, it is unsatisfactory that there exists such a mass difference between these fields, the electromagnetic field being massless.

The assumption that the electromagnetic field were described by a gauge field as seen in Chapter II and the intermediate vector boson field by some gauge field of the Yang-Mills type would not be acceptable since the latter has to be massless if the lagrangean of the theory is invariant under the SU(2) gauge group.

The discovery of the Higgs mechanism was the key which made it finally possible the formulation of unified models of weak and electromagnetic interactions based on the theory of gauge fields : after a lagrangean is constructed with massless fermi and vector boson fields so as to be invariant under a gauge group G, a spontaneous breakdown of the symmetry corresponding to a sub-group g of G is introduced which generates masses for fermions and bosons with the exception of the photon and neutrinos. The Higgs mechanism accomplishes this task without introducing Goldstone bosons. The final physical lagrangean is still invariant under the U(1) gauge sub-group associated to the massless electromagnetic field.

A model proposed by Georgi and Glashow postulated three vector gauge fields associated to the three generators of the SO(3) group ; the theory was arranged so as to identify the neutral field with the electromagnetic field and the two other fields with the intermediate vector boson fields after mass generation by the Higgs mechanism. This model was abandoned since experimentally it was found that besides the two charged boson fields, W_μ, W_μ^+ and the electromagnetic field A_μ, there most likely exists a fourth massive boson neutral field Z_μ, which mediates weak interactions between neutral weak currents ; the latter have been observed in laboratory (Musset et al.[143]).

The simplest, or most economical, unification model of the weak and electromagnetic forces is the model which was independently developed by Weinberg and Salam. Experimental research in the last decade has confirmed so far some of the predictions of this model. The model assumes the SU(2) ⊗ U(1) group as the fundamental gauge group. This is because in correspondence to the four generators of this group, τ_1, τ_2, τ_3 for SU(2), the identify I for U(1), four vector fields are introduced $\mathcal{a}^u{}_a(x)$, a = 1, 2, 3 and $B_\mu(x)$ in connection with the definition of the covariant derivatives needed for the construction of an invariant lagrangean. From these four fields the electromagnetic field A_μ, the charged boson fields W_μ and W_μ^+ and the neutral boson field Z_μ are derived. The simplest basic matter field is the leptonic field formed of an electron and its associated neutrino. They are assumed to be massless together with the above gauge fields in the invariant lagrangean. The mass of the physical particles (electron, charged and neutral bosons W, W^+, Z) is generated by the Higgs mechanism ; therefore a doublet of scalar Higgs fields with mass and quartic self-interaction is introduced in a gauge-invariant way in the lagrangean so as to form an invariant with the matter fields. And the vacuum expectation value of the Higgs field is chosen in such a way as to give a mass-term for the electron field $e(x)$, proportional to $\bar{e}(x)\, e(x)$, and no mass term for the neutrino. As the gauge fields enter the covariant derivatives of the Higgs field, it will turn out that a convenient linear combination of the first two components of the vector field $\mathcal{a}^u{}_a$ will display a mass term. This will then correspond to the physical field W_μ. A similar phenomenon occurs for one of two possible orthogonal linear combinations of $\mathcal{a}^{\mu}_{3}$ and B_{μ} ; it will correspond to the massive neutral boson field Z_{μ}. The remaining, massless, linear combination of $\mathcal{a}^{\mu}_{3}$, B_{μ} is the electromagnetic field. At the same time, the interactions of these fields turn out to be the known electromagnetic and charged weak interactions. A new term, corresponding to an interaction between a neutral weak lepton current and the neutral vector boson field appears in the lagragean. This prediction has been verified experimentally[143)] and a parameter, the Weinberg angle, which defines the electromagnetic and Z-fields in terms of the initial gauge fields has been experimentally measured. At the time of writing, however, (November 1980) no experimental observation has been made of the vector-bosons W, W^{+} and Z, nor of the scalar Higgs bosons.

The unification accomplished by the Salam-Weinberg model is therefore expressed in the fact that the electromagnetic field and the vector boson field are special linear combinations of the components of the vector gauge fields $\mathcal{a}^{\mu}_{a}$ and B^{μ} ; they are thus of the same nature and come out so to say from the same underlying field. The duality still remaining in this theory resides in the occurrence of two coupling constants g, g' in terms of which the charge e and the Weinberg angle are expressed.

VIII. 2 - THE SU(2) ⊗ U(1) GAUGE INVARIANT LAGRANGEAN

We shall now develop the main ideas of the Salam-Weinberg theory of unified weak and electromagnetic interactions[80-85, 94-108)].

Let us consider the simples case of the interactions among electrons and electronic neutrinos. As the neutrino is known to be completely polarized and left-handed the simplest way of combining a neutrino and an electron into an isospin multiplet is to consider a left-handed doublet :

$$L(x) = \begin{pmatrix} \nu_L(x) \\ e_L(x) \end{pmatrix} \qquad \text{(VIII. 1)}$$

where :

$$e_L(x) = \frac{1}{2}\,(I - \gamma^5)\; e(x)$$

$$\nu_L(x) = \frac{1}{2}\,(I - \gamma^5)\; \nu(x)$$

However, the electron is massive, therefore there must exist a right-handed component of the electron field operator which we sall assume to be an isospin singlet (invariant under the SU(2) group) :

$$R(x) = e_R(x) = \frac{1}{2}\,(I + \gamma^5)\; e(x) \qquad \text{(VIII. 1a)}$$

So the basic matter field is represented in table VIII. 2

Basic matter fields in the simplest S-W Theory	
Isospin doublet	Isospin singlet
$L(x) = \begin{pmatrix} \nu_L(x) \\ e_L(x) \end{pmatrix}$	$R(x) = e_R(x)$

Table VIII. 2

However, invariance under the SU(2) group requires that these fields have the same mass. Thus, in a first stage, we assume that the fields (VIII. 1) and (VIII. 1a) are massless.

A lagrangean, invariant under the SU(2) group, associated with these fields is :

$$\mathscr{L} = \bar{L}\, i\, \gamma^{\alpha}\, \partial_{\alpha}\, L + \bar{R}\, i\, \gamma^{\alpha}\, \partial_{\alpha}\, R$$

which describes free massless neutrinos and electrons :

$$i\, \gamma^{\alpha}\, \partial_{\alpha}\, \nu(x) = 0$$

$$i\, \gamma^{\alpha}\, \partial_{\alpha}\, e(x) = 0$$

We want to be able to use the Higgs mechanism to generate the electron mass and for this purpose introduce scalar fields. Thus if we can have, instead of the above equation for the electron, an equation of the form :

$$\left\{ i\, \gamma^{\alpha}\, \partial_{\alpha} - g\, \varphi(x) \right\} e(x) = 0 \qquad \text{(VIII. 1b)}$$

then a change of scalar field variable around its ground state or vacuum expectation value <u>a</u> :

$$\varphi'(x) = \varphi(x) - a \qquad \text{(VIII. 1c)}$$

would give the following equation

$$\left\{ i\, \gamma^{\alpha}\, \partial_{\alpha} - m - g\, \varphi'(x) \right\} e(x) = 0 \qquad \text{(VIII.1d)}$$

with

$$m = g\, a \qquad \text{(VIII.1e)}$$

indicating a generation of the electron mass by the vacuum expectation a of $\varphi(x)$.

In order to get such a mechanism, we look then for a possible interaction of the fields $L(x)$ and $R(x)$ with scalar fields. This interaction must be invariant under the $SU(2)$ group therefore we need an isodoublet of scalar fields :

$$\phi(x) = \begin{pmatrix} \phi_I(x) \\ \phi_{II}(x) \end{pmatrix} \qquad \text{(VIII. 2)}$$

so that the following lagrangean is possible

$$\mathscr{L} = \bar{L}\, i\, \gamma^\alpha \partial_\alpha L + + \bar{R}\, i\, \gamma^\alpha \partial_\alpha R - G\,(\bar{L}\,\phi\, R + \bar{R}\,\phi^+ L) +$$

$$+ \partial^\mu \phi^+ \partial_\mu \phi - M^2 \phi^+ \phi - \frac{\lambda}{4!} (\phi^+ \phi)^2 \qquad \text{(VIII. 3)}$$

The last three terms correspond to a self-interacting ϕ-field and the terms in the coupling constant G define the interaction between ϕ and L and R.

This lagrangean is invariant under the global group $SU(2)$:

$$\begin{aligned} L &\longrightarrow L' = e^{i \vec{\Lambda}\cdot \frac{\vec{\tau}}{2}} L \\ \phi &\longrightarrow \phi' = e^{i \vec{\Lambda}\cdot \frac{\vec{\tau}}{2}} \phi \\ R &\longrightarrow R' = R \end{aligned} \qquad \text{(VIII. 4)}$$

but it is also invariant under the global $U(1)$ group :

$$\begin{aligned} L &\longrightarrow \tilde{L} = e^{-\frac{i}{2}\theta} L \\ \phi &\longrightarrow \tilde{\phi} = e^{\frac{i}{2}\theta} \phi \\ R &\dashrightarrow \tilde{R} = e^{-i\theta} R \end{aligned} \qquad \text{(VIII. 5)}$$

Now the above transformations for constant phases are represented by transformations in Hilbert space, of the type

$$\Omega' = e^{-i\vec{\Lambda}\cdot\vec{T}} \; \Omega \; e^{i\vec{\Lambda}\cdot\vec{T}} \quad ; \quad \Omega = L, \phi, R \qquad \text{(VIII. 4a)}$$

for the group SU(2) (VIII. 4) and of the type :

$$\Omega' = e^{-i\theta \frac{Y}{2}} \; \Omega \; e^{i\theta \frac{Y}{2}} \qquad \text{(VIII. 5a)}$$

for the U(1) group (VIII. 5).

So that the operators $\vec{T}$ (isospin) and Y (hypercharge) in Hilbert space satisfy the commutation rules :

$$\begin{aligned}
\left[\vec{T}, L\right] &= -\frac{\vec{\tau}}{2} L \\
\left[\vec{T}, \phi\right] &= -\frac{\vec{\tau}}{2} \phi \\
\left[\vec{T}, R\right] &= 0 \\
\left[Y, L\right] &= L \\
\left[Y, \phi\right] &= -\phi \\
\left[Y, R\right] &= 2R
\end{aligned} \qquad \text{(VIII. 4b)}$$

Therefore the following are the eigenvalues of the isospin T_3 and hypercharge Y for these fieds (table VIII. 3) :

	T_3	Y	Q
ν_1	1/2	-1	0
e_L	-1/2	-1	-1
e_R	0	-2	-1
ϕ_I	1/2	1	1
ϕ_{II}	-1/2	1	0

Table VIII. 3

Isospin, hypercharge and charge of the matter and Higgs fields

where the charge is given by :

$$Q = (T_3 + \frac{Y}{2})e \qquad \text{(VIII. 5b)}$$

Now in order that an invariant lagrangean under the local SU(2) ⊗ U(1) group be constructed we introduce the vector gauge fields $\mathcal{A}^{\mu}{}_a(x)$, a = 1,2,3 ; and $B^{\mu}(x)$ corresponding to the transformations (we take $g > 0$, $g' < 0$) :

$$\begin{aligned} L &\longrightarrow L' = \exp\left\{ i(g\,\vec{\Lambda}(x)\cdot\frac{\vec{\tau}}{2} + \frac{g'}{2}\theta(x))\right\} L \\ \phi &\longrightarrow \phi' = \exp\left\{ i(g\,\vec{\Lambda}(x)\cdot\frac{\vec{\tau}}{2} - \frac{g'}{2}\theta(x))\right\} \phi \\ R &\longrightarrow R' = \exp\left\{ i\, g'\,\theta\,(x)\right\} R \end{aligned} \qquad \text{(VIII. 5c)}$$

The covariant derivatives applied to the matter and Higgs fields will be :

$$\begin{aligned} D_\mu L\,(x) &= (\partial_\mu + i\,g\,\mathcal{A}_{\mu a}\,(x)\,\frac{\tau_a}{2} + i\,\frac{g'}{2}B_\mu(x))\,L(x) \\ D_\mu \phi\,(x) &= (\partial_\mu + i\,g\,\mathcal{A}_{\mu a}\,(x)\,\frac{\tau_a}{2} - i\,\frac{g'}{2}B_\mu(x))\,\phi(x) \end{aligned} \qquad \text{(VIII. 6)}$$

$$D_\mu R(x) = (\partial_\mu + i g' B_\mu(x)) R(x)$$

according to the rules developed in Chapters II and IV.

The field tensors of the vector fields $\mathcal{a}_{\mu a}$ and B_μ are then :

$$\mathcal{G}^{\mu\nu}{}_a = \partial^\nu \mathcal{a}^\mu{}_a - \partial^\mu \mathcal{a}^\nu{}_a + g\, \varepsilon_{abc}\, \mathcal{a}^\mu{}_b\, \mathcal{a}^\nu{}_c$$

$$B^{\mu\nu} = \partial^\nu B^\mu - \partial^\mu B^\nu \qquad \text{(VIII. 7)}$$

as seen in sections IV. 4 and II. 2 respectively.

The total lagrangean can now be written

$$\mathcal{L} = -\frac{1}{4} \mathcal{G}^{\mu\nu}{}_a\, \mathcal{G}_{\mu\nu,a} - \frac{1}{4} B^{\mu\nu} B_{\mu\nu} + \bar{L}\, i\, \gamma^\alpha D_\alpha L +$$

$$+ \bar{R}\, i\, \gamma^\alpha D_\alpha R + (D_\mu \phi)^+ (D^\mu \phi) - M^2 \phi^+ \phi - \qquad \text{(VIII. 8)}$$

$$- \frac{\lambda}{3!} (\phi^+ \phi)^2 - G(\bar{L}\, \phi\, R + \bar{R}\, \phi^+ L)$$

with the derivatives as defined in (VIII. 6). It is gauge invariant, that is, invariant under the local $SU(2) \otimes U(1)$ group (VIII. 5c) (the coefficient of the term in $(\phi^+ \phi)^2$ is the one we used in formula (VII. 7a), section VII. 3).

VIII. 3 - GENERATION OF THE ELECTRON MASS

As suggested in (VIII. 1b), (VIII. 1e) we now want to break the gauge symmetry by assuming

$$M^2 < 0$$

and shifting the scalar field in a convenient way around a chosen ground state.

Let us consider the terms in G in (VIII. 8) :

$$G(\bar{L}\,\phi\,R + \bar{R}\,\phi^{+}\,L) = G\left\{(\bar{\nu}_L\, e_R)\,\phi_I + (\bar{e}_R\,\nu_L)\,\phi_I^{+} + (\bar{e}_L\, e_R)\,\phi_{II} + (\bar{e}_R\, e_L)\,\phi_{II}^{+}\right\}$$

As the terms in $(\bar{e}_L\, e_R)$ and $(\bar{e}_R\, e_L)$ contain the field ϕ_{II}, ϕ^{+}_{II} as factors respectively, we need to shift only the field ϕ_{II} around the chosen value which gives the spontaneous symmetry breakdown.

We choose to shift the field ϕ_{II} around its vacuum expectation value a :

$$\phi_{II}(x) = \frac{a}{\sqrt{2}} + \frac{1}{\sqrt{2}}\,(\chi_1(x) + i\;\chi_2(x))$$

$$\phi_{II}^{+}(x) = \frac{a}{\sqrt{2}} + \frac{1}{\sqrt{2}}\,(\chi_1(x) - i\;\chi_2(x)) \quad ; \quad a = \langle 0|\phi_{II}(x)|0\rangle \qquad \text{(VIII. 9)}$$

The terms in G become, with this choice :

$$G\,(\bar{L}\,\phi\,R + \bar{R}\,\phi^{+}\,L =$$

$$= G\left\{(\bar{\nu}_L\, e_R)\phi_I + (\bar{e}_R\,\nu_L)\phi_I^{+} + \frac{a}{\sqrt{2}}(\bar{e}e) + \frac{1}{\sqrt{2}}\,(\bar{e}e)\chi_1 + \frac{i}{\sqrt{2}}\,(\bar{e}\,\gamma^5\,e)\chi_2\right\}$$

Therefore if we group these terms together with the kinematic terms in L and R in the lagrangean (VIII. 8) we obtain :

$$\bar{L}\; i\,\gamma^{\alpha}\,\partial_{\alpha}\,L + \bar{R}\; i\,\gamma^{\alpha}\,\partial_{\alpha}\,R - G\,(\bar{L}\,\phi\,R + \bar{R}\,\phi^{+}\,L) =$$

$$= \bar{e}\,(i\,\gamma^{\alpha}\,\partial_{\alpha} - m_e)\,e + \bar{\nu}_L\; i\,\gamma^{\alpha}\,\partial_{\alpha}\,\nu_L - G\left\{(\bar{\nu}_L\, e_R)\,\phi_I + (\bar{e}_R\,\nu_L)\,\phi_I^{+} + \right.$$

$$\left. + \frac{1}{\sqrt{2}}\,(e\bar{e})\,\chi_1 + \frac{i}{\sqrt{2}}\,(\bar{e}\,\gamma^5\,e)\,\chi_2\right\} \qquad \text{(VIII. 10)}$$

where

$$m_e = \frac{1}{\sqrt{2}} \; a \, G \qquad \text{(VIII. 11)}$$

is precisely the electron mass which is thus generated by the vacuum expectation value of the field $\phi_{II}(x)$.

Now, the lagrangean (VIII. 8) is gauge invariant, that is, it is invariant under the transformations (VIII. 4), (VIII. 5) and the transformations of the vector fields :

$$\mathcal{a}'^{\mu}_{a} = \mathcal{a}^{\mu}_{a} - \partial^{\mu} \Lambda_a - g \, \varepsilon_{abc} \, \Lambda_b \, \mathcal{a}^{\mu}_{c}$$

$$B'^{\mu} = B^{\mu} - \partial^{\mu} \theta \qquad \text{(VIII. 12)}$$

As the field $\phi(x)$, given by the doublet (VIII. 2) is equivalent to four real fields we can express it in the following way :

$$\phi(x) = e^{ig\Lambda_a(x) \frac{\tau_a}{2}} \begin{pmatrix} 0 \\ \varphi_1(x) \end{pmatrix} \qquad \text{(VIII. 13)}$$

and four real fields are now the three Λ_a's and φ_1.

The gauge invariance makes $\mathcal{L}$ independent of Λ_a and therefore the above choice is a transformation away of the three fields Λ_a ; so we can choose for the scalar doublet :

$$\phi(x) = \begin{pmatrix} 0 \\ \frac{a}{\sqrt{2}} + \frac{1}{\sqrt{2}} \chi_1(x) \end{pmatrix} \qquad \text{(VIII. 13a)}$$

More precisely, we may fix a SU(2) gauge of a given field $\phi(x)$ such that for the transformed field $\phi'(x)$ one has $\phi'_I(x) = 0$, Im $\phi'_{II}(x) = 0$.

The terms (VIII. 10) are thus :

$$\bar{L}\, i\, \gamma^{\alpha}\, \partial_{\alpha}\, L + \bar{R}\, i\, \gamma^{\alpha}\, \partial_{\alpha}\, R - G\, (\bar{L}\, \phi\, R + \bar{R}\, \phi^{+}\, L) =$$

$$= \bar{e}(i\, \gamma^{\alpha}\, \partial_{\alpha} - m_e)\, e + \bar{\nu}_L\, i\, \gamma^{\alpha}\, \partial_{\alpha}\, \nu_L - \frac{G}{\sqrt{2}}\, (\bar{e}e)\chi_1 \qquad \text{(VIII. 14)}$$

where m_e is given by (VIII. 11).

The Higgs field has thus generated the electron mass and given rise to a scalar field interaction with the electrons, as expressed in the term in χ_1.

VIII. 4 - THE MASS OF THE PHYSICAL HIGGS FIELD

Let us now consider the terms in M^2 and λ in the lagrangean (VIII. 8). The value of ϕ corresponding to a minimum of the potential energy of the self-interacting scalar field is

$$a^2 = \frac{-\ 6\ M^2}{\lambda} \qquad \text{(VIII. 15)}$$

and this value transforms away the terms linear in χ_1 obtained by replacing $\phi(x)$ by (VIII. 13a) in the terms in M^2 and λ of (VIII. 8).

The resulting term in χ^2_1 will be

$$\left(- \frac{1}{2}\, M^2 - \frac{6\lambda}{4!}\, a^2 \right) \chi_1{}^2 = M^2\, \chi_1{}^2$$

As M^2 is negative we see that the field χ_1 has a positive mass :

$$m_H = \sqrt{-2\,M^2} \quad , \qquad M^2 < 0 \tag{VIII. 16}$$

a free parameter of the theory.

VIII. 5 - THE MASSIVE VECTOR BOSONS

Therefore the last six terms in (VIII. 8) are (apart from the interaction of L and R with the gauge fields) :

$$\begin{aligned}
&\bar{e}(i\,\gamma^\alpha\,\partial_\alpha - m_e)e + \bar{\nu}_L\; i\,\gamma^\alpha\,\partial_\alpha\,\nu_L - \frac{G}{\sqrt{2}}\,(\bar{e}e)\chi_1 + \\
&+ \frac{1}{2}\left\{(\partial^\mu\,\chi_1)(\partial_\mu\,\chi_1) - m_H{}^2\,\chi_1{}^2\right\} + \frac{g^2}{4}\;W^{\mu+}\,W_\mu\left\{a + \chi_1\right\}^2 \\
&+ \frac{1}{4\varepsilon^2}\,Z^\mu\,Z_\mu\left\{\frac{a}{\sqrt{2}} + \frac{1}{\sqrt{2}}\,\chi_1\right\}^2 - \frac{\lambda a}{6}\,\chi_1{}^3 - \frac{\lambda}{4!}\,\chi_1{}^4
\end{aligned} \tag{VIII. 17}$$

where the terms in W^μ and Z^μ came from the expression :

$$\left\{(\tfrac{g}{2})^2\,(\mathcal{a}^\mu{}_1\,\mathcal{a}_{\mu 1} + \mathcal{a}^\mu{}_2\,\mathcal{a}_{\mu 2}) + (\tfrac{g}{2})^2\,\mathcal{a}^\mu{}_3\,\mathcal{a}_{\mu 3} + (\tfrac{g'}{2})^2\,B^\mu\,B_\mu + \frac{gg'}{2}\,\mathcal{a}^\mu{}_3\,B_\mu\right\}\varphi_1{}^2$$

with the definition of the fields :

$$\begin{aligned}
W^\mu &= \frac{1}{\sqrt{2}}\left\{\mathcal{a}^\mu{}_1 + i\,\mathcal{a}^\mu{}_2\right\} \\
Z^\mu &= \varepsilon\left\{g\,\mathcal{a}^\mu{}_3 + g'\,B^\mu\right\}.
\end{aligned} \tag{VIII. 18}$$

To see the significance of the latter fields we consider the first two terms in (VIII. 8)

$$-\frac{1}{4}\mathcal{G}^{\mu\nu}{}_a\ \mathcal{G}_{\mu\nu,a} - \frac{1}{4}B^{\mu\nu}B_{\mu\nu}$$

which may be written :

$$-\frac{1}{4}\mathcal{G}^{\mu\nu}{}_a\ \mathcal{G}_{\mu\nu,a} - \frac{1}{4}B^{\mu\nu}B_{\mu\nu} =$$

$$= -\frac{1}{4}\left\{\frac{1}{\sqrt{2}}(\mathcal{G}^{\mu\nu}{}_1 + i\mathcal{G}^{\mu\nu}{}_2)\frac{1}{\sqrt{2}}(\mathcal{G}_{\mu\nu,1} - i\mathcal{G}_{\mu\nu,2}) + \right.$$

$$\left. + \frac{1}{\sqrt{2}}(\mathcal{G}^{\mu\nu}{}_1 - i\mathcal{G}^{\mu\nu}{}_2)\frac{1}{\sqrt{2}}(\mathcal{G}_{\mu\nu,1} + i\mathcal{G}_{\mu\nu,2})\right\} - \qquad \text{(VIII. 19)}$$

$$-\frac{1}{4}\mathcal{G}^{\mu\nu}{}_3\ \mathcal{G}_{\mu\nu,3} - \frac{1}{4}B^{\mu\nu}B_{\mu\nu}$$

In view of the definitions (VIII. 18) and (VIII. 7) we have :

$$\frac{1}{\sqrt{2}}(\mathcal{G}^{\mu\nu}{}_1 + i\mathcal{G}^{\mu\nu}{}_2) = W^{\mu\nu} + i\,g\,(\boldsymbol{a}^{\mu}{}_3 W^{\nu} - \boldsymbol{a}^{\nu}{}_3 W^{\mu})$$

where :

$$W^{\mu\nu} = \partial^{\nu}W^{\mu} - \partial^{\mu}W^{\nu}$$

Therefore :

$$-\frac{1}{4}\mathcal{G}^{\mu\nu}{}_a\ \mathcal{G}_{\mu\nu,a} - \frac{1}{4}B^{\mu\nu}B_{\mu\nu} =$$

$$= -\frac{1}{2}W^{\mu\nu+}W_{\mu\nu} - i\,g\,\boldsymbol{a}_{\mu 3}(W^{\mu\nu+}W_{\nu} - W^{\mu\nu}W_{\nu}{}^{+}) \qquad \text{(VIII. 20)}$$

$$-g^2\left\{\boldsymbol{a}^{\mu}{}_3\,\boldsymbol{a}_{\mu 3}W^{\nu+}W_{\nu} - \boldsymbol{a}^{\mu}{}_3 W_{\mu}W^{\nu+}\boldsymbol{a}_{\nu 3}\right\} - \frac{1}{4}\mathcal{G}^{\mu\nu}{}_3\ \mathcal{G}_{\mu\nu,3} - \frac{1}{4}B^{\mu\nu}B_{\mu\nu}$$

Now :

$$-\frac{1}{4}\mathcal{G}^{\mu\nu}{}_{3}\,\mathcal{G}_{\mu\nu,3} - \frac{1}{4}B^{\mu\nu}B_{\mu\nu} =$$

$$= -\frac{1}{4}\left\{(\partial^{\nu}a^{\mu}{}_{3} - \partial^{\mu}a^{\nu}{}_{3}) + g(a^{\mu}{}_{1}a^{\nu}{}_{2} - a^{\mu}{}_{2}a^{\nu}{}_{1})\right\}\left\{\partial_{\nu}a_{\mu,3} - \partial_{\mu}a_{\nu,3} + g(a_{\mu1}a_{\nu2} - a_{\mu2}a_{\nu1})\right\} -$$

$$-\frac{1}{4}(\partial^{\nu}B^{\mu} - \partial^{\mu}B^{\nu})(\partial_{\nu}B_{\mu} - \partial_{\mu}B_{\nu}) =$$

$$= -\frac{1}{4}(\partial^{\nu}a^{\mu}{}_{3} - \partial^{\mu}a^{\nu}{}_{3})(\partial_{\nu}a_{\mu3} - \partial_{\mu}a_{\nu3}) - \frac{1}{4}(\partial^{\nu}B^{\mu} - \partial^{\mu}B^{\nu})(\partial_{\nu}B_{\mu} - \partial_{\mu}B_{\nu}) -$$

$$-\frac{g}{2}(\partial^{\nu}a^{\mu}{}_{3} - \partial^{\mu}a^{\nu}{}_{3})(a_{\mu1}a_{\nu2} - a_{\mu2}a_{\nu1}) -$$

$$-\frac{g^{2}}{4}(a^{\mu}{}_{1}a^{\nu}{}_{2} - a^{\mu}{}_{2}a^{\nu}{}_{1})(a_{\mu1}a_{\nu2} - a_{\mu2}a_{\nu1})$$

If we define :

$$A^{\mu} = \varepsilon\,(-\,g'\,a^{\mu}{}_{3} + g\,B^{\mu}) \qquad \text{(VIII. 20a)}$$

we see, from the definition of Z^{μ} and W^{μ} in (VIII. 18) that :

$$\partial^{\nu}B^{\mu} - \partial^{\mu}B^{\nu} = \frac{1}{\varepsilon(g^{2} + g'^{2})}(g\,F^{\mu\nu} + g'\,Z^{\mu\nu})$$

$$\partial^{\nu}a^{\mu}{}_{3} - \partial^{\mu}a^{\nu}{}_{3} = \frac{1}{\varepsilon(g^{2} + g'^{2})}(g\,Z^{\mu\nu} - g'\,F^{\mu\nu})$$

where :

$$Z^{\mu\nu} = \partial^\nu Z^\mu - \partial^\mu Z^\nu$$

$$F^{\mu\nu} = \partial^\nu A^\mu - \partial^\mu A^\nu$$

Also

$$\mathcal{a}_{\mu 1} \mathcal{a}_{\nu 2} - \mathcal{a}_{\mu 2} \mathcal{a}_{\nu 1} = - i \left\{ W_\mu^+ W_\nu - W_\nu^+ W_\mu \right\}$$

Therefore :

$$- \frac{1}{4} \mathcal{G}^{\mu\nu}{}_3 \, \mathcal{G}_{\mu\nu,3} - \frac{1}{4} B^{\mu\nu} B_{\mu\nu} =$$

$$= - \frac{1}{4} \frac{1}{\varepsilon^2 (g^2 + g'^2)} \left\{ F^{\mu\nu} F_{\mu\nu} + Z^{\mu\nu} Z_{\mu\nu} \right\} +$$

$$+ \frac{i\, g}{2\varepsilon (g^2 + g'^2)} (g\, Z^{\mu\nu} - g'\, F^{\mu\nu})(W_\mu^+ W_\nu - W_\nu^+ W_\mu) + \qquad \text{(VIII. 21)}$$

$$+ \frac{1}{4} g^2 (W^{\mu +} W^\nu - W^{\nu +} W^\mu)(W_\mu^+ W_\nu - W_\nu^+ W_\mu)$$

We see that if

$$\varepsilon = (g^2 + g'^2)^{-1/2} \qquad \text{(VIII. 22)}$$

The first term on the second hand side is

$$- \frac{1}{4} F^{\mu\nu} F_{\mu\nu} - \frac{1}{4} Z^{\mu\nu} Z_{\mu\nu} \qquad \text{(VIII. 24)}$$

Let us collect the terms in (VIII. 17), (VIII. 20), (VIII. 21), (VIII. 22) :

$$\bar{e}(i\gamma^{\alpha}\partial_{\alpha} - m_e)e + \bar{\nu}_L i\gamma^{\alpha}\partial_{\alpha}\nu_L + \frac{1}{2}\left\{(\partial^{\mu}\chi_1)(\partial_{\mu}\chi_1) - m_H{}^2\chi_1{}^2\right\}$$

$$- \frac{G}{\sqrt{2}}(\bar{e}e)\chi_1 - \frac{\lambda\, a}{3!}\chi_1{}^3 - \frac{\lambda}{4!}\chi_1{}^4 - W^{\mu\nu +} W_{\mu\nu} +$$

$$+ \frac{1}{4}(g^2a^2)\, W^{\mu +} W_{\mu} - \frac{1}{4}F^{\mu\nu}F_{\mu\nu} - \frac{1}{4}Z^{\mu\nu}Z_{\mu\nu} +$$

$$+ \frac{1}{2}\left\{\frac{1}{4}a^2(g^2 + g'^2)\right\}Z^{\mu}Z_{\mu} + \frac{g^2}{2}\left\{\frac{1}{2}\chi_1{}^2 + a\chi_1\right\}W^{\mu +}W_{\mu} +$$

$$+ \frac{1}{4}(g^2 + g'^2)\left\{\frac{1}{2}\chi_1{}^2 + a\chi_1\right\}Z^{\mu}Z_{\mu} -$$

$$- \frac{i\, g}{(g^2 + g'^2)^{1/2}}(g\, Z^{\mu} - g'A^{\mu})(W^{\mu\nu +}W_{\nu} - W^{\mu\nu}W_{\nu}{}^{+}) -$$

$$- \frac{g^2}{(g^2 + g'^2)}(g\, Z^{\alpha} - g'A^{\alpha})(g\, Z^{\beta} - g'A^{\beta})W_{\mu}{}^{+}W_{\nu}(g_{\alpha\beta}g^{\mu\nu} - \delta_{\alpha}{}^{\nu}\delta_{\beta}{}^{\mu}) +$$

$$+ \frac{i\, g}{2(g^2 + g'^2)^{1/2}}(g\, Z^{\mu\nu} - g'F^{\mu\nu})(W_{\mu}{}^{+}W_{\nu} - W_{\nu}{}^{+}W_{\mu}) +$$

$$+ \frac{1}{4}g^2(W^{\mu +}W^{\nu} - W^{\nu +}W^{\mu})(W_{\mu}{}^{+}W_{\nu} - W_{\nu}{}^{+}W_{\mu}) \qquad \text{(VIII. 23)}$$

By inspection of these terms we see that the field W^{μ} has a mass :

$$m_W = \frac{1}{2}g\, a$$

The field Z^μ has a mass :

$$m_Z = \frac{1}{2}\ a(g^2 + g'^2)^{1/2}$$

Thus the Higgs mechanism has also generated masses for the W and Z bosons. The field A^μ is massless and we would like to identify it with the electromagnetic field.

VIII. 6 - THE ELECTROMAGNETIC FIELD AND THE WEINBERG ANGLE

For this purpose we must examine the terms of the lagrangean (VIII. 8) which remain, that is the interaction between L, R and the gauge fields. We have :

$$\bar{L}\ i\ \gamma^\mu\ D_\mu\ L + \bar{R}\ i\ \gamma^\mu\ D_\mu\ R$$

$$= \bar{e}\ i\ \gamma^\alpha \partial_\alpha e + \bar{\nu}_L i \gamma^\alpha \partial_\alpha \nu_L - g\ \bar{L}\gamma^\mu \mathcal{a}_{\mu a}\ \frac{\tau_a}{2}\ L - \frac{g'}{2}\ (\bar{L}\gamma^\mu L)B_\mu - g'(\bar{R}\gamma^\mu R)B_\mu =$$

$$= \bar{e}\ i\ \gamma^\alpha \partial_\alpha e + \bar{\nu}_L i \gamma^\alpha \partial_\alpha \nu_L -$$

$$- g\ (\bar{\nu}\bar{e})\gamma_\mu\ \frac{1}{2}\ (1-\gamma^5) \left\{ \frac{1}{\sqrt{2}}\ \frac{\mathcal{a}^\mu{}_1 + i\mathcal{a}^\mu{}_2}{\sqrt{2}}\ \frac{\tau_1 - i\ \tau_2}{2} + \frac{1}{\sqrt{2}}\ \frac{\mathcal{a}^\mu{}_1 - i\mathcal{a}^\mu{}_2}{\sqrt{2}}\ \frac{\tau_1 + \tau_2}{2} \right\} \begin{pmatrix} \nu \\ e \end{pmatrix} -$$

$$- g\ (\bar{\nu}\bar{e})\gamma_\mu\ \frac{1}{2}(1-\gamma^5)\ \mathcal{a}^\mu{}_3\ \frac{\tau_3}{2} \begin{pmatrix} \nu \\ e \end{pmatrix} - \frac{g'}{2}\ (\bar{\nu}\bar{e})\gamma_\mu\ \frac{1}{2}\ (1-\gamma^5) \begin{pmatrix} \nu \\ e \end{pmatrix}\ B^\mu -$$

$$- g'\ (\bar{e}\ \gamma_\mu\ \frac{1}{2}(1+\gamma^5)e)\ B^\mu$$

So :

$$\bar{L}\, i\, \gamma^{\mu}\, D_{\mu}\, L + \bar{R}\, i\, \gamma^{\mu}\, D_{\mu}\, R =$$

$$= \bar{e}\,(i\,\gamma^{\alpha}\,\partial_{\alpha})e + \bar{\nu}_L(i\,\gamma^{\alpha}\,\partial_{\alpha})\nu_L - \frac{g}{2\sqrt{2}}\left\{\bar{e}\,\gamma^{\mu}(1-\gamma^5)\nu\right\} W_{\mu} -$$

$$- \frac{g}{2\sqrt{2}}\left\{\bar{\nu}\,\gamma^{\mu}(1-\gamma^5)e\right\}W_{\mu}^{+} - \frac{1}{4}\left\{\bar{\nu}\,\gamma_{\mu}(1-\gamma^5)\nu\right\}\left\{g\,a^{\mu}{}_3 + g'\,B^{\mu}\right\} + \qquad \text{(VIII. 25)}$$

$$+ \frac{1}{4}\,(\bar{e}\,\gamma_{\mu}\,e)\left\{g\,a^{\mu}{}_3 - 3\,g'\,B^{\mu}\right\} -$$

$$- \frac{1}{4}\,(\bar{e}\,\gamma_{\mu}\,\gamma^5\,e)\left\{g\,a^{\mu}{}_3 + g'\,B^{\mu}\right\}$$

Now we have the identity :

$$\frac{1}{4}(g\,a^{\mu}{}_3 - 3g'B^{\mu}) = \frac{-\,g\,g'}{g^2 + g'^2}\,(-g'\,a^{\mu}{}_3 + gB^{\mu}) - \frac{3g'^2 - g^2}{g^2 + g'^2}\,\frac{(g\,a^{\mu}{}_3 + g'B^{\mu})}{4}$$

But we saw that (see (VIII. 18), (VIII. 22) and (VIII. 20a)) :

$$Z^{\mu} = \frac{1}{(g^2 + g'^2)^{1/2}}\,(g\,a^{\mu}{}_3 + g'\,B^{\mu})$$

$$\text{(VIII. 26)}$$

$$A^{\mu} = \frac{1}{(g^2 + g'^2)^{1/2}}\,(-\,g'\,a^{\mu}{}_3 + g\,B^{\mu})$$

therefore

$$\frac{1}{4}(\bar{e}\gamma_{\mu}e)(g\,a^{\mu}{}_3 - 3g'B^{\mu}) = -\,\frac{g\,g'}{(g^2 + g'^2)^{1/2}}\,(\bar{e}\gamma_{\mu}e)A^{\mu} - \frac{3g'^2 - g^2}{4(g^2 + g'^2)^{1/2}}\,(\bar{e}\gamma_{\mu}e)Z^{\mu}$$

which gives (VIII. 25) the form :

$$\bar{L}\, i\, \gamma^{\mu} D_{\mu} L + \bar{R}\, i\, \gamma^{\mu} D_{\mu} R =$$

$$= \bar{e}\, i\, \gamma^{\alpha} \partial_{\alpha} e + \bar{\nu}_L\, i\, \gamma^{\alpha} \partial_{\alpha} \nu_L - \frac{g}{2\sqrt{2}} \left\{ \bar{e}\, \gamma^{\mu}(1-\gamma^5)\nu \right\} W_{\mu} -$$

$$- \frac{g}{2\sqrt{2}} \left\{ \bar{\nu}\, \gamma^{\mu}(1-\gamma^5) e \right\} W_{\mu}^{+} - \frac{g\, g'}{(g^2 + g'^2)^{1/2}} (\bar{e}\, \gamma^{\mu} e) A_{\mu} - \quad \text{(VIII. 27)}$$

$$- \frac{3g'^2 - g^2}{4(g^2 + g'^2)^{1/2}} (\bar{e}\, \gamma^{\mu} e) Z_{\mu} -$$

$$- \frac{1}{4} (g^2 + g'^2)^{1/2} \left\{ \bar{\nu}\, \gamma^{\mu}(1-\gamma^5)\, \nu + \bar{e}\, \gamma^{\mu} \gamma^5 e \right\} Z_{\mu}$$

Let us now introduce the Weinberg angle θ_W :

$$\cos \theta_W = \frac{g}{(g^2 + g'^2)^{1/2}} \quad ; \quad \sin \theta_W = \frac{g'}{(g^2 + g'^2)^{1/2}}$$

$$\text{tg}\ \theta_W = \frac{g'}{g} \quad \text{(VIII. 28)}$$

which defines a rotation in the plane $\mathcal{a}^{\mu}_3$, B^{μ} to get the fields A^{μ} and Z^{μ} :

$$Z^{\mu}(x) = \mathcal{a}^{\mu}_3 \cos \theta_W + B^{\mu} \sin \theta_W$$

$$A^{\mu}(x) = -\mathcal{a}^{\mu}_3 \sin \theta_W + B^{\mu} \cos \theta_W \quad \text{(VIII. 29)}$$

From the interaction term of the current $(\bar{e}\gamma^{\mu} e)$ with the field A^{μ} we conclude that this is the electromagnetic field if the charge e is given by

$$e = \frac{g\,g'}{(g^2 + g'^2)^{1/2}} = g \sin\theta_W \qquad \text{(VIII. 30)}$$

The Weinberg angle is a parameter to be determined experimentally.

VIII. 7 - THE EFFECTIVE SALAM-WEINBERG LAGRANGEAN FOR ELECTRONS AND NEUTRINOS

If we collect the terms of the lagrangean (VIII. 8) which have been developed in (VIII. 23), (VIII. 27) we get :

$$\mathcal{L} = \mathcal{L}_{\ell} + \mathcal{L}_{W} + \mathcal{L}_{\gamma} + \mathcal{L}_{Z} + \mathcal{L}_{\chi} + \mathcal{L}_{int} \qquad \text{(VIII. 31)}$$

where :

$$\mathcal{L}_{\ell} = \bar{\nu}_L\, i\,\gamma^{\alpha}\,\partial_{\alpha}\,\nu_L + \bar{e}\,(i\,\gamma^{\alpha}\,\partial_{\alpha} - m_e)e$$

is the free lepton lagrangean with $m_e = \frac{a}{\sqrt{2}}\, G$;

$$\mathcal{L}_{W} = -\frac{1}{2} W^{\mu\nu +} W_{\mu\nu} + m_W^{\ 2} W^{\mu +} W_{\mu} \qquad \text{(VIII. 31a)}$$

is the lagrangean of the free intermediate charged vector boson field with mass :

$$m_W = a\,g\,;$$

$$\mathcal{L}_{\gamma} = -\frac{1}{4} F^{\mu\nu} F_{\mu\nu} \qquad \text{(VIII. 31b)}$$

is the lagrangean of the free electromagnetic field ;

$$\mathcal{L}_{Z} = -\frac{1}{4} Z^{\mu\nu} Z_{\mu\nu} + \frac{1}{2} m_Z^{\ 2} Z^{\mu} Z_{\mu} \qquad \text{(VIII. 31c)}$$

is the lagrangean of the free intermediate neutral vector boson field with mass :

$$m_Z = \frac{1}{2} a \, (g^2 + g'^2)^{1/2} = \frac{m_W}{\cos \theta_W} \quad ; \qquad \text{(VIII. 31d)}$$

$$\mathcal{L}_\chi = \frac{1}{2} \left\{ (\partial^\mu \chi)(\partial_\mu \chi) - m_H^2 \, \chi^2 \right\}$$

is the lagrangean of the free neutral Higgs scalar field with mass :

$$m_H = \sqrt{-2M^2} \quad , \qquad M^2 < 0$$

a free parameter in the model ;

and

$$\begin{aligned} \mathcal{L}_{int} = & \mathcal{L}(\chi, \chi) + \mathcal{L}(\chi, e) + \mathcal{L}(\chi, W) + \\ & + \mathcal{L}(\chi, Z) + \mathcal{L}(W, A, Z) + \mathcal{L}(W, W) + \\ & + \mathcal{L}(e, \nu, W) + \mathcal{L}(e, A) + \mathcal{L}(e, \nu, Z) \end{aligned} \qquad \text{(VIII. 32)}$$

is the interaction lagrangean with the following pieces :

$$\mathcal{L}(\chi, \chi) = -\frac{\lambda \, a}{3!} \chi^3 - \frac{\lambda}{4!} \chi^4 \qquad \text{(VIII. 32a)}$$

is the self-interacting term of Higgs field ;

$$\mathcal{L}(\chi, e) = -\frac{G}{\sqrt{2}}(\bar{e}\, e)\,\chi \qquad \text{(VIII. 32b)}$$

is the interaction between electrons and the Higgs field ;

$$\mathcal{L}(\chi, W) = \frac{g^2}{2}\left(\frac{1}{2}\chi^2 + a\,\chi\right) W^{\mu +} W_\mu \qquad \text{(VIII. 32c)}$$

is the interaction between Higgs and W mesons ;

$$\mathcal{L}(\chi, Z) = \frac{1}{4}(g^2 + g'^2)\left(\frac{1}{2}\chi^2 + a\,\chi\right) Z^{\mu} Z_\mu \qquad \text{(VIII. 32d)}$$

js the interaction between Higgs and Z mesons ;

$$\begin{aligned}\mathcal{L}(W, A, Z) = \cos\theta_W \Big\{ & (g\, Z^\mu - g'\, A^\mu)\, i\, (W^{\mu\nu} W^+_\nu - W^{\mu\nu +} W_\nu) + \\ & + \frac{1}{2}(g\, Z^{\mu\nu} - g'\, F^{\mu\nu})\, i\, (W_\mu^{\ +} W_\nu - W_\nu^{\ +} W_\mu) - \\ & - \cos\theta_W\, (g\, Z^\alpha - g'\, A^\alpha)(g\, Z^\beta - g'\, A^\beta)\, W_\mu^{\ +} W_\nu \cdot \\ & \cdot (g_{\alpha\beta}\, g^{\mu\nu} - \delta_\alpha^{\ \nu}\, \delta_\beta^{\ \mu}) \Big\}\end{aligned} \qquad \text{(VIII. 32e)}$$

is the interaction between W bosons and the electromagnetic and Z fields (note that the electromagnetic coupling constant is $g' \cos\theta_W = e$) ;

$$\mathcal{L}(W, W) = \frac{1}{2} g^2 \left\{ (W^{\nu} W_{\nu})(W^{\mu} W_{\mu})^+ - (W^{\nu +} W_{\nu})^2 \right\} \qquad \text{(VIII. 32f)}$$

is the self-interaction of the W-bosons ;

$$\mathcal{L}(e,\nu,W) = - \frac{g}{2\sqrt{2}} \left\{ (\bar{e}\, \gamma^{\mu}\, (1 - \gamma^5)\, \nu)\, W_{\mu} + (\bar{\nu}\, \gamma^{\mu}(1 - \gamma^5) e)\, W_{\mu}^{+} \right\}$$

(VIII. 32g)

is the weak interaction of the charged weak current with W bosons, from which one derives :

$$\frac{g^2}{8\, m_W^{\ 2}} = \frac{\mathcal{G}_F}{\sqrt{2}} ;$$

$$\mathcal{L}(e, A) = - e\, j_{\mu}\, A^{\mu}$$

$$j_{\mu}^{(\gamma)} = \bar{e}\, \gamma_{\mu}\, e$$

$$e = \frac{g\, g'}{(g^2 + g'^2)^{1/2}} = g' \cos \theta_W = g \sin \theta_W \leq g$$

is the electromagnetic interaction of electrons ; and

$$\mathscr{L}(e,\nu,Z) = -\frac{1}{4}(g^2 + g'^2)^{1/2}(\bar{\nu}\,\gamma^\mu\,(1 - \gamma^5)\nu)\;Z_\mu\; -$$

$$-\left\{\frac{1}{4}(g^2 + g'^2)^{1/2}\,(\bar{e}\,\gamma^\mu\,\gamma^5\,e)\; + \right. \qquad \text{(VIII. 32h)}$$

$$\left. + \;\frac{3\,g'^2 - g^2}{4(g^2 + g'^2)^{1/2}}\,(\bar{e}\,\gamma^\mu\,e)\right\}\;Z_\mu$$

is the interaction of neutrinos and electrons with the Z bosons which may be written :

$$\mathscr{L}(e,\nu,Z) = -\frac{1}{2}(g^2 + g'^2)^{1/2}\; j^\mu{}_{(0)}\;Z_\mu \qquad \text{(VIII. 32i)}$$

where the neutral lepton current is :

$$j^\mu{}_{(0)} = j^\mu{}_3 + 2\sin^2\theta_W\;J^\mu{}_{(\gamma)} \quad ; \qquad j^\mu{}_3 = \frac{1}{2}\left\{\bar{\nu}\,\gamma^\mu(1 - \gamma^5)\nu - \bar{e}\,\gamma^\mu\,(1 - \gamma^5)e\right\}$$

We started with four massless vector gauge fields $\mathcal{a}^\mu$, B^μ needed for the construction of a lagrangean invariant under the $SU(2) \otimes U(1)$ gauge group. The initial matter fields, L, R, are also massless for this invariance to hold.

A scalar massive field is introduced the mass of which is an arbitrary parameter

$$M^2 < 0$$

needed for the generation, by spontaneous symmetry break-down, of the mass of the electron. As a result, four vector fields are defined as a function of the initial massless gauge vector fields and of these new vector fields, three, W, W^+, Z, have acquired a mass, m_W, m_Z, the fourth vector field being massless, the electromagnetic field A^μ ; the theory remains invariant under the U(1) electromagnetic gauge group.

The theory thus describes in a unified fashion the electromagnetic and the weak interactions. It predicts the existence, not only of charged massive vector bosons but also of neutral massive vector bosons. Morever, the charge is intimately related to the weak coupling constant g and as a result the mass of the vector bosons is much higher than the proton mass (results which had been previously predicted in a speculative and intuitive basis, as well as the non-charge-independent nature of the interaction with charged and neutral vector bosons (see refs.[84-90]).

VIII. 8 - PARAMETERS AND PHYSICAL CONSTANTS IN THE SALAM-WEINBERG LEPTON MODEL

Let us collect the parameters and physical constants which appeared in the above derivation of the lagrangean (VIII. 31). They are :

$$m_H^2 = -2M^2 = \frac{\lambda a^2}{3}$$

$$m_e = \frac{a}{\sqrt{2}} G$$

$$m_W = \frac{1}{2} a g \quad , \quad m_Z = \frac{m_W}{\cos\theta_W} \tag{VIII. 33}$$

$$\frac{g^2}{8 m_W^2} = \frac{\mathcal{G}_F}{\sqrt{2}} , \qquad e = \frac{g g'}{(g^2 + g'^2)^{1/2}}$$

$$\sin\theta_W = \frac{g'}{(g^2 + g'^2)^{1/2}} , \quad \cos\theta_W = \frac{g}{(g^2 + g'^2)^{1/2}}$$

The single parameter of the theory is the Weinberg mixing angle, $\sin^2\theta_W$, say. From the relation between e, g, g' and the definition of $\sin\theta_W$ we get :

$$e = g \sin\theta_W$$

So if we determine $\sin\theta_W$ from experiment we obtain the value of g.

From the relation between m_W and $\mathcal{G}_F$ we get the mass of the charged bosons :

$$m_W = \frac{(2)^{1/4}\, g}{2\sqrt{2\mathcal{G}_F}} = \frac{(2)^{1/4}\, e}{2\sqrt{2\mathcal{G}_F}\,\sin\theta_W}$$

and also the mass of the neutral bosons

$$m_Z = \frac{(2)^{1/4}(g^2 + g'^2)^{1/2}}{2\sqrt{2\mathcal{G}_F}} = \frac{m_W}{\cos\theta_W}$$

From

$$m_W = \frac{1}{2}\; a\, g$$

we get (if we compare with the value of m_W above) :

$$a^2 = \frac{\sqrt{2}}{2\mathcal{G}_F}$$

for the vacuum expectation value a of the scalar field. Thus as

$$\mathcal{G}_F \cong 10^{-5}\, m_p^{-2} \quad , \qquad m_p = \text{proton mass}$$

we obtain for the coupling constant G of the Higgs field-electrons interaction :

$$G^2 = \frac{2\, m_e^2}{a^2} \cong 2\sqrt{2} \times 10^{-5} \left(\frac{m_e}{m_p}\right)^2 \cong 10^{-12}$$

which is therefore quite small and negligible as compared to electromagnetic and neutral-boson interactions (it will be about 4×10^4 larger for the case of muons ; 10^7 larger for tau leptons).

We also have for m_W :

$$m_W{}^2 = \frac{\sqrt{2}}{8} \; \frac{e^2}{\mathcal{G}_F \sin^2 \theta_W} = \frac{(37.5 \text{ GeV})^2}{\sin^2 \theta_W}$$

and for m_Z :

$$m_Z{}^2 = \frac{(75 \text{ GeV})^2}{(\sin 2\,\theta_W)^2}$$

For

$$\sin^2 \theta_W = \frac{1}{4}$$

we get

$$m_W \cong 75 \text{ GeV}$$

$$m_Z \cong 90 \text{ GeV}$$

Experimentally, no evidence has yet been found (as of November 1980) of th W and Z bosons directly but for esthetical reasons it is believed that they do exist. Otherwise ... (Remember that pions as well as anti-protons took some time to be detected).

VIII. 9 - THE NEUTRAL LEPTON CURRENTS

Besides justifying the weak V-A charged-current interactions via the W bosons, the Salam-Weinberg model predicts weak neutral-current interactions via the neutral boson Z field. The respective lagrangean interactions are :

$$\mathcal{L}(e,\nu,W) = -\frac{g}{2\sqrt{2}}\left\{(\bar{e}\,\gamma^{\mu}(1-\gamma^5)\nu)W_{\mu}+(\bar{\nu}\,\gamma^{\mu}(1-\gamma^5)e)W_{\mu}^{+}\right\} \qquad \text{(VIII. 34)}$$

for the charged current interaction with the W bosons ;

$$\mathcal{L}(e,\nu,Z) = -\frac{1}{2}(g^2+g'^2)^{1/2}\left\{\frac{1}{2}(\bar{\nu}\gamma^{\mu}(1-\gamma^5)\nu)-\frac{1}{2}(\bar{e}\,\gamma^{\mu}(1-\gamma^5)e)+\right.$$

$$\left.+\,2\sin^2\theta_W(\bar{e}\,\gamma^{\mu}e)\right\}Z_{\mu}$$

for the neutral current interaction with the Z bosons.

The latter can be written :

$$\mathcal{L}(e,\nu,Z) = -\frac{1}{2}\,\frac{g}{\cos\theta_W}\left\{j^{\mu}_{(o)}(\nu)+j^{\mu}_{(o)}(e)\right\}Z_{\mu} \qquad \text{(VIII. 34a)}$$

where

$$j^{\mu}_{(o)}(\nu) = \frac{1}{2}\left\{\bar{\nu}\,\gamma^{\mu}(1-\gamma^5)\,\nu\right\} \qquad \text{(VIII. 34b)}$$

is the neutrino neutral current which is of the V-A form ; and :

$$j^{\mu}_{(o)}(e) = \frac{1}{2}\left\{\bar{e}\,\gamma^{\mu}\,(g_V - g_A\,\gamma^5)\,e\right\} \qquad \text{(VIII. 34c)}$$

where

$$g_V = -1 + 4\sin^2\theta_W$$

$$\text{(VIII. 34d)}$$

$$g_A = -1$$

is the neutral electron current which is therefore not of the V-A form. If we set :

$$\sin^2 \theta_W \cong \frac{1}{4}$$

$j^{\mu}{}_{(o)}(e)$ has indeed rather a predominant axial nature.

Experimentally the Weinberg angle has been determined :

$$\sin^2 \theta_W = 0.23 \pm 0.02 \qquad \text{(Baltay, 1978)}^{145)}$$

VIII. 10 - EXTENSION OF THE MODEL TO THE OTHER LEPTONS

We can easily generalise the theory to the case of the other leptons such as

$$\nu_\mu, \mu \;\; ; \;\; \nu_\tau, \tau \;\; ; \;\; \ldots$$

We only have to consider the left-handed isospinors :

$$L_e = \begin{pmatrix} \nu_L \\ e_L \end{pmatrix} \quad ; \quad L_\mu = \begin{pmatrix} \nu'_L \\ \mu_L \end{pmatrix} \quad ; \quad L_\tau = \begin{pmatrix} \nu''_L \\ \tau_L \end{pmatrix}$$

and the right-handed isoscalars :

$$R_e = e_R \quad ; \quad R_\mu = \mu_R \quad ; \quad R_\tau = \tau_R$$

where we sometimes use the notation

$$\nu' \equiv \nu_\mu \quad ; \quad \nu'' \equiv \nu_\tau \quad ; \ldots$$

In the lagrangean (VIII. 8) the only change is now

$$\bar{L}\, i\, \gamma^{\alpha}\, \partial_{\alpha}\, L + \bar{R}\, i\, \gamma^{\alpha}\, \partial_{\alpha}\, R - G\, (\bar{L}\, \phi\, R + \bar{R}\, \phi^{+}\, L) \longrightarrow$$

$$\longrightarrow \quad \sum_{\ell} \left\{ \bar{L}_{\ell}\, i\, \gamma^{\alpha}\, \partial_{\alpha}\, L_{\ell} + \bar{R}_{\ell}\, i\, \gamma^{\alpha}\, \partial_{\alpha}\, R_{\ell} - G_{\ell}\, (\bar{L}_{\ell}\, \phi\, R_{\ell} + \bar{R}_{\ell}\, \phi^{+}\, L_{\ell}) \right\}$$

where the sum is over $\ell = e, \mu, \tau, \ldots$

The masses of the leptons will be determined by the constants G_{ℓ} :

$$m_e = \frac{a}{\sqrt{2}}\, G_e$$

$$m_{\mu} = \frac{a}{\sqrt{2}}\, G_{\mu}$$

$$m_{\tau} = \frac{a}{\sqrt{2}}\, G_{\tau}$$

The interactions will then be a sum of the interactions of each lepton field with the boson fields (see also paragraph VIII. 12).

VIII. 11 - NEUTRINO-LEPTON SCATTERING AND THE EXPERIMENTAL TESTS OF THE SALAM-WEINBERG MODEL

The interaction of leptons with the W and Z fields is given by (VIII. 34) where we sum over all leptons ℓ, ν_{ℓ}.

The best experimental tests of the Salam-Weinberg theory are provided by the scattering of neutrinos on electrons. The table below gives the diagrams for the different possible types of this scattering

Table VIII. 4

Reactions	First order diagrams		
	charged current interaction		neutral current interaction
$\nu_\mu + e \longrightarrow \nu_e + \mu$	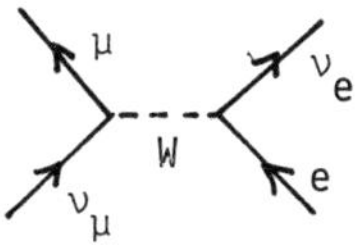		no
$\bar{\nu}_e + e \longrightarrow \bar{\nu}_\mu + \mu$	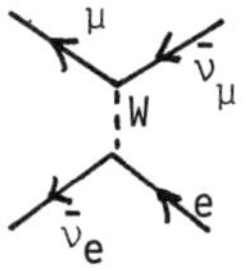		no
$\nu_\mu + e \longrightarrow \nu_\mu + e$	no		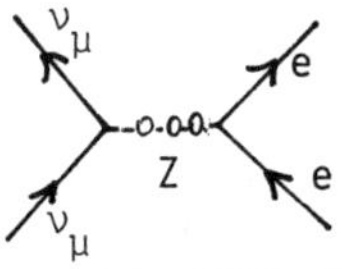
$\bar{\nu}_\mu + e \longrightarrow \bar{\nu}_\mu + e$	no		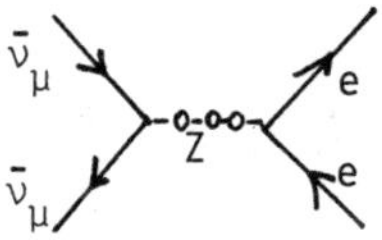
$\nu_e + e \longrightarrow \nu_e + e$	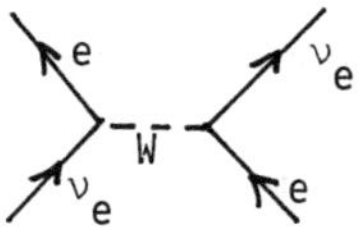	+	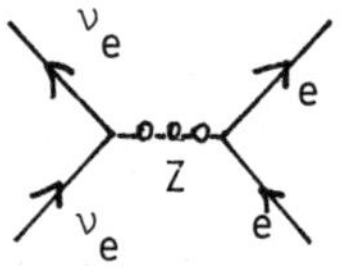
$\bar{\nu}_e + e \longrightarrow \bar{\nu}_e + e$	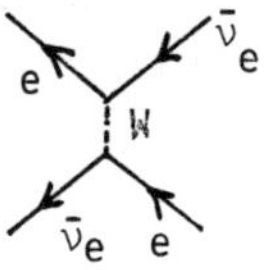	+	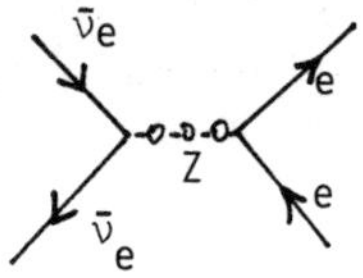

The reactions (as we see from the table VIII. 4) :

$$\nu_\mu + e \longrightarrow \nu_\mu + e \qquad \text{(VIII. 35)}$$

and

$$\bar{\nu}_\mu + e \longrightarrow \bar{\nu}_\mu + e \qquad \text{(VIII. 35a)}$$

provide the good test for the detection of neutral-current interaction 123-130).

From the lagrangean $\mathscr{L}$ (leptons, Z) we deduce that the amplitude for (VIII. 35) is (see (VIII. 34a)) :

$$M = \left(- \frac{g}{4 \cos \theta_W}\right)^2 \left\{ (\nu' \gamma^\mu (1-\gamma^5) \nu') \frac{g_{\mu\nu} - \frac{k_\mu k_\nu}{m_Z^2}}{m_Z^2 - k^2} (\bar{e} \gamma^\nu (g_V - g_A \gamma^5) e) \right\}$$

where

$$k_\mu = p'_\mu(e) - p_\mu(e) = p_\mu(\nu') - p'_\mu(\nu')$$

The contribution of the second term in the propagator vanishes since :

$$\bar{\nu}'_{p'} k_\mu \gamma^\mu (1-\gamma^5) \nu'_p = \bar{\nu}'_{p'} (p_\mu - p'_\mu) \gamma^\mu (1-\gamma^5) \nu'_p = 0$$

as

$$\bar{\nu}'_{p'} \gamma^\alpha p'_\alpha = 0 \quad ; \quad \gamma^\alpha p_\alpha \nu'_p = 0$$

So in the approximation

$$k^2 \ll m_Z^2$$

we get

$$M = \frac{g^2}{16\ m_Z^2 \cos^2 \theta_W} (\bar{\nu}'\ \gamma^\mu(1-\gamma^5)\nu')(\bar{e}\ \gamma_\mu\ (g_V - g_A\ \gamma^5)e)$$

with

$$g_V = -1 + 4 \sin^2 \theta_W$$

$$g_A = -1$$

But by (VIII. 33) we know that

$$m_W = m_Z \cos \theta_W$$

so that the angle disappears from the coupling constant factor :

$$\frac{g^2}{16\ m_Z^2 \cos^2 \theta_W} = \frac{g^2}{16\ m_W^2} = \frac{\mathcal{G}_F}{2\sqrt{2}}$$

Thus :

$$M = \frac{\mathcal{G}_F}{2\sqrt{2}} (\bar{\nu}'\ \gamma^\mu(1-\gamma^5)\nu')(\bar{e}\ \gamma_\mu(g_V - g_A\ \gamma^5)e) \qquad \text{(VIII. 35)}$$

The differential cross-section can be calculated for the reaction :

$$\nu_\mu + e \longrightarrow \nu_\mu + e$$

and has the form :

$$\frac{d\sigma}{dy} = \frac{\mathcal{G}_F^2}{2\pi} \frac{m_e E_\nu}{4} \left\{ (g_V + g_A)^2 + (1-y)^2(g_V - g_A)^2 \right\} \qquad \text{(VIII. 36)}$$

where a small term, $- \frac{m_e y}{E_\nu} (g^2{}_V - g^2{}_A)$ has been neglected and :

$$y = \frac{E_e}{E_\nu}$$

$$g_V = - 1 + 4 \sin^2 \theta_W \qquad \text{(VIII. 37)}$$

$$g_A = - 1$$

E_e is the energy of the electron in the final state, E_ν is the incident neutrino energy in the laboratory. The parameter to be determined is therefore $\sin^2 \theta_W$. To see intuitively the origin of the two terms in (VIII. 36) we shall write the electron neutral current under the following form :

$$\bar{e}\, \gamma_\mu\, (g_V - g_A\, \gamma^5)e =$$

$$= \bar{e}\, \gamma^\mu \left\{ \frac{g_V + g_A}{2} (1 - \gamma^5) + \frac{g_V - g_A}{2} (1 + \gamma^5) \right\} e \qquad \text{(VIII. 38)}$$

If we replace this in (VIII. 35) it is seen that the amplitude will be the sum of two terms

$$M(\nu' + e \rightarrow \nu' + e) = M_{LL} + M_{LR} \qquad \text{(VIII. 39)}$$

where :

$$M_{LL} \equiv M\, (\nu' + e_L \rightarrow \nu' + e_L) =$$

$$= \frac{\mathcal{G}_F}{2\sqrt{2}} (\nu'\, \gamma^\mu(1-\gamma^5)\nu')(\bar{e}\, \gamma_\mu\, \frac{g_V + g_A}{2} (1-\gamma^5)e) \qquad \text{(VIII. 39a)}$$

expresses the interaction between the left-handed neutrino and a left-handed component of the electron field ;

$$M_{LR} \equiv M(\nu' + e_R \rightarrow \nu' + e_R) =$$

$$= \frac{\mathcal{G}_F}{2\sqrt{2}} (\nu' \gamma^\mu (1-\gamma^5)\nu')(\bar{e} \gamma_\mu \frac{g_V - g_A}{2} (1 + \gamma^5) e) \qquad \text{(VIII. 39b)}$$

comes from the interaction of the incoming left-handed neutrino with the right-handed component of the electron field.

If the neutrino initial momentum is taken as the z-axis we have in the center-of-mass system in the initial state

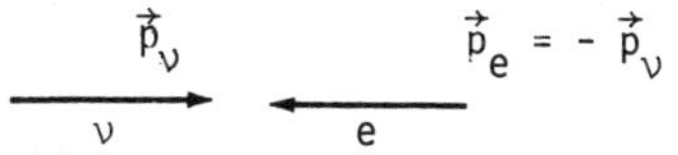

Figure VIII. 1

and as the neutrino is left-handed we have for a left-handed electron (VIII. 39)

initial state

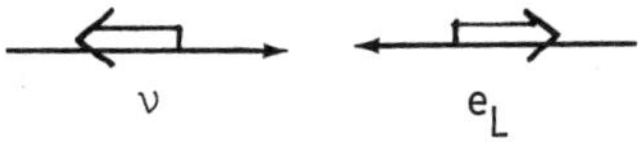

; $J_z = 0$

Figure VIII. 2

so the projection over the z-axis of the angular momentum vanishes

$$J_z = 0$$

and the same is true for the final state for $\theta = 0$ and for $\theta = \pi$:

final state

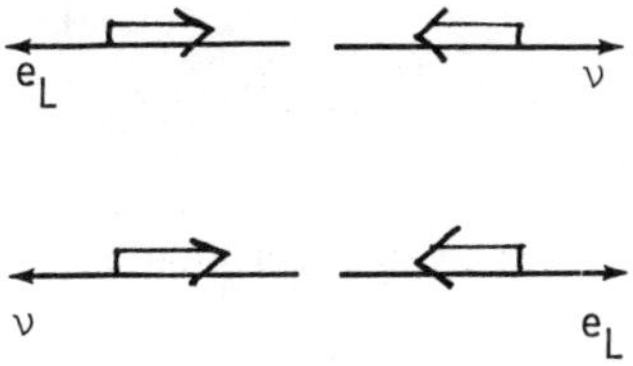

for $\theta = 0$; $J_z = 0$

for $\theta = \pi$; $J_z = 0$

Figure VIII. 3

where θ is the angle in the center-of-mass system :

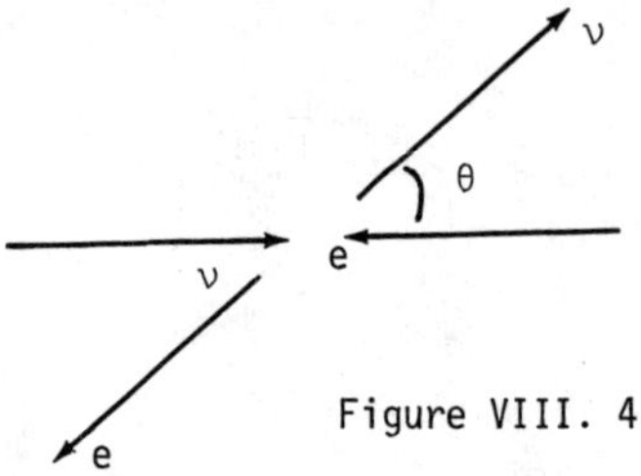

Figure VIII. 4

We expect the cross-section to be independent of θ and

$$\frac{d\sigma_{LL}}{dy} \sim G_F^2\, s \qquad s \gg m_e^2$$

for the left-handed electron (similar to reaction VI. 17).

For the right-handed electron however the situation is as follows :

initial state

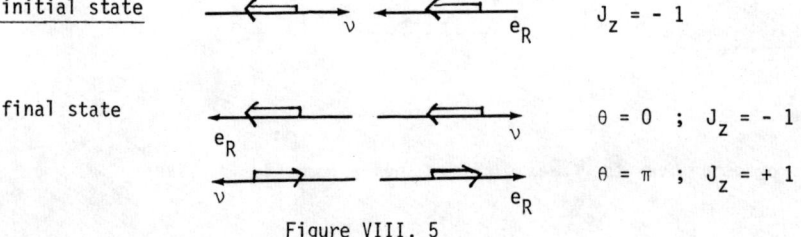

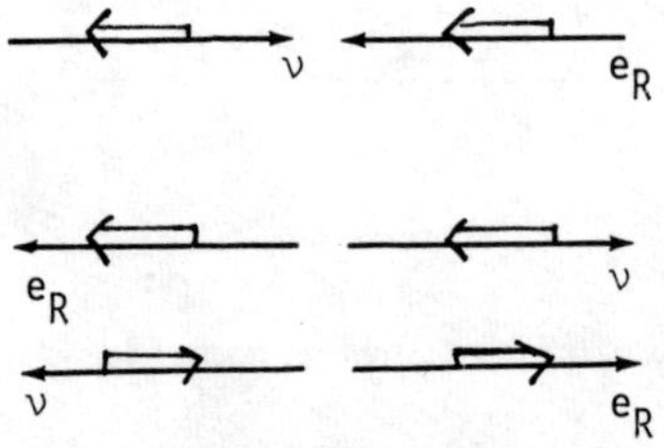

$J_z = -1$

final state

$\theta = 0 \;;\; J_z = -1$

$\theta = \pi \;;\; J_z = +1$

Figure VIII. 5

The state for $J_Z = +1$ is forbidden ; so we expect that the cross-section be of the form (see (VI. 19)) :

$$\frac{d\sigma_{LR}}{dy} \sim G_F^2\, s \left(\frac{1 + \cos\theta}{2} \right)^2$$

We now take into account the coefficients in g_V, g_A, as well as the kinematics of the reaction, namely :

$$\left.\begin{array}{ll} P_e = (m_e, 0, 0, 0) & ; \quad P_\nu = (|\vec{p}_\nu|, \vec{p}_\nu) \\ P'_e = (E_e, \vec{p}_e) & ; \quad P'_\nu = (|\vec{p}'_\nu|, \vec{p}'_\nu) \end{array}\right\} \text{ in the laboratory system ;}$$

$$\left.\begin{array}{ll} P_e = (\tilde{E}_e, \tilde{\vec{p}}) & ; \quad P_\nu = (|\tilde{p}_e|, -\tilde{\vec{p}}_e) \\ P'_e = (\tilde{E}_e, \tilde{\vec{p}}') & ; \quad P'_\nu = (|\tilde{p}_e|, -\tilde{\vec{p}}'_e) \end{array}\right\} \text{ in the center of mass system.}$$

which leads to

$$P_e^{\alpha} \; P'_{e,\alpha} = m_e \; E_e = \tilde{E}_e^{\,2} - \tilde{\vec{p}} \cdot \tilde{\vec{p}}' \simeq \tilde{E}_e^{\,2} \; (1 - \cos\theta)$$

in the approximation :

$$\vec{p}_e^{\,2} \; , \; \vec{p}_\nu^{\,2} \gg m_e^2$$

Then as :

$$s = (P_e + P_\nu)^2_{c.m} = (\tilde{E}_e + |\tilde{p}_e|)^2 \simeq 4 \; \tilde{E}_e^{\,2}$$

we get :

$$m_e \; E_e = \frac{s}{4} \; (1 - \cos\theta)$$

but also :

$$s = (P_e + P_\nu)^2_{lab} = (m_e + |\vec{p}_\nu|)^2 - |\vec{p}_\nu|^2 \cong 2\, m_e\, |\vec{p}_\nu| = 2\, m_e\, E_\nu$$

So that

$$\frac{1 - \cos\theta}{2} = \frac{E_e}{E_\nu}$$

hence :

$$\frac{1 + \cos\theta}{2} = 1 - \frac{E_e}{E_\nu} = 1 - y$$

The origin of the terms in (VIII. 36) is thus intuitively described.

For the reaction with anti-neutrinos

$$\bar{\nu}_\mu + e \rightarrow \bar{\nu}_\mu + e$$

it is seen that the arguments above lead to an exchange of the factors $g_V + g_A$ and $g_V - g_A$ in equation (VIII. 36) since the anti-neutrinos are right-handed.

Having in mind the values of g_V and g_A in (VIII. 37) we thus have :

$$\frac{d\sigma}{dy}(\nu_{\mu e} \rightarrow \nu_{\mu e}) = \frac{\mathcal{G}_F^2}{2\pi}\, m_e\, E_\nu \left\{ (2\sin^2\theta_W - 1)^2 + 4(1-y)^2 \sin^4\theta_W - \frac{m_e}{4\, E_\nu}\, y \left[(4\sin^2\theta_W - 1)^2 - 1 \right] \right\}$$

$$\frac{d\sigma}{dy}(\bar{\nu}_{\mu e} \rightarrow \bar{\nu}_{\mu e}) = \frac{\mathcal{G}_F^2}{2\pi}\, m_e\, E_\nu \left\{ 4\sin^4\theta_W + (1-y)^2 (2\sin^2\theta_W - 1)^2 - \frac{m_e}{4\, E_\nu}\, y \left[(4\sin^2\theta_W - 1)^2 - 1 \right] \right\}$$

where

$y = \frac{E_e}{E_\nu}$, E_ν is the energy of the incident neutrinos, E_e the energy of the outgoing electrons. The above cross-sections allow the determination of sin θ_W.

VIII. 12 - THE SALAM-WEINBERG MODEL FOR HADRONS : THE GIM MECHANISM ; THE QUARK MASSES

In order to extend the Weinberg-Salam model to hadronic matter, we have to consider the quark model so as to be able to describe the weak and electromagnetic interactions of quarks.

As stated in the introduction, there is evidence for the existence of a fifth quark b , and theoretical prejudice would like it to be accompanied by another quark, called t.

We shall, however, consider the Glashow-Illiopoulos-Maiani model[83)] with four quarks, u,d,c,s, which are supposed to exist in three colour states.

All observed hadrons are colour singlets and weak and electromagnetic currents act on flavour space only.

We postulate a lepton - quark symmetry according to which to each lepton there corresponds a quark, in almost a one-to-one correspondence.

Let us consider the two left-handed lepton isospinors :

$$\begin{pmatrix} \nu_e \\ e \end{pmatrix}_L \quad ; \quad \begin{pmatrix} \nu_\mu \\ \mu \end{pmatrix}_L$$

Let us then associate the quark u to ν_e and ν_μ to c :

$$\nu_e \longrightarrow u$$

$$\nu_\mu \longrightarrow c$$

Then we could have either the remaining correspondence :

$$e \longrightarrow d$$

$$\mu \longrightarrow s$$

or the reverse :

$$e \longrightarrow s$$

$$\mu \longrightarrow d$$

From the point of view of charge, d and s have the same charge $-\frac{1}{3}$, so we could take a linear combination for the quark which corresponds to e and the orthogonal one corresponding to μ. We thus write

$$e \longrightarrow d_c = d \cos\theta_c + s \sin\theta_c$$

$$\mu \longrightarrow s_c = s \cos\theta_c - d \sin\theta_c$$

θ_c turns out to be the Cabbibo angle.

So we start by assuming the following operators* :

$$Q_1 \equiv \begin{pmatrix} u \\ d_c \end{pmatrix}_L \quad ; \quad Q_2 \equiv \begin{pmatrix} c \\ s_c \end{pmatrix}_L$$

as isodoublets and

$$u_R \;,\; (d_c)_R \;,\; c_R \;,\; (s_c)_R$$

as isosinglets.

---

* If a third family of quarks, $\binom{t}{b}$, is considered, in correspondence to the leptons $\binom{\nu_\tau}{\tau}$ then, instead of Q_1, Q_2, one will postulate $Q_1 = \binom{u}{d'}_L$, $Q_2 = \binom{c}{s'}_L$ and $Q_3 = \binom{t}{b'}$ where, d', s', b' are connected to d, s, b by a 3 x 3 unitary matrix containing three angles and a phase factor (Kobayashi-Maskawa parameters

As in the case of leptons, we start by writing down a gauge invariant lagrangean which requires that these quarks are massless. The Higgs mechanism requires an interaction between the quarks and the Higgs scalar field. If we call $\phi(x)$ an isodoublet of scalar fields we may assume the SU(2) ⊗ U(1) transformations (with $g > 0$, $g' < 0$) :

$$Q_{1,2} \longrightarrow \exp\left\{ig\,\vec{\Lambda}\cdot\frac{\vec{\tau}}{2}\right\}\exp\left\{i\,\frac{g'}{2}\,(1-y)\,\theta\right\}Q_{1,2}$$

$$u_R \longrightarrow \exp\left\{-\,i\,\frac{g'}{2}\,y\,\theta\right\}u_R$$

$$c_R \longrightarrow \exp\left\{-\,i\,\frac{g'}{2}\,y\,\theta\right\}c_R$$

$$(d_c)_R \longrightarrow \exp\left\{i\,\frac{g'}{2}(2-y)\,\theta\right\}(d_c)_R$$

$$(S_c)_R \longrightarrow \exp\left\{i\,\frac{g'}{2}(2-y)\,\theta\right\}(S_c)_R$$

$$\phi \longrightarrow \exp\left\{ig\,\vec{\Lambda}\cdot\frac{\vec{\tau}}{2}\right\}\exp\left\{-\,ig'\,\frac{\theta}{2}\right\}\phi$$

We shall see that the quark model assumes $y = \frac{4}{3}$.

With the field $\phi(x)$ we can define another field

$$\tilde{\phi}(x) = C\,{}^t\phi^+\,(x)$$

where the matrix C satisfies the relationship :

$${}^t\vec{\tau} = -\,C^{-1}\,\vec{\tau}\,C$$

$${}^tC = -\,C$$

$$C^+C = CC^+ = I$$

(for instance, C may be taken as $i\tau_2$).

Then if under the SU(2) group :

$$\delta \phi = i \vec{\lambda} \cdot \frac{\vec{\tau}}{2} \phi$$

then :

$$\delta \tilde{\phi} = C^t (\delta \phi^+) = i \vec{\lambda} \cdot \frac{\vec{\tau}}{2} \tilde{\phi}$$

that is, $\tilde{\phi}$ is covariant to ϕ under the SU(2) group. But under the U(1) group, if ϕ transforms like

$$\delta \phi = - i g' \frac{\theta}{2} \phi$$

then

$$\delta \tilde{\phi} = i g' \frac{\theta}{2} \tilde{\phi}$$

so $\tilde{\phi}$ is an isospinor like ϕ but has opposite hypercharge :

$$\tilde{\phi} \longrightarrow \exp \left\{ ig \vec{\Lambda} \cdot \frac{\vec{\tau}}{2} \right\} \exp \left\{ i \frac{g'}{2} \theta \right\} \tilde{\phi}$$

Therefore we can construct the following SU(2) ⊗ U(1) invariant interactions with the Higgs field :

$$\mathcal{L}'_{\phi} = \left\{ G_1 \bar{Q}_1 d_{cR} + G_2 \bar{Q}_1 s_{cR} + G_3 \bar{Q}_2 d_{cR} + G_4 \bar{Q}_2 s_{cR} \right\} \phi +$$
$$+ \left\{ G_5 \bar{Q}_1 u_R + G_6 \bar{Q}_1 c_R + G_7 \bar{Q}_2 u_R + G_8 \bar{Q}_2 c_R \right\} \tilde{\phi} + h.c$$

Now if we transform Q_1 and Q_2 and then u_R and c_R (which have the same hypercharge) and d_R and s_R by rotations :

$$Q'_1 = Q_1 \cos\alpha - Q_2 \sin\alpha$$

$$Q'_2 = Q_2 \cos\alpha + Q_1 \sin\alpha$$

$$u'_R = u_R \cos\beta - c_R \sin\beta$$

$$c'_R = c_R \cos\beta + u_R \sin\beta$$

$$d'_R = d_R \cos\gamma - s_R \sin\gamma$$

$$s'_R = s_R \cos\gamma + d_R \sin\gamma$$

then the piece of the lagrangean

$$\mathscr{L}_I = \bar{Q}_1 \, i \, \gamma^\alpha D_\alpha Q_1 + \bar{Q}_2 \, i \, \gamma^\alpha D_\alpha Q_2 +$$
$$+ \bar{u}_R \, i \, \gamma^\alpha D_\alpha u_R + \bar{c}_R \, i \, \gamma^\alpha D_\alpha c_R +$$
$$+ \bar{d}_R \, i \, \gamma^\alpha D_\alpha d_R + \bar{s}_R \, i \, \gamma^\alpha D_\alpha s_R + \mathscr{L}'_\phi$$

where

$$D_\alpha Q_{1,2} = (\partial_\alpha + i g \, \vec{a}_\alpha \cdot \frac{\vec{\tau}}{2} + i \frac{g'}{2}(1 - y) B_\alpha) Q_{1,2}$$

$$D_\alpha u_R = (\partial_\alpha - i \frac{g'}{2} y B_\alpha) u_R$$

$$D_\alpha c_R = (\partial_\alpha - i \frac{g'}{2} y B_\alpha) c_R$$

$$D_\alpha (d_c)_R = (\partial_\alpha + i \frac{g'}{2}(2 - y) B_\alpha)(d_c)_R$$

$$D_\alpha (s_c)_R = (\partial_\alpha + i \frac{g'}{2}(2 - y) B_\alpha)(s_c)_R$$

remains invariant with new constants G'_i as linear functions of the original G_i. One can then chose the angles α, β, γ so as to have

$$G'_6 = G'_7 = 0 \qquad ; \qquad G'_2 = G'_3$$

This choice together with :

$$\phi(x) = \begin{pmatrix} 0 \\ a + \frac{1}{\sqrt{2}} \chi(x) \end{pmatrix}$$

$$\tilde{\phi}(x) = \begin{pmatrix} a + \frac{1}{\sqrt{2}} \chi(x) \\ 0 \end{pmatrix}$$

will give us

$$\mathcal{L}'_\phi = (G_5 a)\, \bar{u}\, u + (G_8 a)\, \bar{c}\, c + (G_1\, a)\, \bar{d}_c\, d_c + (G_4 a)\, \bar{s}_c\, s_c + $$

$$+ (G_2 a)(\bar{d}_c\, s_c + \bar{s}_c\, d_c) \qquad + \text{other terms}$$

As the quark eigenstates of the quantum number operators conserved by strong interactions are u,d,s,c we want this lagrangean to be diagonal in the fields d and s as well as in u and c. As

$$\mathcal{L}'_\phi = (G_5 a)\, \bar{u}\, u + (G_8 a)\, \bar{c}\, c + a \left\{ G_1 \cos^2\theta_c + G_4 \sin^2\theta_c \right\} \bar{d}\, d \; +$$

$$+ a \left\{ G_4 \cos^2\theta_c + G_1 \sin^2\theta_c \right\} \bar{s}\, s + a\,(G_1 - G_4) \sin\theta_c \cos\theta_c (\bar{d}s + \bar{s}d) +$$

$$+ (G_2 a) \cos(2\,\theta_c) \left\{ \bar{d}\, s + \bar{s}\, d + \mathrm{tg}\,(2\,\theta_c)(\bar{s}\, s - \bar{d}\, d) \right\} \qquad + \text{other terms}$$

we see that the terms in $\bar{s}\, d$, $\bar{d}\, s$ are eliminated for

$$t_g\, 2\, \theta_c = \frac{2\, G_2}{G_4 - G_1}$$

In this way the five coupling constants G_1, G_2, G_4, G_5, G_8 gives rise to the Cabbibo angle θ_c and the four quark masses.

VIII. 13 - THE SALAM-WEINBERG QUARK CURRENTS

The generalization of the lepton Salam-Weinberg model to other fermions can best be made if we recall that the interaction between leptons and the vector fields is uniquely determined by the form of the covariant derivative applied to the left and right-handed components of the basic lepton fields (and this form is imposed by the requirement of gauge invariance). Thus, from the derivatives :

$$D_\mu\, L(x) = (\partial_\mu + i\, g\, \mathcal{a}_{\mu a} \frac{\tau_a}{2} + i\, \frac{g'}{2}\, B_\mu)\, L(x)$$

$$D_\mu\, R(x) = (\partial_\mu + i\, g'\, B_\mu)\, R(x)$$

given in (VIII, 6) and from the piece of the lagrangean

$$\mathscr{L}' = \bar{L}\, i\, \gamma^\alpha\, D_\alpha\, L + \bar{R}\, i\, \gamma^\alpha\, D_\alpha\, R$$

we derive for the interaction between leptons and the vector fields :

$$\mathscr{L}_{int} = -\, g\, \mathcal{a}^\mu{}_a\, \bar{L}\, \gamma_\mu\, \frac{\tau_a}{2}\, L - \frac{g'}{2}\, B^\mu\, \bar{L}\, \gamma_\mu\, L - g'\, B^\mu\, \bar{R}\, \gamma_\mu\, R$$

Now we remember that the isospin and the hypercharge operators, $\vec{T}$ and Y are such that :

$$\left[\vec{T}, L\right] = -\frac{\vec{\tau}}{2} L \quad ; \quad \left[\vec{T}, R\right] = 0$$

$$\left[Y, L\right] = L \quad ; \quad \left[Y, R\right] = 2 R$$

so :

$$\mathcal{L}_{int} = g \vec{a}^{\mu} \cdot \bar{L} \gamma_{\mu} \left[\vec{T}, L\right] - \frac{g'}{2} B^{\mu} \left\{ \bar{L} \gamma_{\mu} \left[Y, L\right] - \bar{R} \gamma_{\mu} \left[Y, R\right] \right\}$$

We now consider the relationship between $\vec{a}^{\mu}$, B^{μ} and the electromagnetic field A^{μ} and the neutral boson field Z^{μ}, (VIII. 29) which gives :

$$a^{\mu}_{3} = Z^{\mu} \cos \theta_{W} - A^{\mu} \sin \theta_{W}$$

$$B^{\mu} = Z^{\mu} \sin \theta_{W} + A^{\mu} \cos \theta_{W}$$

and also (by (VIII. 18)) :

$$a^{\mu}_{1} = \frac{1}{\sqrt{2}} (W^{\mu} + W^{\mu+})$$

$$a^{\mu}_{2} = \frac{i}{\sqrt{2}} (W^{\mu+} - W^{\mu})$$

The lagrangean above becomes :

$$\mathcal{L}_{int} = \frac{g}{\sqrt{2}} W^{\mu} \bar{L} \gamma_{\mu} \left[T^{-}, L\right] + \frac{g}{\sqrt{2}} W^{\mu+} \bar{L} \gamma_{\mu} \left[T^{+}, L\right] -$$

$$+ g (Z^{\mu} \cos \theta_{W} - A^{\mu} \sin \theta_{W}) \bar{L} \gamma_{\mu} \left[T_{3}, L\right] +$$

$$- \frac{g'}{2} (Z^{\mu} \sin \theta_{W} + A^{\mu} \cos \theta_{W}) \left\{ \bar{L} \gamma_{\mu} \left[Y, L\right] + \bar{R} \gamma_{\mu} \left[Y, R\right] \right\}$$

where

$$T^- = T_1 - i\,T_2$$

$$T^+ = T_1 + i\,T_2$$

So

$$\mathcal{L}_{int} = \mathcal{L}'_F + \mathcal{L}'_\gamma + \mathcal{L}'_Z$$

where :

$$\mathcal{L}'_F = \frac{g}{\sqrt{2}} \left\{ (\bar{L}\, \gamma_\mu \left[T^-, L\right])\, W^\mu + (\bar{L}\, \gamma_\mu \left[T^+, L\right]) W^{\mu+} \right\}$$

$$\mathcal{L}'_\gamma = -g \sin\theta_W\, A^\mu \left\{ \bar{L}\, \gamma_\mu \left[T_3 + \frac{Y}{2}, L\right] + \bar{R}\, \gamma_\mu \left[\frac{Y}{2}, R\right] \right\}$$

$$\mathcal{L}'_Z = g \cos\theta_W\, Z^\mu \left\{ \bar{L}\, \gamma_\mu \left[T_3 - \frac{Y}{2}\, tg^2\, \theta_W, L\right] + \bar{R}\, \gamma_\mu \left[-\frac{Y}{2}\, tg^2\, \theta_W, R\right] \right\}$$

Now we set

$$Q = T_3 + \frac{Y}{2}$$

then we can write for the electromagnetic interaction

$$\mathcal{L}'_\gamma = -g \sin\theta_W\, A^\mu \left\{ \bar{L}\, \gamma_\mu \left[Q, L\right] + \bar{R}\, \gamma_\mu \left[Q, R\right] \right\}$$

Now

$$\left[Q, L\right] = \left[T_3 + \frac{Y}{2}, L\right] = -\left(\frac{\tau_3}{2} - \frac{1}{2}\right) L = -\frac{1}{2}\left\{\begin{pmatrix}1 & 0\\ 0 & -1\end{pmatrix} - \begin{pmatrix}1 & 0\\ 0 & 1\end{pmatrix}\right\}\begin{pmatrix}\nu_e\\ e\end{pmatrix}_L = \begin{pmatrix}0\\ e\end{pmatrix}_L$$

$$\left[Q, R\right] = \left[\frac{Y}{2}, R\right] = R \equiv e_R$$

So

$$\mathscr{L}'_{\gamma} = - g \sin \theta_W A^{\mu} (\bar{e} \gamma_{\mu} e)$$

and we see again that $g \sin \theta_W = e$.

Thus we get, in a similar procedure

$$\mathscr{L}'_{\gamma} = - e j^{\mu}_{\gamma} A_{\mu} \quad ; \quad e = g \sin \theta_W$$

$$j^{\mu}_{\gamma} = \left\{ \bar{L} \gamma^{\mu} \left[Q, L \right] + \bar{R} \gamma^{\mu} \left[Q, R \right] \right. =$$

$$= \bar{e} \gamma^{\mu} e \quad ;$$

(VIII. 41)

$$\mathscr{L}'_{F} = g_W (j^{\mu}_{W} W_{\mu} + j^{\mu +}_{W} W_{\mu}^{+}) \quad ; \quad g_W = \frac{g}{2\sqrt{2}} \quad ;$$

$$j^{\mu}_{W} = 2 \left\{ \bar{L} \gamma^{\mu} \left[T^{-}, L \right] \right\} =$$

$$= \bar{e} \gamma^{\mu} (1 - \gamma^5) \nu \quad ;$$

$$\mathscr{L}'_{Z} = g_Z Z^{\mu} j_{\mu}(0) \quad ; \quad g_Z = \frac{g}{2 \cos \theta_W} = \frac{(g^2 + g'^2)^{1/2}}{2} \quad ;$$

$$j_{\mu(o)} = 2 \left\{ \bar{L} \gamma_\mu \left[T_3 \cos^2 \theta_W - \frac{Y}{2} \sin^2 \theta_W, L \right] + \right.$$

$$\left. + \bar{R} \gamma_\mu \left[- \frac{Y}{2} \sin^2 \theta_W, R \right] \right\} =$$

$$= 2 \left\{ \bar{L} \gamma_\mu \left[T_3 - \sin^2 \theta_W Q, L \right] + \right.$$

$$\left. + \bar{R} \gamma_\mu \left[T_3 - \sin^2 \theta_W Q, R \right] \right\} =$$

$$= \frac{1}{2} \left\{ \bar{\nu} \gamma_\mu (1 - \gamma^5)\nu - \bar{e} \gamma_\mu (1 - \gamma^5)_e + \right.$$

$$\left. + 4 \sin^2 \theta_W (\bar{e} \gamma_\mu e) \right\} \equiv$$

$$\equiv 2 \left\{ \bar{L} \gamma_\mu \left[T_3, L \right] - \sin^2 \theta_W j_{\mu(\gamma)} \right\}$$

$$= j_{\mu 3} - 2 \sin^2 \theta_W j_\mu(\gamma)$$

Also :

$$\mathscr{L}'_Z = \frac{e}{\sin \theta_W \cos \theta_W} \left\{ \bar{L} \gamma^\mu \frac{\tau_3}{2} L - \sin^2 \theta_W j^\mu{}_{(\gamma)} \right\} Z_\mu$$

Now we are ready to take the quarks into account. We generalise the above formulae for the currents to any fermion field ψ, L being replaced by ψ_L and R by ψ_R. Then we have for the basic quarks fields

$$Q_1 = \begin{pmatrix} u_L \\ (d_c)_L \end{pmatrix} \quad ; \quad Q_2 = \begin{pmatrix} c_L \\ (s_c)_L \end{pmatrix} ; u_R ; (d_c)_R ; (s_c)_R ; c_R ,$$

$$- \left[\vec{T}, Q_1 \right] = \frac{\vec{\tau}}{2} Q_1 \quad ; \quad - \left[\vec{T}, Q_2 \right] = \frac{\vec{\tau}}{2} Q_2 \quad ;$$

$$\left[\vec{T}, u_R \right] = \left[\vec{T}, (d_c)_R \right] = \left[\vec{T}, (s_c)_R \right] = \left[\vec{T}, c_R \right] = 0$$

Also as we want that the charges Q of u,d,c,s be $\frac{2}{3}, -\frac{1}{3}, \frac{2}{3}, -\frac{1}{3}$ respectively we have :

$$- \left[Q, u_{L,R} \right] = \frac{2}{3} u_{L,R} \quad ;$$

$$- \left[Q, d_{L,R} \right] = -\frac{1}{3} d_{L,R} \quad ;$$

$$- \left[Q, c_{L,R} \right] = \frac{2}{3} c_{L,R} \quad ;$$

$$- \left[Q, s_{L,R} \right] = -\frac{1}{3} s_{L,R} \quad .$$

Then for the Cabibbo mixtures

$$d_c = d \cos\theta + s \sin\theta \quad ;$$

$$s_c = s \cos\theta - d \sin\theta \qquad \text{(VIII. 41a)}$$

we get

$$- \left[Q, (d_c)_{L,R} \right] = -\frac{1}{3} (d_c)_{L,R} \quad ;$$

$$- \left[Q, (s_c)_{L,R} \right] = -\frac{1}{3} (s_c)_{L,R}$$

Therefore we have for the hypercharge :

$$- \left[Y, \begin{pmatrix} u_L \\ (d_c)_L \end{pmatrix} \right] \equiv - \left[Y, Q_1 \right] = - \left[2(Q-T_3), Q_1 \right] = \begin{pmatrix} \frac{4}{3} - 1 & 0 \\ 0 & -\frac{2}{3} + 1 \end{pmatrix} \begin{pmatrix} u_L \\ (d_c)_L \end{pmatrix} =$$

$$= \frac{1}{3} \begin{pmatrix} u_L \\ (d_c)_L \end{pmatrix}$$

also :

$$\left[Y, Q_2 \right] = \frac{1}{3} \; Q_2$$

That is :

$$\left[Y, \begin{pmatrix} c_L \\ (s_c)_L \end{pmatrix} \right] = \frac{1}{3} \begin{pmatrix} c_L \\ (s_c)_L \end{pmatrix}$$

Then also :

$$\left[Y, u_R \right] = \left[2Q - T_3, u_R \right] = \frac{4}{3} u_R \quad ;$$

$$\left[Y, (d_c)_R \right] = \left[2Q - T_3, (d_c)_R \right] = -\frac{2}{3} (d_c)_R \quad ;$$

$$\left[Y, (s_c)_R \right] = -\frac{2}{3} (s_c)_R \quad ;$$

$$\left[Y, c_R \right] = \frac{4}{3} c_R \quad .$$

We thus obtain for the quark currents :

$$j^\mu{}_{(\gamma)} = \frac{2}{3} (\bar{u} \gamma^\mu u + \bar{c} \gamma^\mu c) - \frac{1}{3} (\bar{d} \gamma^\mu d + \bar{s} \gamma^\mu s) \; ;$$

$$j^\mu{}_{(W)} = \bar{u} \gamma^\mu (1 - \gamma^5) d_c + \bar{c} \gamma^\mu (1 - \gamma^5) s_c \; ;$$

$$j^\mu{}_{(0)} = \frac{1}{2} \Big\{ \bar{u} \gamma^\mu (1 - \gamma^5) u + \bar{c} \gamma^\mu (1 - \gamma^5) c - \bar{d} \gamma^\mu (1 - \gamma^5) d - \bar{s} \gamma^\mu (1 - \gamma^5) s \Big\} - 2 \sin^2 \theta_W \, j^\mu{}_{(\gamma)} = j^\mu{}_3 - 2 \sin^2 \theta_W \, j^\mu{}_{(\gamma)} \qquad \text{(VIII. 42)}$$

VIII. 14 - THE SUPPRESSION OF THE STRANGENESS-CHANGING NEUTRAL CURRENT

We note that the neutral current $j^{\mu}(0)$ above, which is coupled to the field Z^{μ}, does not contain terms of the form $\bar{d}\,\gamma^{\mu}(1-\gamma^5)\,s$ which would give rise to a strangeness changing neutral current. This is because, besides the terms $\bar{u}\,\gamma^{\mu}(1-\gamma^5)\,u$ and $\bar{c}\,\gamma^{\mu}(1-\gamma^5)\,c$, there occur the terms in d_c and s_c which are such that

$$\bar{d}_c\,\gamma^{\mu}(1-\gamma^5)\,d_c + \bar{s}_c\,\gamma^{\mu}(1-\gamma^5)\,s_c =$$

$$= (\bar{d}\cos\theta_c + \bar{s}\sin\theta_c)\,\gamma^{\mu}\,(1-\gamma^5)(d\cos\theta_c + s\sin\theta_c) +$$

$$+ (\bar{s}\cos\theta_c - \bar{d}\sin\theta_c)\,\gamma^{\mu}\,(1-\gamma^5)(s\cos\theta_c - d\sin\theta_c)$$

and we see that the strangeness changing terms $\bar{d}\,\gamma^{\mu}(1-\gamma^5)\,s\cos\theta_c\sin\theta_c$ and $\bar{s}\,\gamma^{\mu}(1-\gamma^5)\,d\cos\theta_c\sin\theta_c$ cancel out :

$$\bar{d}_c\,\gamma^{\mu}(1-\gamma^5)d_c + \bar{s}_c\,\gamma^{\mu}(1-\gamma^5)\,s_c = \bar{d}\,\gamma^{\mu}(1-\gamma^5)\,d + \bar{s}\,\gamma^{\mu}(1-\gamma^5)\,s$$

It was in order to have such a suppression mechanism that Glashow, Iliopoulos and Maiani proposed the charmed quark and the Cabibbo combination (VIII. 41a).

Thus, to lowest order, the reaction

$$K_L \rightarrow \mu^+\,\mu^-$$

cannot occur through exchange of a Z boson.

However, as shown in the figures below, the K_L being thought of as composed of a pair quark-antiquark $(d\,\bar{s})$, there can be virtual emission of a W^- and a W^+ which then give rise to the creation of a pair μ^+, μ^-. As the d quark emits a W^- boson it can go over into a u quark with a coupling $-\frac{g}{2\sqrt{2}}\cos\theta_c$ and the u quark annihilates $\bar{s}$ with a coupling $-\frac{g}{2\sqrt{2}}\sin\theta_c$ and emission of a W^+ (according to (VIII. 41), (VIII. 41a), (VIII. 42) :

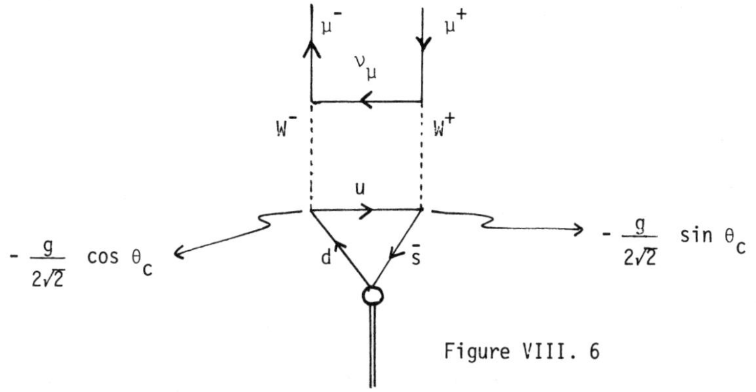

Figure VIII. 6

But the d quark can also go over into a c quark

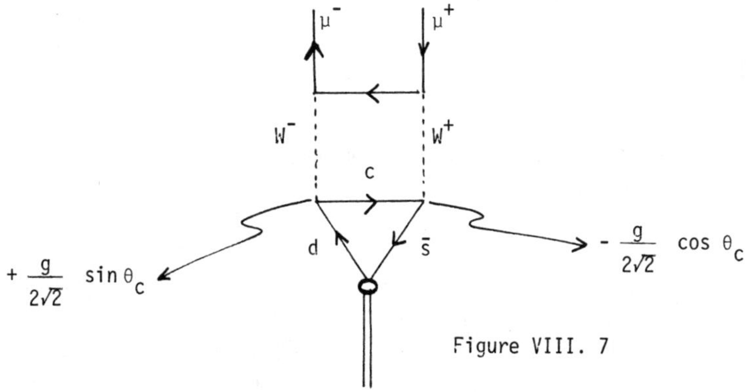

Figure VIII. 7

with coupling $\frac{g}{2\sqrt{2}}\sin\theta_c$ and the c quark turns over into a s quark with coupling $-\frac{g}{2\sqrt{2}}\cos\theta_c$.

To these diagrams with four vertices we must add the following ones of the same order :

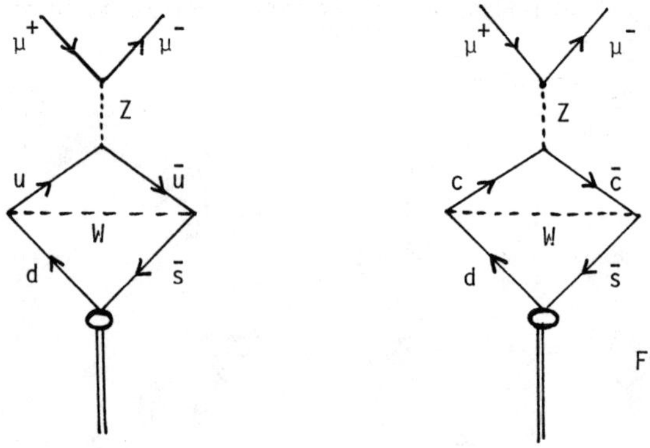

Figure VIII. 8

which involve intermediate W and Z bosons.

The amplitude for these diagrams would be cancelled if the masses of the u and c quarks were equal, as a result of the sign difference of the coupling constant in the diagram with u and c respectively. This amplitude will have a term of the form, assuming $m^2_W \gg m^2_c \gg m^2_u$:

$$\frac{g^4}{8m_W^4} (m^2_c - m^2_u) \cos\theta_c \sin\theta_c \cong \frac{G_F\alpha}{m^2_W} (m^2_c - m^2_u) \cos\theta_c \sin\theta_c$$

which will have the role of an effective coupling constant. From the study of this reaction and of the reaction $\bar{K}_0 \to K_0$ an estimate has been made of the mass of the charmed quark

$$m_c \cong 1.5 \text{ GeV}$$

VIII. 15 - ESTIMATES OF THE QUARK MASSES

Here are the estimates of the masses of quarks (those of the u and d quarks are estimated from hadron masses and the baryon magnetic moments) :

$$m_u \approx m_d \approx 350 \text{ MeV}$$

$$m_s \approx 500 \text{ MeV} \cong m_u + 150 \text{ MeV}$$

$$m_c \approx 1{,}5 \text{ GeV} \cong m_u + 1{,}2 \text{ GeV}$$

$$m_b \approx 5 \text{ GeV}$$

The last estimate results from the observation of the upsilon particles :

$$\Upsilon(9.4), \qquad \Upsilon'(10.0), \qquad \Upsilon''(10.4)$$

which are particle states with masses 9.4 GeV, 10 GeV and 10.4 GeV respectively. They are bosons with spin 1 and are believed to be 1^- - s wave bound states of a b quark and $\bar{b}$ antiquark. They are analogous to the charmonium particle states $\psi(3.1)$ and $\psi(3.7)$ which are assumed to be bound states of a charmed c quark and an antiquark $\bar{c}$.

Now the D-mesons have a mass of 1.86 GeV and they are structures with a charm quantum number C = 1 and zero spin, zero strangeness and isospin $\frac{1}{2}$:

$$C = 1 \; : \quad D^+ \sim c\,\bar{d}$$

$$D^0 \sim c\,\bar{u}$$

We may thus write for the mass of the lowest state :

$$m_D \sim 1.86 \text{ GeV} \simeq \frac{1}{2} m_\psi + 300 \text{ MeV} \sim m_c + m_{u,d}$$

whence

$$m_c \sim 1{,}5 \text{ Gev}$$

Similary for a B meson with a composition of the type $b\bar{d}$ or $b\bar{u}$ one would have, with evidence at CERN for such a meson and :

$$m_B \sim 5.3 \text{ GeV}$$

then :

$$m_B \sim 5.3 \text{ GeV} \simeq \frac{1}{2} m_\Upsilon + 300 \text{ MeV} \sim m_b + m_{u,d}$$

so

$$m_b \sim 5 \text{ GeV}$$

VIII. 16 - THE PARTON-QUARK MODEL

As we saw in § VIII. 10, the Salam-Weinberg model describes well the neutrino-lepton scattering experiments and predicts, for example, a cross-section

$$\sigma = 1.3 \times 10^{-42} \ E_\nu \text{ cm}^2$$

for the reaction

$$\bar{\nu}_\mu + e \rightarrow \bar{\nu}_\mu + e$$

for

$$\sin^2 \theta_W = 0.23$$

In order to check this model for hadrons, as the interactions of photons as well as of the W and Z bosons with quarks are theoretically known, one has to study how to obtain cross-sections for hadron-electron and for hadron-neutrino collisions from the knowledge of the above interactions. In other words, one needs to know how the quarks inside hadrons will contribute to the hadron-neutrino collisions or hadron-electron or hadron-photon interaction processes.

Consider the reaction neutrino-proton :

$$\nu_\mu + p \rightarrow \mu^- + h$$

where h indicates hadron particles.

Usually, experimentalists measure only the energy and the angle of the outgoing muon μ^- as well as the missing mass $\mathcal{M}^2$ as defined by :

$$\mathcal{M}^2 \equiv P^2_h$$

where P_h is the momentum of the hadron system h. If p is the momentum of the proton p and k_ν, k_μ those of the incident ν_μ and of the outgoing μ^- then :

$$\mathcal{M}^2 \equiv (p + k_\nu - k_\mu)^2$$

Such a reaction where the system h is not detected is called an <u>inclussive</u> reaction.

The proton is formed of two u quarks and one d quark (an anti-symmetric combination of such quarks of different colours) :

$$p \sim u\,u\,d$$

If the interaction between quarks inside the proton during the above collision process is such that the interaction of the neutrino takes place with each quark as if the latter were free then we may consider the cross-section for the reaction as the sum of the cross-sections for the interaction of the neutrino with each quark (Fig. VIII. 9)

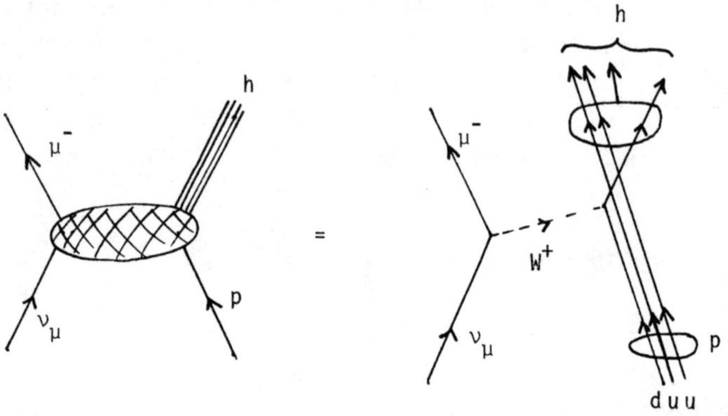

Figure VIII. 9

In this reaction only the quark u can absorb a virtual W^+ and give the final hadronic system h with the remaining spectator quarks.

For a reaction with antineutrinos

$$\bar{\nu}_\mu + p \rightarrow \mu^+ + h$$

the above hypothesis of <u>incoherence</u> of the scattering in the deep inelastic region gives the following picture (Fig. VIII. 10)

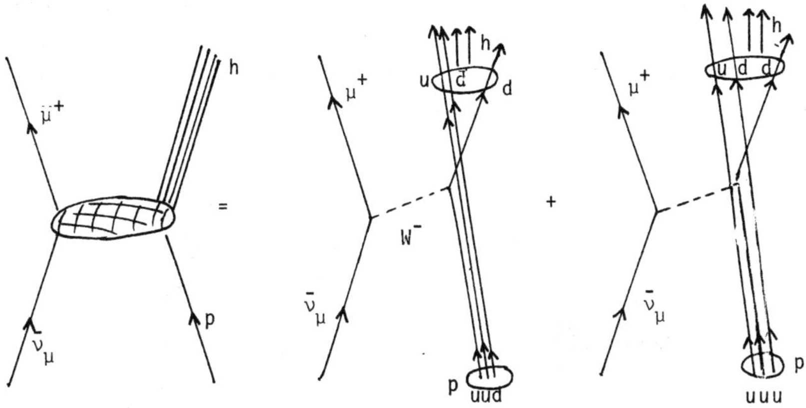

Figure VIII. 10

The quark inside the proton which absorbs the virtual boson W receives a large energy-momentum transfer from the incident neutrinos. A hypothesis in this picture is that this excited quark will recombine with the other quarks to form the final hadronic system h.

In the above figure, if

$$h \sim u\ d\ d$$

we have the reaction :

$$\bar{\nu}_\mu + p \rightarrow \mu^+ + n$$

But we could as well have the reaction

$$\bar{\nu}_\mu + p \rightarrow \mu^+ + \pi^- + p$$

in which case the diagrams would be of the type of Fig. VIII. 11

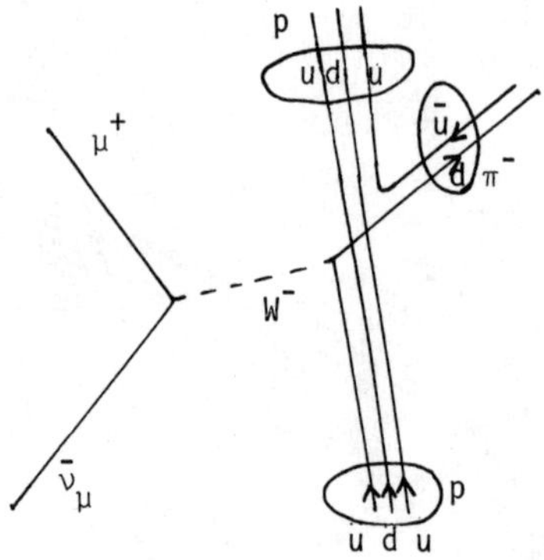

Figure VIII. 11

In this case we see that a pair $u\,\bar{u}$ is created of which $\bar{u}$ combines with d to form the pion π^- and the u quark of the pair combines with u and d to give the final proton.

Thus besides the structure quarks of the hadrons, called valence quarks there are also virtual quark-antiquark pairs which are said to constitute the quark sea. Such pairs will contribute to the production of the many hadron sytem in the final state h for high energy collisions.

As we have said previously, the structure quarks of baryons are of different colours and in such a combination that the baryon ground state wave function is colour antisymmetric : hadrons are, by hypothesis, colourless. It is also assumed that all observed particles are colourless, that is why (coloured) quarks must not be observed in the state of a free particle. The last statement leads us also to the so-called jet-hypothesis : quarks (as well as all other components of hadrons, called partons) endowed with large momentum, produce a jet of hadrons (fragmentation).

The parton model [59-68], developed mainly by Feynman, was invented to describe the deep inelastic collisions of hadrons with photons, electrons or neutrinos ; in these collisions the hadrons are shown to behave as formed of small parts -the partons- which scatter the incident particles, as if they were free point-like particles.

It is convenient, in the study of such deep inelastic processes, to introduce certain kinematic variables.

Consider, in general, the Fig. VIII. 12

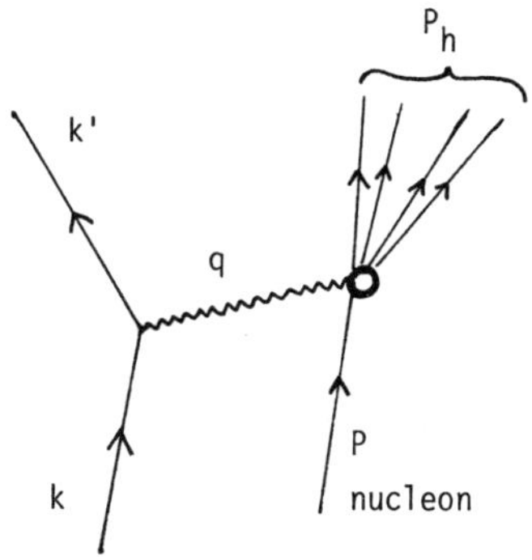

Figure VIII. 12

A light particle with momentum k (an electron, a neutrino or antineutrino) interacts with a nucleon with momentum P through the virtual particle (wavy line : a photon or a W boson or a Z boson) to give an outgoing light particle with momentum k' and a system of hadrons h.

The momentum transfer is

$$q_\mu = k_\mu - k'_\mu = P_h - P$$

We define the variable Q by :

$$Q^2 = - q^2 = - t$$

The variable s is

$$s = (k + P)^2$$

Another variable, denoted by ν, is defined as (M designates the proton mass) :

$$\nu = P.q / M$$

whereas the so called scaled parameters x and y are

$$x = \frac{Q^2}{2 M \nu}$$

$$y = \frac{2 M \nu}{s} \qquad \text{(VIII. 43)}$$

In the laboratory system the nucleon is initially at rest so :

$$P = (M, 0, 0, 0)$$

If we neglect, at high energies, the lepton mass, then :

$$k = (E, E, 0, 0)$$

$$k' = (E', E' \cos \theta, E' \sin \theta, 0)$$

The direction of $\vec{k}$ is taken as the $0\,x$ axis and the $0\,z$ axis is perpendicular to the plane $\vec{k}\,0\,\vec{k}'$ (Fig. VIII. 13).

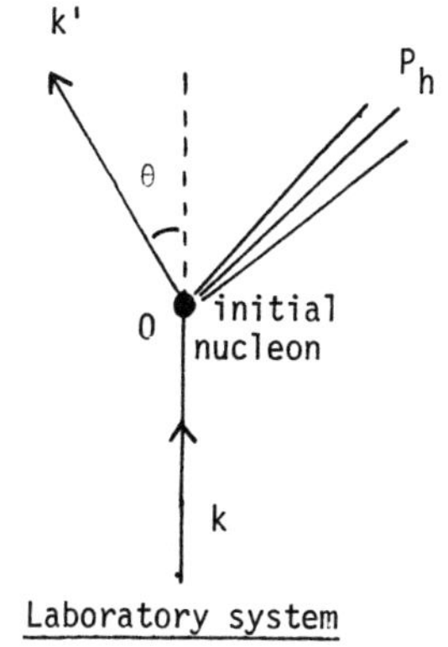

Figure VIII. 13

In this system therefore :

$$s = (k + P)^2 = (E + M)^2 - E^2 = 2\,M\,E + M^2$$
$$\cong 2\,M\,E$$

thus $\frac{s}{2M}$ in the laboratory system is the lepton initial energy ;

$$Q^2 = -\,(k - k')^2 = -\,(E' - E)^2 + (E - E'\cos\theta)^2 + (E'\sin\theta)^2 =$$

$$= 2\,E\,E'\,(1 - \cos\theta)$$

so

$$Q^2 = 4\,E\,E'\,\sin^2\frac{\theta}{2} > 0$$

we see that the momentum transfer q_μ is a space-like vector $q^2 < 0$.

The variable ν, in the laboratory system is such that :

$$\nu = E - E'$$

ν is the energy transfer in the collision.

Also :

$$x = \frac{Q^2}{2 M \nu} = \frac{E'}{M} \frac{2 \sin^2 \frac{\theta}{2}}{1 - \frac{E'}{E}}$$

$$y = \frac{2 M \nu}{s} = 1 - \frac{E'}{E} = \frac{\nu}{E}$$

How do we use these variables in the collision cross-section for the deep inelastic reaction above ?

The conventional calculation[69-78)] gives the differential cross-section for an outgoing lepton with energy between E' and $E' + d E'$ and three-dimensional momentum inside a solid angle $d \Omega$ with angle θ, $\frac{d^2 \sigma}{d E' d \Omega}$ where :

$$d \Omega = 2 \pi |d \cos \theta|$$

With the variables above we have :

$$\frac{d^2 \sigma}{d Q^2 d \nu} = \frac{\pi}{E E'} \frac{d^2 \sigma}{d E' d \Omega} = \frac{1}{\nu s} \frac{d^2 \sigma}{dx \, dy}$$

For the cross-section for electron-nucleon deep inelastic scattering it is found, in the laboratory system :

$$\frac{d^2\sigma}{dE'\,d\Omega} = \frac{\alpha^2}{4E^2(\sin\frac{\theta}{2})^4}\left\{2\sin^2\frac{\theta}{2}W_1 + \cos^2\frac{\theta}{2}W_2\right\}$$

for $e + p \rightarrow p + h$

where $\alpha = \frac{e^2}{4\pi}$ and :

$$W_1 = W_1\,(Q^2, \nu)$$

$$W_2 = W_2\,(Q^2, \nu)$$

are structure functions associated with the proton vertex. For the neutrino-nucleon scattering the corresponding cross-section is

$$\frac{d^2\sigma}{dE'\,d\Omega} = \frac{G_F^2}{2\pi^2}E'^2\left\{2\sin^2\frac{\theta}{2}W_1 + \cos^2\frac{\theta}{2}W_2 - \frac{E+E'}{M}\sin^2\frac{\theta}{2}W_3\right\}$$

for $\nu + p \rightarrow \mu^- + h$

In term of the variables x and y we have (with neglect of terms of order $\frac{M}{E}$) :

$$\frac{d^2\sigma}{dx\,dy} = \frac{G_F^2}{2\pi}\,s\left\{x\,y^2\,M\,W_1(Q^2, x) + (1-y)\nu\,W_2(Q^2, x) - (1-\frac{y}{2})x\,y\,\nu\,W_3(Q^2, x)\right\}$$

(VIII. 43a)

Experimentally it is found that these structure functions have the following behaviour : in the high energy region known as the Bjorken limit :

$$Q^2 \to \infty , \quad \nu \to \infty$$

with

$$x = \frac{Q^2}{2M\nu} \quad \text{fixed}$$

the structure functions have the following limiting behaviour :

$$M\, W_1\, (Q^2, x) \to F_1(x) ,$$

$$\nu\, W_2\, (Q^2, x) \to F_2(x) , \qquad \text{(VIII. 43b)}$$

$$\nu\, W_3\, (Q^2, x) \to F_3(x)$$

where the functions F_1, F_2, F_3 do not depend on Q^2 but only on x.

This means that in this reaction a structure determined by a mass or length paramenter is essentially absent.

The parton model describes these scaling properties. According to this model a high energy nucleon with large momentum $\vec{P}$ is regarded as incoherently formed of massless point-like constituents, the partons with momentum $\vec{p}_i$, the sum of which gives the momentum $\vec{P}$:

$$\vec{p}_i = x_i\, \vec{P} + \vec{P}_i^{\,t}$$

where x_i is the fraction of the total momentum $\vec{P}$ shared by the parton i along $\vec{P}$ and $\vec{P}_i^{\,t}$ is the transverse component of $\vec{p}_i$

$$\vec{P}.\, \vec{P}_i^{\,t} = 0$$

$$\sum_i x_i = 1$$

If the transverse component $|\vec{P}_i{}^t|$ is bounded then in the limit $\vec{P}^2 \to \infty$ we have for the massless i-parton energy :

$$E_i = \sqrt{x_i{}^2(\vec{P}^2 + M^2) + (\vec{P}_i{}^t)^2 - x_i{}^2 M^2} \cong$$

$$\cong x_i E + \frac{1}{2 x_i E} \left[(\vec{P}_i{}^t)^2 - x_i{}^2 M^2 \right] + \dots$$

where

$$E = \sqrt{\vec{P}^2 + M^2} \text{ is the proton energy}$$

And so in the limit $\vec{P}^2 \to \infty$, if $|\vec{P}_i{}^t|$ is bounded we have

$$E_i \cong x_i E$$

Thus :

$$p_i \cong x_i P$$

The so-called quark-parton model identifies the partons with quarks and antiquarks.

Let us now suppose that the deep inelastic scattering of Fig. VIII. takes place by elastic scattering from each parton

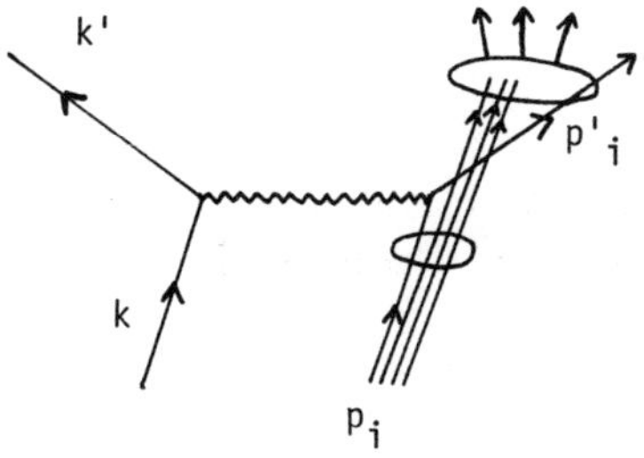

Figure VIII. 14

Then the virtual quantum is absorbed by the parton i and the momentum transfer in this process is :

$$q = p'_i - p_i$$

we have

$$Q^2 = - q^2 = - (p'_i - p_i)^2 = 2\, p_i\, p'_i$$

since partons are massless (at high energies). Now

$$p_i\, q = p'_i\, p_i - p_i^{\,2} = p'_i\, p_i$$

therefore ;

$$Q^2 = 2\, p_i\, q = 2\, x_i\, P\, q$$

and as

$$\nu = \frac{P.q}{M}$$

we obtain :

$$\frac{Q^2}{2\nu M} = x_i$$

we see by the formula (VIII. 43) that the scaling variable x is the fraction x_i of the proton momentum shared by parton i. This means that the deep inelastic collision results from the interaction between the virtual quantum emitted by the lepton and a quark with momentum

$$p = x\, P \quad , \qquad |\vec{P}|^2 \rightarrow \infty \; .$$

The deep inelastic scattering

$$\nu_\mu + p \rightarrow \mu^- + h$$

will then be described, in first approximation, by the scattering with the valence quark d in the proton. The cross-section for the latter is the same as for an electron, namely for the reaction :

$$\nu_\mu + e \rightarrow \mu^- + \nu_e$$

which is :

$$\frac{d\sigma}{dt}(\nu_\mu\, e \rightarrow \mu^-\, \nu_e) = \frac{G_F^{\,2}}{\pi}$$

Therefore, if we designate with $d(x)\,dx$ the probability that a d quark is inside the proton with momentum fraction between x and $x + dx$ we may write :

$$\frac{d^2\,\sigma}{dx\,dt}(\nu_\mu\, p \rightarrow \mu^-\, h) = d(x)\,\frac{d\sigma}{dt}(\nu_\mu\, d \rightarrow \mu^-\, u)$$

However, if we take into account the existence of the quark sea, the reaction may take place not only as indicated in Fig. VIII. 9 but also by the transition of an antiquark $\bar{u}$ of the sea into a $\bar{d}$ (Fig. VIII. 15)

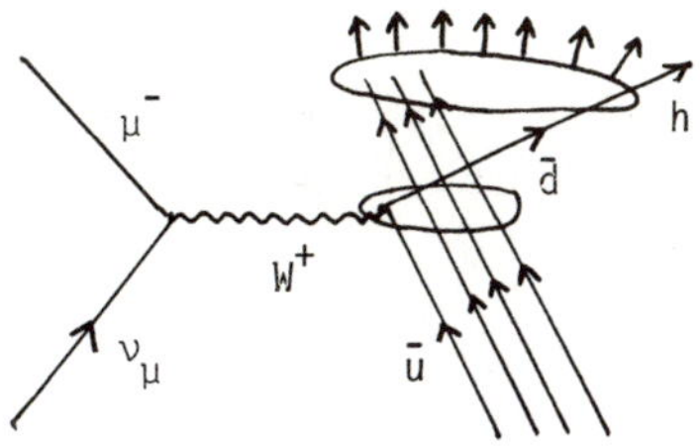

Figure VIII. 15

Thus, if $\bar{u}(x)\,dx$ is the probability of finding an antiquark $\bar{u}$ with momentum fraction between x and $x + dx$ in the high-energy hadron we must have :

$$\frac{d^2\sigma}{dx\,dt}(\nu_\mu\, p \to \mu^-\, h) = d(x)\,\frac{d\sigma}{dt}(\nu_\mu\, d \to \mu^-\, u) + \bar{u}(x)\,\frac{d\sigma}{dt}(\nu_\mu\, \bar{u} \to \mu^-\, \bar{d})$$

Now it is known that :

$$\frac{d\sigma}{dt}(\nu_\mu\, d \to \mu^-\, u) = \frac{G_F^{\,2}}{\pi}\cos^2\theta_c$$

$$\frac{d\sigma}{dt}(\nu_\mu\, \bar{u} \to \mu^-\, \bar{d}) = \frac{G_F^{\,2}}{\pi}(1 - y)^2\cos^2\theta_c$$

Therefore, as :

$$dx\,dt = xs\,dx\,dy$$

we shall have :

$$\frac{d^2\sigma}{dx\,dy}(\nu_\mu\, p \to \mu^-\, h) = \frac{G_F^{\,2}}{\pi}\, s \left\{ x\,d(x) + x\,\bar{u}(x)(1 - y)^2 \right\} \cos^2\theta_c \qquad \text{(VIII. 44)}$$

The functions $d(x)$, $\bar{d}(x)$, $u(x)$ and $\bar{u}(x)$ satisfy the equations :

$$\int_0^1 dx(u(x) - \bar{u}(x)) = 2$$

$$\int_0^1 dx(d(x) - \bar{d}(x)) = 1$$

for a proton since this particle has two u quarks and one d quark.

Besides these relations we impose :

$$Q = \int_0^1 dx \left\{ \frac{2}{3}(u(x) - \bar{u}(x)) + \frac{2}{3}(c(x) - \bar{c}(x)) - \frac{1}{3}(d(x) - \bar{d}(x)) - \frac{1}{3}(s(x) - \bar{s}(x) \right\} = 1$$

for the proton charge ;

$$T_Z = \int_0^1 dx \, \frac{1}{2} \left\{ u(x) - \bar{u}(x) - (d(x) - \bar{d}(x)) \right\} = \frac{1}{2}$$

for the proton isospin z component ;

$$S = \int_0^1 dx \, (s(x) - \bar{s}(x)) = 0$$

for the proton strangeness ; and

$$C = \int_0^1 dx \, (c(x) - \bar{c}(x)) = 0$$

for the proton charm.

As the proton has two valence u quarks and one d quark and the neutron has two valence d quarks and one u quark, distribution functions for the neutron are obtained from those of the proton by the substitution

$$\begin{aligned} &u(x) \text{ (for proton)} \rightarrow d(x) \text{ (for neutron)} \\ &d(x) \text{ (for proton)} \rightarrow u(x) \text{ (for neutron)} \end{aligned} \qquad \text{(VIII. 45)}$$

and

$$s(x) \text{ (proton)} \rightarrow s(x) \text{ (neutron)}$$

$$c(x) \text{ (proton)} \rightarrow c(x) \text{ (neutron)}$$

and similarly for the antiquark distributions.

One thus obtains

$$\frac{d^2\sigma}{dx\,dy}(\nu_\mu\, n \to \mu^-\, h) = \frac{G_F{}^2 sx}{\pi}\left\{u(x) + \bar{d}(x)(1-y)^2\right\}\cos^2\theta_c \qquad \text{(VIII. 46)}$$

This expression as well as (VIII. 44) hold true if we neglect the contribution of the quark sea with s and c quarks.

If one takes all quarks and antiquarks of the sea into account, namely :

$$\frac{d\sigma}{dt}(\nu_\mu\, d \to \mu^-\, c) = \frac{d\sigma}{dt}(\nu_\mu\, s \to \mu^-\, u) = \frac{G_F{}^2}{\pi}\sin^2\theta_c$$

$$\frac{d\sigma}{dt}(\nu_\mu\, s \to \mu^-\, c) = \frac{d\sigma}{dt}(\nu_\mu\, d \to \mu^-\, u) = \frac{G_F{}^2}{\pi}\cos^2\theta_c$$

and

$$\frac{d\sigma}{dt}(\nu_\mu\, \bar{u} \to \mu^-\, \bar{s}) = \frac{d\sigma}{dt}(\nu_\mu\, \bar{c} \to \mu^-\, \bar{d}) = \frac{G_F{}^2}{\pi}(1-y)^2\sin^2\theta_c$$

$$\frac{d\sigma}{dt}(\nu_\mu\, \bar{c} \to \mu^-\, \bar{s}) = \frac{d\sigma}{dt}(\nu_\mu\, \bar{u} \to \mu^-\, \bar{d}) = \frac{G_F{}^2}{\pi}(1-y)^2\cos^2\theta_c$$

we get

$$\frac{d^2\sigma}{dx\,dy}(\nu_\mu p \to \mu^- h) = \frac{G_F{}^2}{\pi}\, s\, x\left\{\left[d(x) + s(x)\right] + \left[\bar{u}(x) + \bar{c}(x)\right](1-y)^2\right\}$$

$$\frac{d^2\sigma}{dx\,dy}(\nu_\mu n \to \mu^- h) = \frac{G_F{}^2}{\pi}\, s\, x\left\{\left[u(x) + s(x)\right] + \left[\bar{d}(x) + \bar{c}(x)\right](1-y)^2\right\}$$

For antineutrinos one finds :

$$\frac{d^2\sigma}{dx\,dy}(\bar{\nu}_\mu p \to \mu^+ h) = \frac{G_F^{\,2}}{\pi}\, s\, x \left\{ \left[u(x) + c(x) \right] (1-y)^2 + \left[\bar{d}(x) + \bar{s}(x) \right] \right\}$$

$$\frac{d^2\sigma}{dx\,dy}(\bar{\nu}_\mu n \to \mu^+ h) = \frac{G_F^{\,2}}{\pi}\, s\, x \left\{ \left[d(x) + c(x) \right] (1-y)^2 + \left[\bar{u}(x) + \bar{s}(x) \right] \right\}$$

A comparison with the formula (VIII. 43a) taken in the Bjorken limit (VIII. 43b) namely :

$$\frac{d^2\sigma(\nu\, p)}{dx\,dy} = \frac{G_F^{\,2}}{2\pi}\, s \left\{ x \left[F_1(x) - \frac{F_3}{2}(x) \right] + x \left[F_1(x) + \frac{F_3}{2}(x) \right] (1-y)^2 \right.$$
$$\left. + \left(F_2(x) - 2\,x\,F_1(x)\right)(1-y) \right\}$$

$$\frac{d^2\sigma(\bar{\nu}\, p)}{dx\,dy} = \frac{G_F^{\,2}}{2\pi}\, s \left\{ x \left[F_1(x) + \frac{F_3}{2}(x) \right] + x \left[F_1(x) - \frac{F_3}{2}(x) \right] (1-y)^2 \right.$$
$$\left. + \left(F_2(x) - 2\,x\,F_1(x)\right)(1-y) \right\}$$

gives for the structure functions :

$$2\,x\,F_1^{\nu p}(x) = F_2^{\nu p}(x) = 2\,x \left[d(x) + s(x) + \bar{u}(x) + \bar{c}(x) \right]$$

$$F_3^{\nu p}(x) = 2 \left[\bar{u}(x) + \bar{c}(x) - d(x) - s(x) \right]$$

$$2\,x\,F_1^{\bar{\nu} p}(x) = F_2^{\bar{\nu} p}(x) = 2\,x \left[\bar{d}(x) + \bar{s}(x) + u(x) + c(x) \right]$$

$$F_3^{\bar{\nu} p}(x) = -\,2 \left[u(x) + c(x) - \bar{d}(x) - \bar{s}(x) \right]$$

The substitution rule (VIII. 45) gives us immediately the stucture functions for the case $\nu\, n \to \mu^- h$ and $\bar{\nu}\, n \to \mu^+ h$.

It is of interest to have the cross-sections for isoscalar targets, that is formed of nuclei which have the same number of neutrons and protons :

$$Z \cong \frac{A}{2}$$

Then the measured quantity is the average cross-section for reaction with p and with n :

$$\frac{d^2\,\sigma_{iso}(\nu)}{dx\,dt} = \frac{1}{2}\left\{\frac{d^2\,\sigma}{dx\,dy}\,(\nu_\mu\, p \to \mu^- h) + \frac{d^2\,\sigma}{dx\,dy}\,(\nu_\mu\, n \to \mu^-\, h)\right\}$$

The total cross-section as defined by :

$$\sigma_{iso} = \int_0^1 dx\,dy\,\frac{d^2\,\sigma_{iso}}{dx\,dy}$$

is :

$$\sigma_{iso}(\nu) = \frac{G_F^{\,2}}{\pi}\,s\left\{\int_0^1 dx(u(x) + d(x) + 2s(x)) + \frac{1}{3}\int_0^1 dx\;x(\bar{u}(x) + \bar{d}(x) + 2\bar{c}(x))\right\}$$

Therefore the ratio

$$R = \frac{\sigma_{iso}(\bar{\nu})}{\sigma_{iso}(\nu)} = \frac{\frac{1}{3}\,M + \bar{M}}{M + \frac{1}{3}\,\bar{M}}$$

where, if we neglect the contribution of the s, $\bar{s}$ and c, $\bar{c}$ quarks of the sea, M and $\bar{M}$ are :

$$M = \int_0^1 dx\;x\;(u(x) + d(x))$$

$$\bar{M} = \int_0^1 dx\;x\;(\bar{u}(x) + \bar{d}(x))$$

If the contribution of $\bar{M}$ is negligible we obtain

$$R \cong \frac{1}{3}$$

Experiment confirms this small contribution of $\bar{u}$, $\bar{d}$ and of $s, c, \bar{s}, \bar{c}$ since the experimental value of R is :

$$R = 0.38 \pm 0.02 \quad \text{(Gargamelle)}$$

$$= 0.35 \pm 0.04 \quad \text{(NAL)}$$

VIII. 17 - THE VALUE OF THE WEINBERG ANGLE FOR THE NEUTRINO-NUCLEON SCATTERING

The calculation of the cross-section for the scattering reactions

$$\nu_\mu + p \rightarrow \nu_\mu + \chi$$

$$\nu_\mu + n \rightarrow \nu_\mu + \chi'$$

with neutral current interaction is carried out having in mind the expression of the neutral current $j^\mu(o)$ for quarks as given in equation (VIII. 42).

Let us call :

$$R^{<\nu N>}\left(\frac{n}{c}\right) = \frac{\sigma(\nu_\mu + p \rightarrow \nu_\mu x) + \sigma(\nu_\mu + n \rightarrow \nu_\mu + x')}{\sigma(\nu_\mu + p \rightarrow \mu^- + h) + \sigma(\nu_\mu + n \rightarrow \mu^- + h')}$$

$$R^{<\bar{\nu} N>}\left(\frac{n}{c}\right) = \frac{\sigma(\bar{\nu}_\mu + p \rightarrow \bar{\nu}_\mu + x) + \sigma(\bar{\nu}_\mu + n \rightarrow \bar{\nu}_\mu + x')}{\sigma(\bar{\nu}_\mu + p \rightarrow \mu^+ + h) + \sigma(\bar{\nu}_\mu + n \rightarrow \mu^+ + h')}$$

These are ratios of neutral to charged current reactions.

Then one may write with Sehgal[140]:

$$R^{<\nu N>}\left(\frac{n}{c}\right) = \left(\frac{1}{2} - \frac{2}{3}\sin^2\theta_W\right)^2 + \left(\frac{1}{2} - \frac{1}{3}\sin^2\theta_W\right)^2 +$$

$$+ \frac{1}{3}\left[\left(\frac{2}{3}\sin^2\theta_W\right)^2 + \left(\frac{1}{3}\sin^2\theta_W\right)^2\right]$$

$$= \frac{1}{2} - \sin^2\theta_W + \frac{20}{27}\left(\sin^2\theta_W\right)^2$$

$$R^{<\bar{\nu} N>}\left(\frac{n}{c}\right) = \left(\frac{1}{2} - \frac{2}{3}\sin^2\theta_W\right)^2 + \left(\frac{1}{2} - \frac{1}{3}\sin^2\theta_W\right)^2 +$$

$$+ 3\left[\left(\frac{2}{3}\sin^2\theta_W\right)^2 + \left(\frac{1}{3}\sin^2\theta_W\right)^2\right] =$$

$$= \frac{1}{2} - \sin^2\theta_W + \frac{20}{9}\left(\sin^2\theta_W\right)^2$$

In these calculations one keeps the contributions of the u and d quarks only. These expressions are shown to be the lower bounds for $R^{<\nu N>}$ and $R^{<\bar{\nu} N>}$ respectively when this restriction is relaxed under certain additional assumptions

Experimentally it is found for the weighted average of results from different laboratories :

$$R^{<\nu N>}\left(\frac{n}{c}\right) = 0.29 \pm 0.01$$

(Baltay, 1978)[145]

$$R^{<\bar{\nu} N>}\left(\frac{n}{c}\right) = 0.35 \pm 0.025$$

implying a value for θ_W :

$$\sin^2\theta_W = 0.24 \pm 0.02$$

According to Paschos and Wolfenstein [63] the quark-parton model dependence in the previous calculations is eliminated if one computes the ratio :

$$S = \frac{\sigma^{<\nu N>}_{\text{neutral current}} - \sigma^{<\bar{\nu} N>}_{\text{neutral current}}}{\sigma^{<\nu N>}_{\text{charg. current}} - \sigma^{<\bar{\nu} N>}_{\text{charg. current}}}$$

to obtain :

$$S = \frac{1}{2} - \sin^2 \theta_W$$

The data indicate agreement for

$$\sin^2 \theta_W = 0.22 \pm 0.05$$

All data so far available point to a consistently unique value for the Weinberg angle to describe different experiments. The table VIII. 5 gives a summary of this excellent verification of the Salam-Weinberg gauge field theory of the electro - weak interactions.

Let me finally point out that chromodynamics gives a satisfactory justification of the assumption of the parton model, that partons behave as free particles in the infinite momentum frame of reference. This theory, which describes strong interactions by the coupling between quarks and the massless colour gluon fields (see § VIII. 10), has the property that these interactions become very small for very large momentum transfers and therefore, for small distances. This property is called asymptotic freedom and follows from the fact that the effective coupling constant in chromodynamics is a decreasing function of Q^2. For large Q^2, perturbation theory can thus be used and calculations can be compared with experiment. For small Q^2 or large distances, the coupling constant increases and becomes larger and larger and calculations are more difficult to be made by present techniques [124-133].

Process	Experimental results	$\sin^2\theta$	W-S Prediction with $\sin^2\theta=0.23$
1. Purely leptonic			
$\bar{\nu}_e + e^- \rightarrow \bar{\nu}_e + e^-$	$(5.7 \pm 1.2) \times 10^{-42}\ E_\nu\ cm^2$	0.29 ± 0.05	5.0
$\nu_\mu + e^- \rightarrow \nu_\mu + e^-$	$(1.7 \pm 0.5) \times 10^{-42}\ E_\nu\ cm^2$	$0.21^{+0.09}_{-0.06}$	1.5
$\bar{\nu}_\mu + e^- \rightarrow \bar{\nu}_\mu + e^-$	$(1.8 \pm 0.9) \times 10^{-42}\ E_\nu\ cm^2$	$0.30^{+0.10}_{-0.30}$	1.3
2. Elastic scattering			
$\nu_\mu + p \rightarrow \nu_\mu + p$	$(0.11 \pm 0.02) \times \sigma(\nu_\mu + n \rightarrow \mu^- + p)$	0.26 ± 0.06	0.12
$\bar{\nu}_\mu + p \rightarrow \bar{\nu}_\mu + p$	$(0.19 \pm 0.08) \times \sigma(\bar{\nu}_\mu + p \rightarrow \mu^+ + n)$	≤ 0.5	0.11
3. Single pion production			
$\nu_\mu + N \rightarrow \nu_\mu + N + \pi^0$	$(0.45 \pm 0.08) \times \sigma(\nu_\mu + N \rightarrow \mu^- + N + \pi^0)$	0.22 ± 0.09	0.42
$\bar{\nu}_\mu + N \rightarrow \bar{\nu}_\mu + N + \pi^0$	$(0.57 \pm 0.11) \times \sigma(\bar{\nu}_\mu + N \rightarrow \mu^+ + N + \pi^0)$	0.15 - 0.52	0.60
4. Inclusive			
$\nu_\mu + N \rightarrow \nu_\mu + \ldots$	$(0.29 \pm 0.01) \times \sigma(\nu_\mu + N \rightarrow \mu^- + \ldots)$	0.24 ± 0.02	0.30
$\bar{\nu}_\mu + N \rightarrow \bar{\nu}_\mu + \ldots$	$(0.35 \pm 0.025) \times \sigma(\bar{\nu}_\mu + N \rightarrow \mu^+ + \ldots)$	0.3 ± 0.1	0.38

From C. Baltay, Neutrino interactions, Proc. 19 th International Conf. on High Energy Phys., Tokyo (1979)

TABLE VIII. 5 - Neutral currents comparison with the Salam-Weinberg model

PROBLEMS

VIII - 1. Establish the equations (VIII. 4b) for the corresponding infinitesimal transformations.

VIII - 2. Given the lagrangean of a Yang-Mills field in the absence of matter :

$$\mathscr{L} = -\frac{1}{4} F^{\mu\nu}{}_a F_{\mu\nu i a}$$

a) Write the field equations and specify which are the equations of motion and which are the constraint equations.
b) Give the canonical equal time commutation rules and the hamiltonian in the Coulomb gauge $A^o{}_a = 0$, $\partial_k A^k{}_o = 0$. Read Abers and Lee, ref. 104, especially § 13.

VIII - 3. Show that the charges associated with the currents $j^\mu{}_a(x)$ in gauge field theories satisfy the same commutation rules as the generators of the gauge group.

VIII - 4. From Problem VIII. 3 it follows that if $Q = Q_1 - i\,Q_2$ is the charge associated with the current

$$j^\mu{}_a = \bar{L}\,\gamma^\mu\,(1 - \gamma^5)\,\frac{\tau_a}{2}\,L \;;\quad L = \begin{pmatrix} \nu \\ e \end{pmatrix}_L$$

which is coupled to the boson field W_μ, then Q_3, given by

$$Q_3 = \frac{1}{2}\,[Q, Q^+]$$

is also a gauge charge and must be associated with a current $j^\mu{}_3(x)$ coupled to a gauge field, call it $W_{\mu 3}(x)$, with the same coupling constant.
If $j^\mu{}_3(x)$ is not the electromagnetic current, it is a new neutral current coupled to a new field Z_μ and this is what is predicted by the Salam-Weinberg SU(2) ⊗ U(1) model.

Another possibility is to assume that $j^{\mu}{}_{3}(x)$ is the electromagnetic current and that therefore $Q_3 \equiv Q_e$, Q, Q^+ form an SU(2) algebra.

Discuss this case, the SU(2) model proposed by Georgi and Glashow. The fact that Q_e can have zero eigenvalue and is a generator of SU(2) requires an integer spin representation of the latter group such as SO(3) and this implies the need of postulating two heavy leptons, one positively charged and one neutral.
Consult Georgi and Glashow, ref. 135 ; Bjorken and Llewellyn-Smith, ref. 136 ; and L. O'Raifeartaigh, ref. 109, especially pag. 204 and foll.

VIII - 5. Study the method of path integral quantization and formulation of field theory. Consult Coleman, ref. 24, especially § 4 and 5 ; Taylor, ref. 185, especially Chaps. 10 and 11 ; Nash, ref. 186 ; Abers and Lee, ref. 104, especially § 11 and 12.

VIII - 6. Study the dimensional regularization method. Read C. Bollini and J.J. Giambiagi, Nuovo Ciment 12B, 20 (1972) ; Phys. Lett. 40B, 566 (1972) ; G. 't Hooft and M. Veltman, Nucl. Phys. 44B, 189 (1972) ; Nash, ref. 186 ; Taylor, ref. 185, Chaps. 13 and 14.

VIII - 7. For the experimental investigation on the predictions of the SU(2) ⊗ U(1) model, study refs. 140-147, especially Musset and Vialle, ref. 143 ; Baltay, ref. 146 ; Steinberger, ref. 147.

CHAPTER IX

Gauge Theory with Lepton Flavour Non-Conservation

IX. 1 - SU(2) ⊗ U(1) GAUGE THEORY WITH HEAVY LEPTONS

It is well known that the neutrino which accompanies the muon in pi-decay :

$$\pi^- \rightarrow \mu^- + \nu_\mu$$

has its own (leptonic) muon quantum number which is different from the electron quantum number of the neutrino which accompanies the electron in the decay :

$$\pi^- \rightarrow e^- + \nu_e$$

These numbers are given in the table V in the introduction and the separate exact conservation of L_e and L_μ is invoked to explain the lack of mixture of the two types of neutrinos as well as the absence of the muon radiative decay :

$$\mu \rightarrow e + \gamma \qquad \text{(IX. 1)}$$

However, one might inquire into whether the conservation of L_e and of L_μ is not exact so that very small rates of reaction (IX. 1) would be possible.

The idea of neutrino mixture was first studied by Pontecorvo[140].

If one tries an analogy with the Cabibbo mixture of the quarks d and s and assumes an interaction of the form

$$L = g\, W^\mu \left\{ (\bar{e}\, \gamma_\mu (1 - \gamma^5) \nu_1) + (\bar{\mu}\, \gamma_\mu (1 - \gamma^5) \nu_2 \right\}$$

where

$$\nu_1 = \nu_e \cos\theta + \nu_\mu \sin\theta$$

$$\nu_2 = \nu_\mu \cos\theta - \nu_e \sin\theta$$

then it is easy to see that the rate for reaction (IX. 1) is zero. This is similar to the GIM mechanism of suppression of strangeness changing neutral currents and results from the cancellation of the diagram amplitudes (Fig. IX. 1)

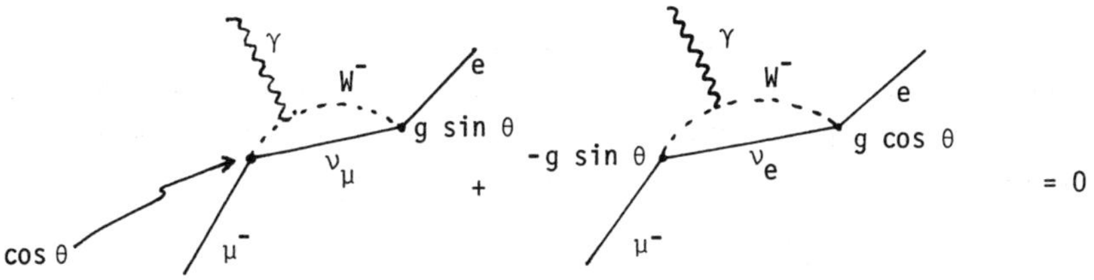

Figure IX. 1

(we omit the diagrams where γ is emitted by μ or e).

This is because both ν_e and ν_μ are both massless and the propagator is the same for both.

An extension of the Salam-Weinberg theory to encompass possible muon number non-conservation was proposed by Ragiadakos and the author[136] in 1976.

For this, one must assume a heavy neutral lepton L_o and mix it with ν_e and ν_μ in the following way. Call

$$\nu_1 \equiv \nu_e, \quad \nu_2 \equiv L_o, \quad \nu_3 \equiv \nu_\mu \qquad \text{(IX. 1a)}$$

and define the three linear combinations

$$\nu'_j = a_{jk}\, \nu_k \qquad \text{(IX. 2)}$$

where the a_{jk} are matrix elements of the SO(3) group

$$a_{jk}\, a_{j\ell} = \delta_{k\ell}$$

The following left-handed doublets are defined :

$$e_L = \frac{1}{2}(1 - \gamma^5)\begin{pmatrix} \nu'_1 \\ e \end{pmatrix}, \quad \mu_L = \frac{1}{2}(1 - \gamma^5)\begin{pmatrix} \nu'_3 \\ \mu \end{pmatrix} \qquad \text{(IX. 3)}$$

and the singlets :

$$\begin{aligned} e_R &= \frac{1}{2}(1 + \gamma^5)\, e, & \mu_R &= \frac{1}{2}(1 - \gamma^5)\mu \\ \mathscr{L}_\ell &= \frac{1}{2}(1 - \gamma^5)\, L_o, & \mathscr{L}_R &= \frac{1}{2}(1 + \gamma^5) L_o \end{aligned} \qquad \text{(IX. 3a)}$$

After application of the Higgs mechanism with the requirement that the masses of ν_e and ν_μ vanish whereas that of the heavy lepton is different from zero, one obtains the effective lagrangean with the following currents :

$$\ell_\alpha(W) = \sum_{k=1}^{3} \left\{ a_{1k}\, \bar{\nu}_k\, \gamma_\alpha(1-\gamma^5)e + a_{3k}\, \bar{\nu}_k\, \gamma_\alpha(1-\gamma^5)\mu \right\}$$

is the charged current which interacts with the W-field ; and

$$\ell_\alpha(o) = \sum_{l=1,2} \sum_{j,k=1}^{3} a_{ik}\, a_{ij}\, \bar{\nu}_k\, \gamma_\alpha(1-\gamma^5)\nu_j +$$

$$+ \bar{e}\, \gamma_\alpha \left(\frac{3\, g'^2 - g^2}{g^2 + g'^2} - \gamma^5 \right) e + \bar{\mu}\, \gamma_\alpha \left(\frac{3\, g'^2 - g^2}{g^2 + g'^2} - \gamma^5 \right) \mu$$

is the neutral current in interaction with the Z-field.

Clearly the fact that the hypothetical neutral lepton L_0 is massive gives rise to a possible gamma-decay of the muon, the diagrams of which are given in Fig. IX. 2 (omitting the diagrams with γ emitted by μ or e) :

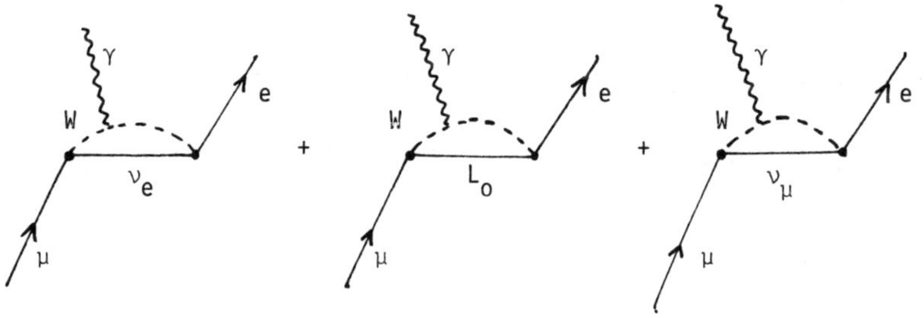

Figure IX. 2

The amplitude for the reaction of Fig. (IX. 2) will be of the form :

$$A(\mu \rightarrow e + \gamma) \sim e\, g^2\, a_{32}\, a_{12}\, m_L^{\ 2}\, I\,(P_e, P_\mu, K, m_L, m_W)$$

where

$$I = \int \frac{d^4k}{(2\pi)^4}\, \bar{u}\,(p_e)\, \gamma^\alpha(1 - \gamma^5)\, \frac{\gamma.k + \frac{k^2}{m_L}}{k^2(k^2 - m_L^{\ 2})} \quad \cdot$$

$$\cdot \quad \frac{1}{(p_\mu - k)^2 - m_W^2}\, \Gamma_\alpha\, \varepsilon^\alpha(K_\gamma)\, \frac{1}{(p_e - K)^2 - m_W^2}\, \gamma_\alpha(1 - \gamma^5)\, u\,(p_\mu) \quad ,$$

and

$$\Gamma_\alpha = (p_e + p_\mu - 2K)_\alpha \ , \quad K_\gamma = p_\mu - p_e \ , \quad \text{among other terms.}$$

In virtue of the orthogonality of the coefficients a_{jk} one has :

$$a_{31}\, a_{11} + a_{32}\, a_{12} + a_{33}\, a_{13} = 0$$

Therefore the sum of the propagators of the intermediate neutral leptons becomes

$$g^2 \left\{ \frac{\gamma.k}{k^2}\,(a_{31}\, a_{11} + a_{33}\, a_{13}) + \frac{\gamma.k + m_L}{k^2 - m_L^{\ 2}}\, a_{32}\, a_{12} \right\}$$

$$= \ g^2\, a_{32}\, a_{12}\, m_L^{\ 2}\, \frac{\gamma.k + \frac{k^2}{m_L}}{k^2(k^2 - m_L^{\ 2})}$$

If the additional neutral lepton were massless the amplitude $A(\mu \rightarrow e\,\gamma)$ would vanish.

One obtains for the ratio of the decay rates $\mu \to e\gamma$ and $\mu \to \nu_\mu + e + \bar{\nu}_e$:

$$R = \frac{\Lambda(\mu \to e\gamma)}{\Lambda(\mu \to \nu_\mu e \nu_e} \cong C \frac{\alpha}{24\pi} \left(\frac{a_{12}\, a_{32}}{a_{11}\, a_{33}} \right)^2 \left(\frac{m_L}{m_W} \right)^4$$

The coefficients a_{jk} can be estimated in connection with the electron and muon production in the reaction neutrino-nucleon collisions.

After this theory was proposed, experiments to find the reaction (IX. 1) led other physicists to rediscuss the model, as well as to suggest other mechanisms , such as the introduction of more than one doublet of Higgs fields (see Pontecorvo, ref. 142). Experiments, however, indicate that

$$R \lesssim 10.^{-9}$$

IX. 2 - SPECULATIONS ON LEPTON STRUCTURE*

Another theoretical possibility for the non-conservation of lepton flavour is given by the speculative assumption that leptons may interact with hadrons at a primary level. The non-observation of such interactions for leptons at the present energies might result from the existence of some suppression mechanism.

It is probably an unsatisfactory feature of the present theories that they assume that quarks and leptons are on the same theoretical level as fundamental constituents of matter. This assumption has helped to eliminate the triangle Adler anomalies. However, leptons share with hadrons the property of being observable and hadrons are assumed to be quark structures. If quarks

* The reader may omit this section.

are retained as fundamental objects of which matter is composed one might boldly assume that leptons also have a quark structure. In order to have cancellation of the baryon number for leptons these must then be structures composed of a quark-antiquark pair coupled in some manner with hypothetical fundamental particles such as heavy leptons [148-152].

If one assumes that an electron behaves as a structure of the type

$$e \sim (\bar{u}\ d\ L_1)$$

where L_1 is a convenient linear combination of three fundamental heavy leptons of nearly the same mass it may be argued that in this structure the binding forces are due to exchange of gluons between the quarks and of some exotic lepto-baryonic quanta between the heavy lepton and the quarks.

One may assume that the muon has a structure of the types

$$\mu \sim \bar{u}\ s\ L_2$$

and the tauon :

$$\tau \sim \bar{c}\ d\ L_1$$

In this way a correlation is established between leptons and hadronic bosons such as pions, kaons and D-mesons.

An effective lagrangean which makes possible to verify some consequences of these ideas is :

$$\mathcal{L} = g \left\{ (\bar{e}\ (1 - \gamma^5)\ L_1)\ \pi + (\bar{\mu}(1 - \gamma^5)\ L_2)\ K + \right.$$
$$\left. + (\bar{\tau}(1 - \gamma^5)\ L_1)\ D + \ldots \right\} + \text{h.c.}$$

and the choice of linear combinations of heavy leptons is designed to eliminate decays such as

$$K^{+} \rightarrow \pi^{+} + \mu^{+} + e^{-}$$

If one, however, considers that the pions interact with both electrons and muons through transition of the latter to orthogonal heavy lepton combinations L_1 and L_2 then it is possible to have a contribution to the gamma-decay of muons.

These speculations on lepton structure lead to the difficult question of the nature of the neutrinos.

But if quarks are not a part of leptons, one still probably needs to go down a step further in the question of lepton structure in order to achieve a satisfactory unification of leptons and hadrons. This question is, however, still very speculative and no experimental evidence brings so far any support at the presently known energies, for the existence of a quark content of leptons 148-162).

PROBLEMS

IX - 1. Given Dirac's equation for a fermion $\psi(x)$ with charge e in interaction with an electromagnetic field show that the field :

$$\psi^{c}(x) = C\, {}^{t}\bar{\psi}(x)$$

where C is a 4×4 antisymmetric unitary matrix such that :

$${}^{t}\gamma^{\mu} = - C^{-1} \gamma^{\mu} C$$

satisfies Dirac's equation for a fermion with opposite charge $-e$.

IX - 2. A global phase transformation on field operators $\psi(x)$ induces a transformation in Hilbert space $U(\alpha)$ with the charge Q as generator :

$$\psi \rightarrow \psi'(x) = e^{-i\alpha Q}\,\psi(x)\,e^{i\alpha Q} = e^{ie\alpha}\,\psi(x)$$

If $\mathscr{C}$ is the transformation in Hilbert space defining charge conjugation, show that :

$$\mathscr{C}^{-1}\,U(\alpha)\,\mathscr{C} = U^{-1}(\alpha)$$

IX - 3. Dirac's equation for a free fermion can have the form :

$$\left\{ \gamma^5 \left(-\,i\,\vec{\Sigma}\cdot\vec{\nabla}\right) + \beta\,m \right\} \psi(x) = i\,\frac{\partial\psi(x)}{\partial\,t}$$

from which one deduces :

$$-\,i\,(\vec{\Sigma}\cdot\vec{\nabla})\,\psi_L(x) - \beta\,m\,\psi_R(x) = -\,i\,\frac{\partial\,\psi_L(x)}{\partial\,t}$$

$$-\,i\,(\vec{\Sigma}\cdot\vec{\nabla})\,\psi_R(x) + \beta\,m\,\psi_L(x) = i\,\frac{\partial\,\psi_R(x)}{\partial\,t}$$

where

$$\vec{\Sigma} = \begin{pmatrix} \vec{\sigma} & 0 \\ 0 & \vec{\sigma} \end{pmatrix},$$

$$\psi_L = \frac{1}{2}\,(1-\gamma^5)\,\psi \;;\; \psi_R = \frac{1}{2}\,(1+\gamma^5)\,\psi$$

Consider the development :

$$\psi(x) = \frac{1}{(2\pi)^{3/2}} \int \frac{d^3p}{2p^0} \sum_{s=1,-1} \left\{ a(p,\,s)\,u(p,\,s)\,e^{-ipx} + \right.$$

$$\left. +\,b^+(p,\,s)\,v\,(p,\,s)\,e^{ipx} \right\}$$

a) Obtain the equations for $u(p,\,s)$ and $v(p,\,s)$ for $m = 0$ in terms of $\vec{\Sigma}\cdot\vec{p}$ and γ^5, $p^0 = +\,|\vec{p}|$.

b) If s is the eigenvalue of the helicity operator $\dfrac{\vec{\Sigma}\cdot\vec{p}}{|\vec{p}|}$ what are the

the values of s for $\frac{1}{2}(1 - \gamma^5)\, u\,(p, s)$ and $\frac{1}{2}(1 - \gamma^5)\, v\,(p, s)$?

c) Show as a result of b), that a zero mass particle described by ψ_L is either a left-handed particle (with negative helicity) or a righ-handed antiparticle (with positive helicity).

d) What are the charge conjugates of ψ_L and ψ_R ?

IX - 4. Show that the two-component neutrino theory is equivalent to the theory of Majorana [a Majorana field being defined by :

$$\psi^c = \eta\, \psi$$

where $\eta = \pm 1$; a two-component neutrino is described by :

$$\psi_L = \frac{1}{2}(1 - \gamma^5)\, \psi \quad \text{with} \quad \gamma^5 = \begin{pmatrix} -I & 0 \\ 0 & I \end{pmatrix}$$]

IX - 5. Work out the Salam-Weinberg model with the fermions (IX. 1), (IX. 2), (IX. 3), (IX. 3a).

IX - 6. Concerning possible new leptons, read M.L. Perl, $e^+ e^-$ physics today and tomorrow, SLAC-Pub-2615 (1980) and J. Leite Lopes, J.A. Martins Simoẽs and D. Spehler, ref. 162 and literature quoted there.

CHAPTER X

Attempts at a "Grand" Unification: The SU(5) Model

X. 1 - THE SU(5) GAUGE FIELDS AND GENERATORS

If the strong interactions are correctly described by chromodynamics, which is a theory based on the exact colour SU(3) gauge symmetry, and if weak and electromagnetic interactions are correctly incorporated in the SU(2) ⊗ U(1) gauge theory, it is natural to ask if and how one can have a unifying theory of strong, electromagnetic and weak interactions. This is the so-called grand unification.

In chromodynamics one single coupling constant is introduced, call it g_s, in the definition of the covariant derivative acting on the quark operator (IV. 32)

$$D_\mu\, q(x) = (\partial_\mu + i\, g_s\, A_{\mu,k}(x)\, \frac{\lambda_k}{2}\,)\; q(x)$$

Eight massless vector gauge fields $A_{\mu,k}(x)$ are thereby introduced associated to the eight generators of the SU(3) group.

In the Salam-Weinberg theory we need two coupling constants, call them g_1 and g_2, in the definition of the covariant derivative which introduces a set of three gauge vector fields $\mathcal{a}_{\mu,a}$ and another gauge field B_μ (see VIII. 6). Here, therefore, there is no real unification in the sense that one puts from the beginning two coupling constants in the model and four massless gauge fields associated to the four generators of the semi-simple SU(2) ⊗ U(1) group. Spontaneous symmetry breaking develops mass for three of these fields and the result is the theory of weak and electromagnetic forces with U(1) gauge invariance.

Clearly for a "grand" unification we want a group G which will be larger than the product of those separately associated to the strong and to the weak and electromagnetic interactions :

$$SU(3)_c \otimes SU(2) \otimes U(1) \subset G$$

The direct product is of rank 4, i.e, it will have four diagonal matrices, then the rank of G will be at least four. The group considered by Georgi and Glashow SU(5), has rank four and twenty-four generators and it is the minimal to satisfy the requirements, as we shall see. To these, therefore, it is possible to associate twenty-four massless gauge vector fields and one single coupling constant, so that if ψ represents a multiplet belonging to a SU(5) representation one will have :

$$D_\mu \psi = (\partial_\mu + i\, g_0\, \mathcal{G}_{\mu,k}\, T_k)\, \psi \qquad (X.\ 1)$$

for the covariant derivative, where T_k are the generators in this representation. The twenty-four fields must comprise the ones we have seen before namely the gluon fields and W^+, W^-, Z and γ fields. As there exist (or we assume to exist) (at least) three families of basic fermions, namely

$$u_1, d_1 \;;\; u_2, d_2 \;;\; u_3, d_3 \;;\; \nu_e, e$$

$$c_1, s_1 \;;\; c_2, s_2 \;;\; c_3, s_3 \;;\; \nu_\mu, \mu$$

$$t_1, b_1 \;;\; t_2, b_2 \;;\; t_3, b_3 \;;\; \nu_\tau, \tau$$

we study only the generation of one family after which the lagrangean will be taken as the sum of terms of all families.

The SU(5) matrices will act on a basic quintuplet ; we will take this to be formed of the three right-handed colour components of a given quark flavour of the first family, call them q_1, q_2, q_3 plus the right-handed positron and electronic-antineutrino :

$$\psi_2 = \begin{pmatrix} q_1 \\ q_2 \\ q_3 \\ e^c \\ -\nu_e^c \end{pmatrix}_R \qquad (X.\ 2)$$

This is because in the Salam-Weinberg model we postulated the left-handed doublet $\begin{pmatrix} \nu_e \\ e \end{pmatrix}_L$ since neutrinos are left-hand polarized. So antineutrinos are right-handed and therefore its lepton companion must be e^c_R, c designates charge conjugate.

The negative sign in ν^e_c comes from the following : as seen in § VIII. 11 for the Higgs doublet, given a doublet like

$$\begin{pmatrix} \nu_e \\ e \end{pmatrix}_L$$

then the charge conjugate of this doublet which transforms in the same manner under the SU(2) group is :

$$\hat{C} \begin{pmatrix} \nu_e^{\,c} \\ e^c \end{pmatrix}_R$$

where C is a 2 x 2 matrix such that :

$${}^t\vec{\tau} = - C^{-1} \, \vec{\tau} \, C$$

$${}^tC = - C$$

$$C^+ = C^{-1}$$

If one chooses

$$C = i \, \tau_2$$

then we have :

$$C \begin{pmatrix} \nu_e^{\,c} \\ e^c \end{pmatrix}_R = \begin{pmatrix} 0 & 1 \\ -1 & 0 \end{pmatrix} \begin{pmatrix} \nu_e^{\,c} \\ e^c \end{pmatrix} = \begin{pmatrix} e^c \\ -\nu_e^{\,c} \end{pmatrix}$$

Now the covariant derivative acting on (X. 2) will be of the form (X. 1) where $\mathcal{G}_{\mu,k}(x)$, k = 1, ... 24 are 24 gauge vector fields and the T_k are the generators in the five dimensional representation. These are of the form :

$$T_k = \frac{1}{2} t_k \qquad (X.\ 3)$$

where :

$$t_a = \left(\begin{array}{ccc|cc} & & & 0 & 0 \\ & \lambda_a & & 0 & 0 \\ & & & 0 & 0 \\ \hline 0 & 0 & 0 & 0 & 0 \\ 0 & 0 & 0 & 0 & 0 \end{array}\right) \qquad \text{for} \quad a = 1, \dots 8$$

and λ_a are the SU(3) matrices given in (IV. 28) :

$$t_b = \left(\begin{array}{cccc|c} & & & & 0 \\ & & & & 0 \\ & \eta_b & & & 0 \\ & & & & 0 \\ \hline 0 & 0 & 0 & 0 & 0 \end{array}\right) \qquad \text{for} \quad b = 9, \dots 14$$

and η_b are the SU(4) matrices :

$$\eta_9 = \begin{pmatrix} 0 & 0 & 0 & 1 \\ 0 & 0 & 0 & 0 \\ 0 & 0 & 0 & 0 \\ 1 & 0 & 0 & 0 \end{pmatrix}, \ \eta_{10} = \begin{pmatrix} 0 & 0 & 0 & -i \\ 0 & 0 & 0 & 0 \\ 0 & 0 & 0 & 0 \\ i & 0 & 0 & 0 \end{pmatrix}, \ \eta_{11} = \begin{pmatrix} 0 & 0 & 0 & 0 \\ 0 & 0 & 0 & 1 \\ 0 & 0 & 0 & 0 \\ 0 & 1 & 0 & 0 \end{pmatrix}$$

$$\eta_{12} = \begin{pmatrix} 0 & 0 & 0 & 0 \\ 0 & 0 & 0 & -i \\ 0 & 0 & 0 & 0 \\ 0 & i & 0 & 0 \end{pmatrix}, \ \eta_{13} = \begin{pmatrix} 0 & 0 & 0 & 0 \\ 0 & 0 & 0 & 0 \\ 0 & 0 & 0 & 1 \\ 0 & 0 & 1 & 0 \end{pmatrix}, \ \eta_{14} = \begin{pmatrix} 0 & 0 & 0 & 0 \\ 0 & 0 & 0 & 0 \\ 0 & 0 & 0 & -i \\ 0 & 0 & i & 0 \end{pmatrix} \ ;$$

and then :

$$t_{15}=\frac{1}{\sqrt{6}}\begin{pmatrix}1&0&0&0&0\\0&1&0&0&0\\0&0&1&0&0\\0&0&0&-3&0\\0&0&0&0&0\end{pmatrix},\ t_{16}=\begin{pmatrix}0&0&0&0&1\\0&0&0&0&0\\0&0&0&0&0\\0&0&0&0&0\\1&0&0&0&0\end{pmatrix},\ t_{17}=\begin{pmatrix}0&0&0&0&-i\\0&0&0&0&0\\0&0&0&0&0\\0&0&0&0&0\\i&0&0&0&0\end{pmatrix}$$

$$t_{18}=\begin{pmatrix}0&0&0&0&0\\0&0&0&0&1\\0&0&0&0&0\\0&0&0&0&0\\0&1&0&0&0\end{pmatrix},\ t_{19}=\begin{pmatrix}0&0&0&0&0\\0&0&0&0&-i\\0&0&0&0&0\\0&0&0&0&0\\0&i&0&0&0\end{pmatrix}$$

$$t_{20}=\begin{pmatrix}0&0&0&0&0\\0&0&0&0&0\\0&0&0&0&1\\0&0&0&0&0\\0&0&1&0&0\end{pmatrix},\ t_{21}=\begin{pmatrix}0&0&0&0&0\\0&0&0&0&0\\0&0&0&0&-i\\0&0&0&0&0\\0&0&i&0&0\end{pmatrix}\qquad (X.\ 3)$$

$$t_{22}=\begin{pmatrix}0&0&0&0&0\\0&0&0&0&0\\0&0&0&0&0\\0&0&0&0&1\\0&0&0&1&0\end{pmatrix},\ t_{23}=\begin{pmatrix}0&0&0&0&0\\0&0&0&0&0\\0&0&0&0&0\\0&0&0&0&-i\\0&0&0&i&0\end{pmatrix}$$

$$t_{24}=\frac{1}{\sqrt{10}}\begin{pmatrix}1&0&0&0&0\\0&1&0&0&0\\0&0&1&0&0\\0&0&0&1&0\\0&0&0&0&-4\end{pmatrix}$$

There are four diagonal matrices, t_3, t_8, t_{14} and t_{24} and all the 24 matrices are traceless.

Clearly, we want that out of the twenty four gauge fields, eight be identified with the eight gluon massless fields, one must be the massless electromagnetic field and, after spontaneous symmetry breaking, three fields must give the weak W and Z fields. There remain twelve other fields which are related to very weak lepton-quark transitions predicted in this theory.

But first let us identify the quark flavor in q_1, q_2, q_3. As the electromagnetic field must be one of the fields $\mathcal{G}_{\mu,k}(x)$ then the generator T_k associated to it will be the charge Q which then will also be traceless. Applied to the quintuplet (X. 2) the diagonal elements of Q will give :

$$3\,Q_q + Q_{e^c} = 0$$

hence :

$$Q_q = -\frac{1}{3}$$

Thus the flavor of the quark q is that of the d-quark :

$$\psi_R = \begin{pmatrix} d_1 \\ d_2 \\ d_3 \\ e^c \\ -\nu^c \end{pmatrix}_R \qquad \text{(X. 2a)}$$

As we want that Higgs fields generate a mass for d_1, d_2, d_3 and for the electron, we shall have to consider besides these right-handed components also left-handed components for these fields. We must also introduce the u fields which also acquire a mass. Therefore we must consider 15 operators namely :

$$d_{1R},\ d_{2R},\ d_{3R},\ d_{1L},\ d_{2L},\ d_{3L}$$

$$u_{1R},\ u_{2R},\ u_{3R},\ u_{1L},\ u_{2L},\ u_{3L}$$

e_R, e_L and ν_L (or alternatively $e^c{}_R$, $e^c{}_L$, $\nu^c{}_R$)

As five of these operators are in the right-handed quintuplet (X. 2a) we need a left-handed decuplet to account for the remaining fermi fields.

With the quintuplet we can construct an irreducible ten-dimensional representation of SU(5) obtained by making the antisymmetrized product of two quintuplets. Thus if we call

$$\psi = \begin{pmatrix} \psi_1 \\ \psi_2 \\ \psi_3 \\ \psi_4 \\ \psi_5 \end{pmatrix} \qquad \text{(X. 4)}$$

then this antisymmetric product will be :

$$\psi_{ab} = \frac{1}{\sqrt{2}} (\psi_a(1)\ \psi_b(2) - \psi_a(2)\ \psi_b(1)) \ ; \ a,\ b = 1,\ \ldots\ 5 \qquad \text{(X. 5)}$$

and will have ten independent components. The first three ψ_a, $a = 1, 2, 3$ are a colour triplet, therefore ψ_{ab} for $a, b = 1, 2, 3$ represents a product of two SU(3) triplets which decompose as (see Kokedee, ref. 51) :

$$3 \otimes 3 = 6 + \bar{3}$$

and the sextuplet is symmetric and the antitriplet antisymmetric. Thus the ψ_{ab} for a, b = 1, 2, 3 constitute an anti-colour triplet. These also have charge $-\frac{2}{3}$ therefore we may identify them with the anti u-quarks. We shall put :

$$\psi_{ab} = \varepsilon_{abk} \, (u^c{}_k)_L \qquad ; \qquad a, b = 1, 2, 3$$

since the charge conjugate of a right-handed spinor is left-handed : $(\psi_R)_c = (\psi_c)_L$ where c means charge-conjugate.

Also if we take ψ_{a4} these will represent a colour triplet with charge $\frac{2}{3}$ whereas ψ_{a5} will represent a colour triplet with charge $-\frac{1}{3}$. Therefore we identify :

$$\psi_{a4} = (u_a)_L \qquad ; \qquad a = 1, 2, 3$$

$$\psi_{a5} = (d_a)_L \qquad ; \qquad a = 1, 2, 3$$

ψ_{45} will be colour-singlet and isospin singlet, therefore as its charge is + 1 we identify it with $e^c{}_L$:

$$\psi_{45} = e^c{}_L$$

Hence, besides the quintuplet (X. 2) we postulate the following decuplet (left-handed) :

$$\psi_L = \frac{1}{\sqrt{2}} \begin{pmatrix} 0 & u^c{}_3 & -u^c{}_2 & u_1 & d_1 \\ -u^c{}_3 & 0 & u^c{}_1 & u_2 & d_2 \\ u^c{}_2 & -u^c{}_1 & 0 & u_3 & d_3 \\ -u_1 & -u_2 & -u_3 & 0 & e^c \\ -d_1 & -d_2 & -d_3 & -e^c & 0 \end{pmatrix}_L$$

Now with these objects we construct immediately the corresponding lagrangean piece which is gauge invariant namely :

$$\mathscr{L}_o = (\bar{\psi}_R)_a \, i\, \gamma^\mu (D_\mu)_{aa'} (\psi_R)_{a'} + (\bar{\psi}_L)_{ab} \, i\, \gamma^\mu (D_\mu)_{ab;a'b'} (\psi_L)_{a'b} =$$

$$= (\bar{\psi}_R)_a \, i\, \gamma^\mu \left[\partial_\mu \, \delta_{aa'} + i\, g_o \, \mathscr{G}_{\mu,k} \, (T_k)_{aa'} \right] (\psi_R)_{a'} \; +$$

$$+ (\bar{\psi}_L)_{ab} \, i\, \gamma^\mu \left[\partial_\mu \, \delta_{ab;a'b'} + i\, g_o \, \mathscr{G}_{\mu,k} \, (T_k)_{ab;a'b'} \right] (\psi_L)_{a'b'}$$

Here the first term contains the generator matrices $(T_k)_{aa'}$ in the five-dimensional representation which are the ones given in (X. 3). The second term, involving ψ_L contains the generator matrices in the ten-dimensional representation $(T_k)_{ab;a'b'}$. Now an infinitesimal gauge transformation on the quintuplet (X. 4) is given by :

$$\psi'_a = \left[\delta_{ab} + i\, g_o \, \Lambda_k (T_k)_{ab} \right] \psi_b \; ; \; T_k = \frac{t_k}{2}$$

therefore this transformation induces the following one on the decuplet (X. 5)

$$\psi'_{ab} = \frac{1}{2} \Big\{ \left[\delta_{aa'} + i\, g_o \, \Lambda_k (T_k)_{aa'} \right] \left[\delta_{bb'} + i\, g_o \, \Lambda_k (T_k)_{bb'} \right] -$$

$$- \left[\delta_{ba'} + i\, g_o \, \Lambda_k (T_k)_{ba'} \right] \left[\delta_{ab'} + i\, g_o \, \Lambda_k (T_k)_{ab'} \right] \Big\} \psi_{a'b'}$$

If we compare this with the transformation :

$$\psi'_{ab} = \left[\delta_{aa'} \, \delta_{bb'} + i\, g_o \, \Lambda_k (T_k)_{ab;a'b'} \right] \psi_{a'b'}$$

we find :

$$(T_k)_{ab,a'b'} = \frac{1}{2}\Big\{ \delta_{aa'}(T_k)_{bb'} + \delta_{bb'}(T_k)_{aa'} -$$

$$- \delta_{ab'}(T_k)_{ba'} - \delta_{ba'}(T_k)_{ab'} \Big\}$$

and :

$$\bar{\psi}_{ab}(T_k)_{ab;a'b'}\ \psi_{a'b'} = 2\ \bar{\psi}_{ab}(T_k)_{aa'}\ \psi_{a'b} = \bar{\psi}_{ab}(t_k)_{aa'}\ \psi_{ab'}$$

This then indicates how to operate with the ten-dimensional representation of the generators.

The interaction of the fermion fields with the gauge fields is contained in the covariant derivative terms ; it is

$$\mathscr{L}_{int} = - g_0 \Big\{ (\bar{\psi}_R)_a\ \gamma^\mu\ (\frac{t_k}{2})_{aa'}(\psi_R)_{a'} +$$

$$+ (\bar{\psi}_L)_{ab}\ \gamma^\mu (T_k)_{ab;a'b'}\ (\psi_L)_{a'b'} \Big\}\ \mathscr{G}_{\mu,k}$$

We expect that the fields from $k = 1$ to $k = 8$ describe the gluon fields.

Given the expression of T_{15} in (X. 3) in the 5-dimensional representation we see that

$$- g_0(\bar{\psi}_R\ \gamma^\mu\ T_{15}\ \psi_R)\ \mathscr{G}_{\mu,15} = - \frac{g_0}{2\sqrt{6}}\left[\bar{\psi}_R\ \gamma^\mu \begin{pmatrix} 1 & 0 & 0 & 0 & 0 \\ 0 & 1 & 0 & 0 & 0 \\ 0 & 0 & 1 & 0 & 0 \\ 0 & 0 & 0 & -3 & 0 \\ 0 & 0 & 0 & 0 & 0 \end{pmatrix} \psi_R\right] \mathscr{G}_{\mu,15}$$

Now if we identify the field $\mathcal{G}_{\mu,15}$ with the electromagnetic field :

$$\mathcal{G}_{\mu,15} = A_\mu$$

then we should have :

$$\frac{g_0}{2\sqrt{6}} = \frac{e}{3}$$

that is :

$$\frac{1}{2}\, g_0 = \sqrt{\frac{2}{3}}\; e$$

so that :

$$- g_0(\bar{\psi}_R\, \gamma^\mu\, T_{15}\, \psi_R)\, \mathcal{G}_{\mu,1} = e\left[\bar{\psi}_R\, \gamma^\mu \begin{pmatrix} -1/3 & 0 & 0 & 0 & 0 \\ 0 & -1/3 & 0 & 0 & 0 \\ 0 & 0 & -1/3 & 0 & 0 \\ 0 & 0 & 0 & 1 & 0 \\ 0 & 0 & 0 & 0 & 0 \end{pmatrix} \psi_R\right] A_\mu$$

where ψ_R is given by (X. 2a).

The constant g_0 is thus determined.

The decuplet term will contribute with the left-handed part of the current so as to give the correct electromagnetic interaction of the quarks u, d and the electron. Given the term corresponding to k = 24 we identify $\mathcal{G}_{\mu,24}$ with the neutral meson field Z_μ. For the sector electron-neutrino this must give a term of the form (VIII. 41) :

$$- \frac{e}{\sin\theta_W \cos\theta_W} \left\{ \bar{L}\, \gamma^\mu\, \frac{\tau_3}{2}\, L - \sin^2\theta_W\, j^\mu_{(\gamma)} \right\} Z_\mu$$

If we call

$$T_{3L} = \int \bar{L} \gamma^o \frac{\tau_3}{2} L \, d^3x$$

we have for the neutral charge T^Z :

$$T^Z = T_{3L} - \sin^2 \theta_W Q$$

Thus as

$$Tr (T_k T_{k'}) = \frac{1}{2} \delta_{kk'}$$

we have that :

$$Tr (T^Z Q) = Tr (T_{3L} Q - \sin^2 \theta_W Q^2) = 0$$

whence

$$since^2 \theta_W = \frac{Tr(T_{3L}{}^2)}{Tr(Q^2)}$$

where :

$$Q = T_{3L} + \frac{Y}{2}$$

As the trace is taken over the quintuplet we have (see (X.3)) :

$$\sin^2 \theta_W = \frac{\frac{1}{2}}{1 + 3 \times \frac{1}{9}} = \frac{3}{8}$$

The value of the Weinberg angle is thus fixed in the region of exact symmetry by the condition of grand unification. This value is different from the one obtained in the Salam-Weinberg model because in the latter case spontaneous symmetry breaking has been introduced and the final effective lagrangean is used in determining the value indicated at the end of paragraph VIII. 8.

Clearly, in obtaining the value $\frac{3}{8}$ for $\sin^2 \theta_W$ we should remember that we are still in the exact symmetry limit where all constants, like g_1, g_2 and g_s are equal to g_0 and therefore $\sin^2 \theta_W$ has this limiting value, and this happens at very high energies ($\sim 10^{15}$ GeV).

X. 2 - HIERARCHY OF SPONTANEOUSLY BROKEN SYMMETRIES - LEPTO QUARK BOSONS

The next step is to try and introduce Higgs fields in order to implement spontaneous symmetry breaking procedures which will give rise to masses and to the final physical particles and coupling constants.

Let us first look at the couplings of the twenty-four gauge fields with the fermions, for instance with the quintuplet (X. 2). For the transition of a d quark from a colour state a to a colour state b we have an interaction with the gluon fields (Fig. X. 1) :

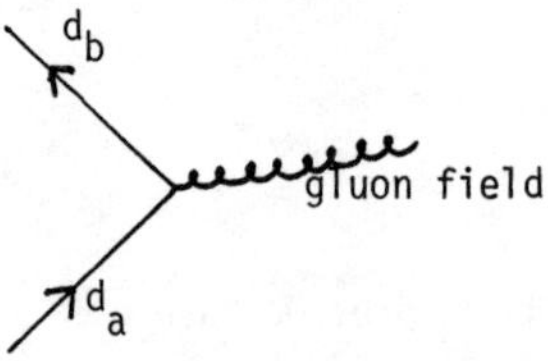

Figure X. 1

for a transition between the same colour states we have an additional interaction with photons or with Z^0 bosons (Fig. X. 2) :

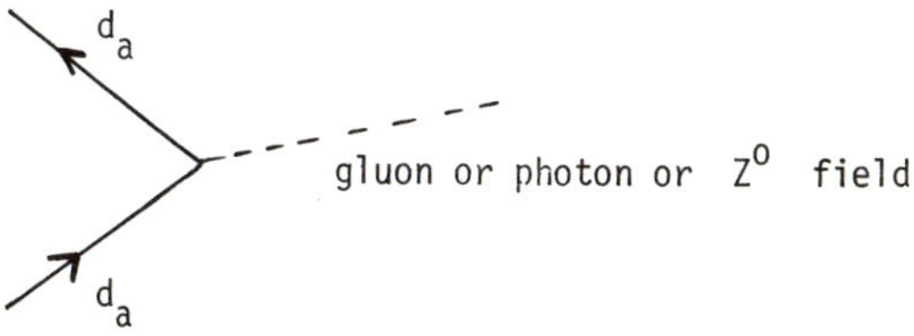

Figure X. 2

We also have the usual transitions between lepton states which involve interactions with the W, Z and γ fields, namely (Fig. X. 3):

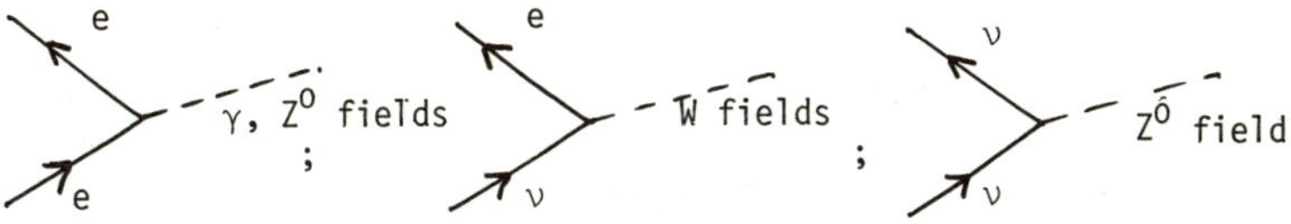

Figure X. 3

But the lagrangean also gives rise to a transition from a quark to a lepton with interaction with vector fields which one may call $X_a(\frac{4}{3})$ and $Y_a(\frac{1}{3})$, namely (Fig. X. 4):

Figure X. 4

that is, an antineutrino may emit a vector field $Y_a(\frac{1}{3})$ with colour a and charge $\frac{1}{3}$ to give a coloured quark d_a(with charge $-\frac{1}{3}$). And a positron may emit a vector field $X_a(\frac{4}{3})$ with colour a and charge $\frac{4}{3}$ to give a quark d_a. These vector fields, required by the unifying SU(5) group, carry colour, electric charge as well as leptonic and baryonic number. We still have the following possible interactions

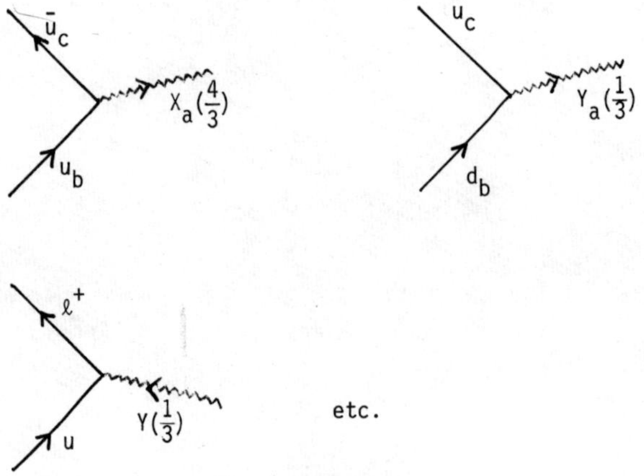

Figure X. 5

These interactions clearly imply <u>instability of the proton</u> since the following decays would be possible (note the conservation of B-L in the SU(5) model although $\Delta B \neq 0$, $\Delta L \neq 0$) :

$$p \rightarrow e^+ + \pi^0$$

$$p \rightarrow \bar{\nu}_e + \pi^+$$

$$p \rightarrow \mu^+ + K^0$$

$$n \rightarrow \bar{\nu}_e + \pi^0$$

etc.

As we must require that the lifetime of the proton be extremely large, we see that the mass of the mesons X and Y must be extremely large. Thus, possible diagrams for the proton decay are the following (Fig. X. 6)

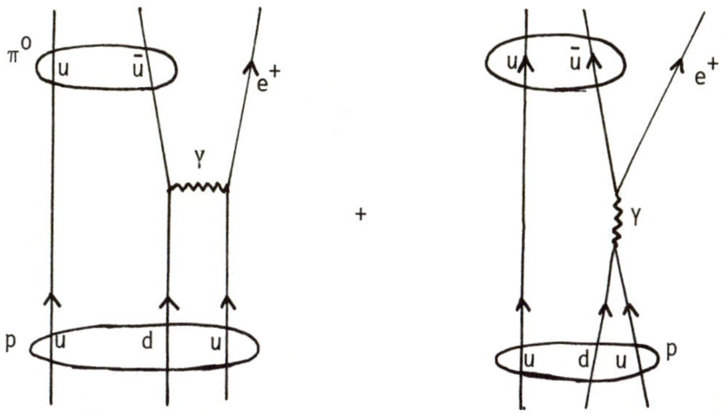

Figure X. 6

and it is seen that the Y-meson propagator enters these diagrams, therefore the decay rate is expected to be proportional to $(M_y)^{-4}$.

So the Higgs mechanism must give very large masses to the X and Y mesons in order to suppress this decay ; these have been estimated to be of the order of 10^{15} GeV -these masses give rise to a proton lifetime of the order 10^{31} years. This mechanism then should be carried out in two steps. The first step makes a transition from the group SU(5) to the direct product $SU(3)_c \otimes SU(2) \otimes U(1)$ and conserves the rank of the group. For this procedure, a set of 24 Higgs fields is required, which are postulated to have a non-vanishing vacuum expectation value. This first spontaneous symmetry breaking gives masses to the X and Y mesons but keeps the other twelve fields massless, that is the symmetry $SU(3)_c \otimes SU(2) \otimes U(1)$ is still conserved.

In a second step, another set of five Higgs fields is introduced which will give a mass to the W^+, W^- and Z^0 fields. The energies corresponding to the masses of the X and Y mesons are of the order 10^{15} GeV, those for the W, Z mesons, as we know, are of order 100 GeV.

Let us see briefly how the first step may be carried out. If $\varphi_a(x)$ is the set of twenty-four scalar fields, the covariant derivatives applied to this set are :

$$(D_\mu \varphi)_a = (\partial_\mu \delta_{aa'} + i\, g_0\, \mathscr{G}_{\mu,k} (\mathscr{F}_k)_{aa'}) \cdot \varphi_{a'}$$

where now the $(\mathscr{F}_k)_{aa'}$ are twenty-four matrices, 24 x 24, which act on the twenty-four φ_a.

As φ_a transforms like $\bar{\psi}\, \frac{t_a}{2}\, \psi$, ψ being the quintuplet (X. 4), we deduce from a comparison between :

$$\varphi_a \sim \bar{\psi}\, \frac{t_a}{2}\, \psi \rightarrow \bar{\psi}_\alpha \left[\delta_{\alpha\alpha'} - i\, g_0\, \Lambda_k \left(\frac{t_k}{2}\right)_{\alpha\alpha'}\right] \left(\frac{t_a}{2}\right)_{\alpha'\beta'} \cdot$$

$$\cdot \left[\delta_{\beta'\beta} + i\, g_0\, \Lambda_\ell \left(\frac{t_\ell}{2}\right)_{\beta'\beta}\right] \psi_\beta$$

and

$$\varphi_a \sim (\delta_{ab} + i\, g_0\, \Lambda_n\, (\mathcal{F}_n)_{ab})\, \varphi_b$$

the following reationship between the two representations of T :

$$\bar{\psi}_\alpha \left(\frac{t_a}{2} \right)_{\alpha\beta} \psi_\beta - i\, g_0\, \Lambda_k\, \bar{\psi}_\alpha \left(\frac{t_k}{2} \right)_{\alpha\alpha'} \left(\frac{t_a}{2} \right)_{\alpha'\beta} \psi_\beta \; +$$

$$+\, i\, g_0\, \Lambda_k\, \bar{\psi}_\alpha \left(\frac{t_a}{2} \right)_{\alpha\beta'} \left(\frac{t_k}{2} \right)_{\beta'\beta} \psi_\beta \;\sim\; \varphi_a + i\, g_0\, \Lambda_n\, (\mathcal{F}_n)_{ab}\, \varphi_b$$

that is

$$(\mathcal{F}_n)_{ab}\, \varphi_b \;\approx (\mathcal{F}_n)_{ab}\, \bar{\psi}_\alpha \left(\frac{t_b}{2} \right)_{\alpha\beta} \psi_\beta \sim \bar{\psi}_\alpha \left[\frac{t_a}{2} , \frac{t_n}{2} \right]_{\alpha\beta} \psi_\beta$$

therefore

$$(\mathcal{F}_n)_{ab} \sim \frac{1}{4}\; \left[t_a, t_n \right]_{\alpha\beta}\, (t_b)_{\beta\alpha} = \frac{1}{4}\, \mathrm{Tr}\Big(\left[t_a, t_n \right] t_b \Big)$$

The Higgs fields may be replaced by the 5 x 5 traceless matrices

$$\phi = \sum_{a=1}^{24} \varphi_a\, \frac{t_a}{2}$$

so that a term like

$$\frac{1}{2} \sum_{a=1}^{24} \partial_\mu \varphi_a\, \partial^\mu \varphi_a$$

may represented by :

$$\mathrm{Tr}\; \partial_\mu\, \phi\, \partial^\mu\, \phi$$

From the commutation relations between the matrices t which define the structure constants of the group, f_{abc} :

$$\left[\frac{t_a}{2}\ ,\ \frac{t_b}{2}\right] = i\ f_{abc}\ \frac{t_c}{2}$$

we deduce that

$$(\mathcal{F}_n)_{ab} \sim i\ f_{anb}$$

The kinetic gauge invariant term of the Higgs fields will be

$$\mathrm{Tr}\ (D^{\mu}\ \phi)^{+}\ \ (D_{\mu}\ \phi)$$

where

$$D_{\mu}\ \phi = \partial_{\mu}\ \phi + i\ g_0\left[\mathcal{G}_{\mu k}\frac{t_k}{2}\ ,\ \phi\right].$$

We see that when we make ϕ to acquire a non-vanishing vacuum expectation value

$$<0\,|\ \phi\ |\,0> \equiv a$$

the mass matrix of the vector fields will be of the type :

$$g_0{}^2\mathrm{Tr}\left(\left[\mathcal{G}_{\mu,k}\ \frac{t_k}{2}\ , a\right]\left[\ \mathcal{G}^{\mu}{}_{\ell}\ \frac{t_\ell}{2}\ , a\ \right]\right) = (m^2)_{k\ell}\ \mathcal{G}^{\mu}{}_{k}\,\mathcal{G}^{\mu}{}_{\ell}$$

The mass acquired by the vector fields thus depends on the commutator between the 5 x 5 matrix $a \equiv <0\,|\ \phi\ |\,0>$ and the generators t_k.

As we want that the gluon fields remain massless we want that

$$\left[t_{\ell}, a \right] = 0 \quad \text{for} \quad \ell = 1, \ldots 8$$

This will be ensured if $\underline{a}$ has the unit matrix as its first 3×3 sub-matrix :

$$a = \left(\begin{array}{ccc|ccc} 1 & 0 & 0 & & & \\ 0 & 1 & 0 & & & \\ 0 & 0 & 1 & & & \\ \hline & & & & & \\ & & & & & \\ & & & & & \end{array} \right)$$

since

$$t_j = \left(\begin{array}{ccc|cc} & & & 0 & 0 \\ & \lambda_j & & 0 & 0 \\ & & & 0 & 0 \\ \hline 0 & 0 & 0 & 0 & 0 \\ 0 & 0 & 0 & 0 & 0 \end{array} \right) , \quad j = 1, \ldots 8$$

Also, as the electromagnetic field was identified as $\mathcal{G}_{\mu,15}$ we require

$$\left[t_{15}, a \right] = 0$$

for the photon mass to vanish.

And if also the mass of Z^0 is zero at this stage of spontaneous symmetry breaking then :

$$\left[t_{24}, a \right] = 0$$

We conclude that $\mathcal{a}$ must be of the form :

$$\mathcal{a} = \mu\, t_{15} + \nu\, t_{24}$$

where μ and ν are constants that is :

$$\mathcal{a} = \begin{pmatrix} \frac{\mu}{\sqrt{6}} + \frac{\nu}{\sqrt{10}} & 0 & 0 & 0 & 0 \\ 0 & \frac{\mu}{\sqrt{6}} + \frac{\nu}{\sqrt{10}} & 0 & 0 & 0 \\ 0 & 0 & \frac{\mu}{\sqrt{6}} + \frac{\nu}{\sqrt{10}} & 0 & 0 \\ 0 & 0 & 0 & -\frac{3\mu}{\sqrt{6}} + \frac{\nu}{\sqrt{10}} & 0 \\ 0 & 0 & 0 & 0 & -\frac{4\nu}{\sqrt{10}} \end{pmatrix}$$

If we call

$$-\frac{3\mu}{\sqrt{6}} + \frac{\nu}{\sqrt{10}} = \alpha + \beta$$

$$-\frac{4\nu}{\sqrt{10}} = \alpha - \beta$$

then we can write :

$$\mathcal{a} = \begin{pmatrix} -\frac{2\alpha}{3} & 0 & 0 & 0 & 0 \\ 0 & -\frac{2\alpha}{3} & 0 & 0 & 0 \\ 0 & 0 & -\frac{2\alpha}{3} & 0 & 0 \\ 0 & 0 & 0 & \alpha + \beta & 0 \\ 0 & 0 & 0 & 0 & \alpha - \beta \end{pmatrix}$$

The sub-matrix

$$\begin{pmatrix} \alpha + \beta & 0 \\ 0 & \alpha - \beta \end{pmatrix}$$

however would imply a mass for the W mesons. As we do not want this to happen at this stage we assume

$$\beta = 0$$

The masses of the X and Y mesons are then proportional to $g_0^2 \alpha^2$ and we postulate this mass to be extremely large

$$m_X \sim 10^{15} \text{ GeV}$$

the vacuum expectation value of the Higgs fields at this stage must be tremendously large.

The second step of spontaneous symmetry breaking will be achieved with the introduction of five other Higgs fields and will give rise to the smaller-masses, of the order of 100 GeV, of the W and Z bosons.

X. 3 - CONCLUDING REMARKS

We shall not proceed further with the discussion of the SU(5) model, still under study at present and urge the interested reader to consult the literature on the subject [163-172].

As we have seen, in this model, as in all models so far, one assumes as fundamental objects a certain number of fermions -quarks and leptons- which are grouped, for no known reasons, into families. These objects interact with basic gauge vector fields and with Higgs scalar fields, according to the prescription of a "grand" gauge group. This group is SU(5) in the model proposed by Georgi and Glashow ; it is SU(4) ⊗ SU(4) in the unifying model of Pati and Salam (in this model twelve integrally charged quarks, four flavors with three colours each, and four leptons are placed in a $(4, \bar{4})$ representation of the group SU(4) ⊗ SU(4)).

The spontaneous symmetry breaking is carried out as if there existed a hierarchy of broken symmetries. The "grand" group is broken down to another one by Higgs fields which acquire an extremely large vacuum expectation value and give rise to tremendously heavy vector mesons. Then the latter group is broken down to a succeeding one with another set of Higgs fields which give rise to less heavy vector mesons.

A prediction of these unifying theories is the instability of the proton since they put quarks and leptons in the same multiplet (this is also true in the unifying model proposed by the author on lepton structure [148]). The superheavy vector mesons help to suppress this decay.

As we saw for the case of the SU(5) model, there is an enormous gap between the masses of the X and Y mesons and those of the W and Z mesons. Are there no other meson masses in between ?

Of course, the question of the occurrence of quarks and leptons into families is still unexplained and the total number of supposedly basic fermions is now twenty-four (if t quarks exist). The question of a possible structure for these basic fermions is now the subject of investigation [148-184].

There also remains the question of the unification of the strong, electromagnetic and weak interactions with the gravitational field. Supersymmetry [179-] and supergravity theories which started from the attempt at deeply correlating and unifying fermion fields and boson fields, are being developed and point out to a possible super unification of the forces in nature, bringing into the picture gravitons and possibly massless spin - 3/2 particles (Table X.1).

Such a possibility of spin 3/2 basic particles [162] is also envisaged if leptons as well as quarks have a structure.

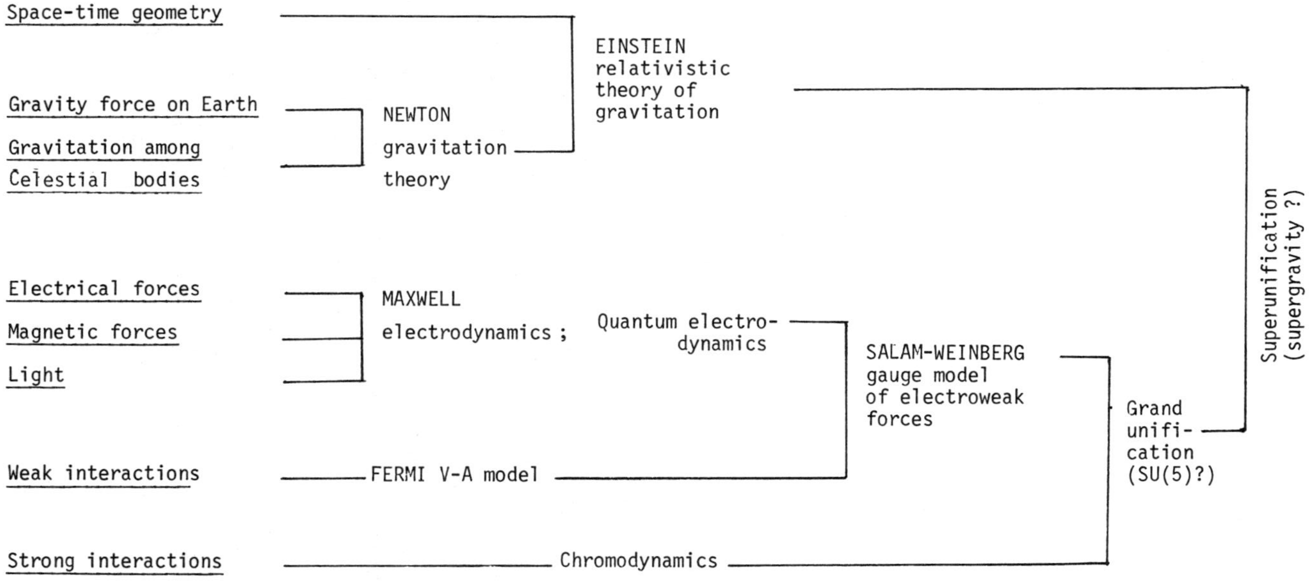

TABLE X. 1 - Grand unification and superunification are still non-tested conjectures, as of March 1980

PROBLEMS

X - 1. Two operators a_A, a_B satisfy the following commutation rules with their hermitian conjugates :

$$\left[a_A, a^+_A \right]_- = \left[a_B, a^+_B \right]_- = 1$$

and the other commutators vanish.

a) What are the commutation rules between a_A, a_B and the operator number of objects N :

$$N = N_A + N_B,$$

$$N_A = a^+_A\, a_A, \quad N_B = a^+_B\, a_B.$$

b) If $|n_A, n_B >$ is an eigenvecor of N with eigenvalues $n_A + n_B$ what are the eigenvalues of $a_A|n_A, n_B >, a_B|n_A, n_B >, a^+_A|n_A, n_B >, a^+_B|n_A, n_B >$?

What is the expression of the latter vectors ?

c) From these expressions obtain the matrix elements of the operator Ω in the following representation :

$$\Omega = \begin{pmatrix} < 1, 0 \mid \Omega \mid 1, 0 > & < 1, 0 \mid \Omega \mid 0, 1 > \\ < 0, 1 \mid \Omega \mid 1, 0 > & < 0, 1 \mid \Omega \mid 0, 1 > \end{pmatrix}$$

where Ω is one of the four possible operators which do not change the total number of objects : $a^+_A\, a_B,\ a^+_B\, a_A,\ a^+_A\, a_A - a^+_B\, a_B,\ a^+_A\, a_A + a^+_B\, a_B$.

d) What linear combinationsof these matrices give the usual Pauli matrices $\tau_1, \tau_2\, \tau_3$, the generators of SU(2) ?

X - 2. Apply the method of the preceding problem to determine the form of :

a) the eight generators λ_k of the SU(3) group.

b) The fifteen generators η_ℓ of the SU(4) group.

c) The twenty-four generators t_ℓ of the SU(5) group, as given in (X. 3).

d) How many diagonal matrices are there among the generators for each of these groups ?

X - 3. Complete the treatment of the SU(5) model. Consult Jarlskog, ref. 165 ; Ellis, ref. 166 ; Paul Langacker, Grand unified theories and proton decay, SLAC-Pub-2544 (1980).

X - 4. Whereas the postulate of an SU(2) singlet right-handed neutrino, ν_R, will give rise, by coupling with a Higgs doublet in the Salam-Weinberg theory, to a mass term $m_D\,(\bar{\nu}_L\,\nu_R + \bar{\nu}_R\,\nu_L)$, (Called Dirac mass), another mass term, called Majorana mass, may be introduced by coupling the singlet ν_R with the singlet $(\nu_R)^c = (\nu^c)_L$:

$$m_M \left\{ \overline{(\nu^c)_L}\;\nu_R + \text{h.c.} \right\}$$

a) express the Majorana mass term as a function of ${}^t\nu_R$ and the charge conjugation matrix C.

b) Consult P. Langacker, Grand unified theories and proton decay on Dirac and Majorana masses.

c) Read Bilenky and Pontecorvo, ref. 140 ; L.Maiani, Neutrino oscillations, CERN preprint-TH-2846 (1980).

X - 5. Study the SO(10) model. Consult Langacker, ref. in the preceding problem.

Solutions of Problems

I - 1

a) $D^{-1}(\ell)\ \gamma^{\mu}\ D(\ell) = \ell^{\mu}{}_{\nu}\ \gamma^{\nu}$

b) The choice of hermitian conjugate of γ^{μ} being adopted as :

$$(\gamma^{\mu})^{+} = \gamma^{0}\ \gamma^{\mu}\ \gamma^{0}$$

one obtains :

$$\gamma^{0}\ D^{+}(\ell)\ \gamma^{0} = D^{-1}\ (\ell)$$

The 26 equations are then :

$$\det D^{+} = \det D^{-1}$$

$$\text{Trace } D^{+} = \text{Trace } D^{-1}$$

$$\frac{1}{4}\ \text{Trace}\ (D^{-1}\ \gamma^{\mu}\ D\ \gamma^{\alpha}) = \ell^{\mu}{}_{\nu}\ g^{\nu\alpha}$$

$$\frac{1}{4}\ \text{Trace}\ (D^{-1}\ \gamma^{\mu}\ D\ \frac{1}{2}\ (\gamma^{\alpha}\ \gamma^{\beta} - \gamma^{\beta}\ \gamma^{\alpha})\ \gamma^{\lambda}) = \ell^{\mu}{}_{\nu}\ (g^{\nu\alpha}\ g^{\lambda\beta} - g^{\nu\beta}\ g^{\lambda\alpha})$$

I - 2

A - a) $\ell^{\mu}{}_{m}\ \ell^{\nu}{}_{n}\ \tilde{F}'_{\mu\nu}(x') = (\det \ell)\ \tilde{F}_{mn}\ (x)$

b) $S'(x') = S(x)$

$V^{\mu'}(x') = \ell^{\mu}{}_{\nu}\ V^{\nu}(x)$

$A^{\mu'}(x') = (\det \ell)\ \ell^{\mu}{}_{\nu}\ A^{\nu}(x)$

$S'^{\mu\nu}(x') = \ell^{\mu}{}_{m}\ \ell^{\nu}{}_{n}\ S^{mn}\ (x)$

$P'(x') = (\det \ell)\ P(x).$

B - a) $D(\ell) = I - \frac{i}{2}\ \frac{\Sigma_{\mu\nu}}{2}\ \alpha^{\mu\nu}$, infinitesimal

$D(\ell) = \exp\left(-\frac{i}{2}\ \frac{\Sigma_{\mu\nu}}{2}\ \alpha^{\mu\nu}\right)$, finite

$$\Sigma_{\mu\nu} = \frac{i}{2}\left[\gamma_\mu, \gamma_\nu\right]$$

b) $$\gamma^0 = \begin{pmatrix} 0 & \sigma_2 \\ \sigma_2 & 0 \end{pmatrix} = \begin{pmatrix} 0 & 0 & 0 & -i \\ 0 & 0 & i & 0 \\ 0 & -i & 0 & 0 \\ i & 0 & 0 & 0 \end{pmatrix}$$

$$\gamma^1 = \begin{pmatrix} -i\,\sigma_3 & 0 \\ 0 & -i\,\sigma_3 \end{pmatrix}$$

$$\gamma^2 = \begin{pmatrix} 0 & -\sigma_2 \\ \sigma_2 & 0 \end{pmatrix}$$

$$\gamma^3 = \begin{pmatrix} i\,\sigma_1 & 0 \\ 0 & i\,\sigma_1 \end{pmatrix}$$

c) $$\Sigma_{23} = i\begin{pmatrix} 0 & -\sigma_3 \\ \sigma_3 & 0 \end{pmatrix};\ \Sigma_{31} = \begin{pmatrix} \sigma_2 & 0 \\ 0 & \sigma_2 \end{pmatrix};\ \Sigma_{12} = i\begin{pmatrix} 0 & \sigma_1 \\ -\sigma_1 & 0 \end{pmatrix}$$

$$\Sigma_{01} = i\ \alpha_1 = i\begin{pmatrix} 0 & \sigma_1 \\ \sigma_1 & 0 \end{pmatrix}$$

$$\Sigma_{02} = i\ \alpha_2 = i\begin{pmatrix} I & 0 \\ 0 & -I \end{pmatrix}$$

$$\Sigma_{03} = i\ \alpha_3 = i\begin{pmatrix} 0 & \sigma_3 \\ \sigma_3 & 0 \end{pmatrix}$$

hence $D(\ell)$ is real in this representation.

I - 3

Multiply Rarita-Scwinger's equation by γ^μ to the left, to obtain

$$2\, i\, (\partial_\alpha\, \psi^\alpha) + m\, (\gamma_\alpha\, \psi^\alpha) = 0$$

Multiply by the derivative operator ∂_μ so that

$$2\, i\, (\gamma_\alpha\, \partial^\alpha)(\partial_\mu\, \psi^\mu) + m\, (\gamma_\alpha\, \partial^\alpha)(\gamma_\mu\, \psi^\mu) - 3\, m\, (\partial_\mu\, \psi^\mu) = 0$$

These two equations give the Dirac's equation and the subsidiary conditions for $m \neq 0$.

I - 4

$$i\, \frac{\partial\, \varphi}{\partial x^0} = -\, i\, \sigma_1\, \frac{\partial\, \varphi}{\partial x} + \sigma_2\, m\, \varphi$$

where $\varphi = \begin{pmatrix} \varphi_1 \\ \varphi_2 \end{pmatrix}$ is a two component spinor σ_1 and σ_2 are Pauli matrices

The above equation is of the form :

$$H\, \varphi = i\, \frac{\partial\, \varphi}{\partial x_0}$$

where :

$$H = \alpha_x\, p + \beta\, m$$

and

$$\alpha_x \rightarrow \sigma_1$$

$$\beta \rightarrow \sigma_2$$

Under the form :

$$(i\, \gamma^\alpha\, \partial_\alpha - m)\, \varphi = 0$$

we have :

$$\gamma^0 \rightarrow \sigma_2$$
$$\gamma_x \rightarrow -i\,\sigma_3$$

so that :

$$(\gamma^0)^2 = I, \ (\gamma_x)^2 = -\ I$$

and

$$\gamma^5 \rightarrow \gamma^0\ \gamma_x \rightarrow \sigma_1$$

Infinitesimal Lorentz transformation :

$$\varphi'(x') = D(\ell)\ \varphi(x), \quad \ell^{\mu}{}_{\nu} = \delta^{\mu}{}_{\nu} + \varepsilon^{\mu}{}_{\nu}$$
$$D = I - \frac{1}{2}\sigma_1\ \varepsilon^{01}$$

Adjoint equation :

$$i\ \partial_0\ \bar{\varphi}\ \sigma_2 + \partial_x\ \bar{\varphi}\ \sigma_3 + m\ \bar{\varphi} = 0$$

where :

$$\bar{\varphi}(x) = \varphi^+(x)\ \sigma_2$$

Conserved current :

$$j^{\mu}(x) = \bar{\varphi}\ \gamma^{\mu}\ \varphi, \quad \gamma^0 = \sigma_2, \ \gamma^1 = -\ i\ \sigma_3$$

Left and right-handed components are :

$$\varphi_L \doteq \frac{1}{2}\ (I - \sigma_1)\ \varphi(x)$$
$$\varphi_R = \frac{1}{2}\ (I + \sigma_1)\ \varphi(x)$$

and satisfy the equations :

$$\left[i\ \partial_0 + i\ (\sigma_1\ \partial_x) \right]\ \varphi_R = m\ \sigma_2\ \varphi_L$$
$$\left[i\ \partial_0 - i\ (\sigma_1\ \partial_x) \right]\ \varphi_L = m\ \sigma_2\ \varphi_R$$

from which equations with $m = 0$ follow. Find the other possible solutions.

I - 5

a) $(\square + m^2)\varphi = g\bar{\psi}\psi \dfrac{1}{(1+K\varphi)^2}$

$(i\gamma^\alpha \partial_\alpha - M)\psi = -g \dfrac{\varphi}{1+K\varphi}\psi$

$i(\partial_\alpha \bar{\psi})\gamma^\alpha + M\bar{\psi} = g\dfrac{\varphi}{1+K\varphi}\bar{\psi}$

b) $T_{\mu\nu} = \partial_\mu \varphi\, \partial_\nu \varphi + \frac{i}{2}\left[\bar{\psi}\gamma_\mu \partial_\nu \psi - (\partial_\nu \bar{\psi})\gamma_\mu \psi\right] +$

$+ g_{\mu\nu}\Big\{ \frac{1}{2}(m^2\varphi^2 - \partial_\alpha\varphi\,\partial^\alpha\varphi) + \frac{1}{2}\bar{\psi}(-i\gamma^\alpha\partial_\alpha + M)\psi +$

$+ \frac{1}{2}(i(\partial_\alpha\bar{\psi})\gamma^\alpha + M\bar{\psi})\psi - g\bar{\psi}\psi\dfrac{\varphi}{1+K\varphi}\Big\}$

c) $H = \int d^3x \Big\{ \pi^2 + (\vec{\nabla}\varphi)^2 + m^2\varphi^2 + \psi^+\left[-i\vec{\alpha}\cdot\vec{\nabla} + M\beta\right]\psi$

$= g\bar{\psi}\psi \sum_n (-1)^n K^n \varphi^{n+1}\Big\}$,

$\pi = \partial_0\varphi$

- 6

H anti-kink $= \dfrac{\sqrt{8}}{3}\dfrac{m^3}{\lambda}$

- 8

a) $A \neq \frac{1}{2}$; $B = \frac{3}{2}A^2 - A + \frac{1}{2}$;

$C = 3A^2 - 3A + 1$

b) $A = B = C = 1$. The equation in this case is the one adopted in supergravity to describe massless Majorana spin 3/2 fields.

c) This relationship allows the following form for the equation in b) :

$\varepsilon^{\alpha\mu\nu\beta}\gamma^5\gamma_\mu(\partial_\nu + \frac{im}{2}\gamma_\nu)\psi_\beta = 0.$

II - 1

a) Let $\delta Z^\mu(s)$ be arbitrary but vanishing at the limits of integration ; the corresponding δS is then after partial integration :

$$\delta S_1 = \int ds \left\{ m_0 c \left(\frac{d}{ds} \frac{dZ^\mu}{ds} \right) \delta Z_\mu + \right.$$

$$\left. + \frac{e}{c} \left(\frac{\partial A_\mu}{\partial Z^\alpha} \frac{dZ^\alpha}{ds} \delta Z^\mu - \frac{\partial A_\mu}{\alpha Z^\alpha} \frac{dZ^\mu}{ds} \delta Z^\alpha \right) \right\}$$

which gives, for $\delta S = 0$ for the above arbitrary δZ_μ :

$$m_0 c^2 \frac{d}{ds} \frac{dZ^\mu}{ds} = e \frac{dZ_\alpha}{ds} F^{\alpha\mu} (Z(s)),$$

$$F^{\alpha\mu} = \partial^\mu A^\alpha - \partial^\alpha A^\mu$$

b) Variation of S corresponding to an arbitrary field variation $\delta A_\mu(x)$ which vanishes on the boundaries of space-time integrations gives :

$$\delta S_2 = - \frac{1}{c} \int d^4x \left\{ \frac{1}{2} F_{\mu\nu}(x) (\partial^\nu \delta A^\mu(x) - \partial^\mu \delta A^\nu(x)) + \right.$$

$$\left. + e j_\mu(x) \delta A^\mu(x) \right\} = \text{after partial integration} =$$

$$= \frac{1}{c} \int d^4x \left\{ \partial_\nu F^{\mu\nu} - e j^\mu \right\} \delta A_\mu(x)$$

which gives, for $\delta S_2 = 0$ for the chosen arbitrary $\delta A_\mu(x)$:

$$\partial_\mu F^{\mu\nu} = e j^\mu$$

c) We have :

$$\partial_\nu T^{\mu\nu}{}_m = m_0 c^2 \int \frac{dZ^\mu}{ds} \frac{dZ^\nu}{ds} \frac{\partial}{\partial x^\nu} \delta^4(x - Z(s))\, ds = \text{(after partial}$$

$$\text{integration)} = m_0 c^2 \int \left(\frac{d}{ds} \frac{dZ^\mu}{ds} \right) \delta^4(x - Z(s))\, ds$$

so that, in view of the particle's equations of motion :

$$\partial_\nu T^{\mu\nu}_{m} = e\, j_\nu F^{\nu\mu}$$

d)

$$L_{\lambda\mu\nu} = x_\mu T_{\lambda\nu} - x_\nu T_{\lambda\mu}$$

$$L_{jk} = \int L_{ojk}\, d^3x = Z_j p_k - Z_k p_j$$

II - 2

a) $$\vec{\nabla} \cdot \vec{E} = e\rho$$

$$\vec{\nabla} \times \vec{B} - \partial_0 E = e\vec{j}$$

$$\vec{\nabla} \times \vec{E} + \partial_0 \vec{B} = 0$$

$$\vec{\nabla} \cdot \vec{B} = 0$$

b) $$F^{\mu\nu} \to \tilde{F}^{\mu\nu} \; ; \; \tilde{F}^{\mu\nu} \to - F^{\mu\nu}$$

or

$$\vec{E} \to \vec{B} \; ; \; \vec{B} \to - \vec{E}$$

c) Let h^μ be the magnetic monopole current. We may postulate :

$$\partial_\nu F^{\mu\nu} = e\, j^\mu$$

$$\partial_\nu \tilde{F}^{\mu\nu} = g\, h^\mu$$

From :

$$\partial'_\nu \tilde{F}'^{\mu\nu}(x') = g'\, h'^\mu(x')$$

obtain :

$$h'^\mu(x') = (\det \ell)\, \ell^\mu{}_\nu\, h^\nu(x)$$

if

$$g' = g$$

II - 3

In a two dimensional Minkowski space there is no magnetic field :

$$F^{01} = E = - \partial^0 A^1 - \partial_1 A^0$$

$$F^{\mu\nu} = \begin{pmatrix} 0 & E \\ -E & 0 \end{pmatrix}$$

The lagrangean is :

$$\mathcal{L} = - \frac{1}{4} F^{\mu\nu} F_{\mu\nu} + \bar{\psi} (i \gamma_\mu D^\mu - m) \psi =$$

$$= \frac{1}{2} E^2 + \bar{\varphi} \left\{ \gamma^0 (i \partial_0 - e A_0) + \gamma^1 (i \partial_1 - e A_1) - m \right\} \varphi$$

where

$$\bar{\varphi} = \varphi^+ \sigma_2, \quad \gamma^0 = \sigma_2, \quad \gamma^1 = - i \sigma_3$$

Maxwell's equations are :

$$\partial_x E (x, x 0) = e j^0 (x, x 0)$$

$$\partial_0 E (x, x 0) = - e j^1 (x, x 0)$$

$$j^0 (x, x 0) = \bar{\varphi} \sigma_2 \varphi = \varphi^+ \varphi$$

$$j^1 (x, x 0) = - i \bar{\varphi} \sigma_3 \varphi$$

The conservation equation is clearly

$$\frac{\partial j^0}{\partial x^0} + \frac{\partial j^1}{\partial x^1} = 0$$

In the absence of matter, E = const. and the energy in each interval $(- \ell, \ell)$ in constant E^2_o if $E = \frac{E_0}{\ell}$

II - 4

a) $\Box A^{\mu} - \partial^{\mu}(\partial_{\nu} A^{\nu}) = e\, j^{\mu}$

gives rise, in the Coulomb gauge :

$$\vec{\nabla} \cdot \vec{A} = 0$$

to the equations :

$$\Box \vec{A} = e\, \vec{j} - \vec{\nabla}\, (\partial_{0} A^{0})$$

$$\nabla^{2} A^{0} = - e\, j^{0}$$

The Green's function of the latter equation is :

$$\mathcal{G}(\vec{x}\, ; \vec{x}\,') = \frac{1}{4\pi} \frac{1}{|\vec{x} - \vec{x}\,'|}$$

so that

$$A^{0}(\vec{x}, t) = \frac{e}{4\pi} \int \frac{j^{0}(\vec{x}\,', t)}{|\vec{x} - \vec{x}\,'|}\, d^{3}x'$$

and

$$j^{\mu}(x) = \bar{\psi}(x)\, \gamma^{\mu}\, \psi(x)$$

The hamiltonian is :

$$H = \int d^{3}x \left\{ \psi^{+}(x) \left[\vec{\alpha} \cdot (-i\, \vec{\nabla} - e\, \vec{A}) + \beta\, m \right] \psi(x) + \frac{1}{2} \left[\vec{E}^{2} + \vec{B}^{2} \right] \right\}$$

after partial integration over a term in $\vec{E} \cdot \vec{\nabla} A^{0}$.

b) The decomposition :

$$\vec{E} = \vec{E}_{\ell} + \vec{E}_{t}$$

into a longitunal and a transverse component of $\vec{E}$ gives rise to :

$$\frac{1}{2}\int d^3x\left[\vec{E}^2 + \vec{B}^2\right] = \frac{1}{2}\int d^3x\, \vec{E}^2_{\ell} + \frac{1}{2}\int d^3x\left[\vec{E}^2_{t} + \vec{B}^2\right]$$

the first term on the right-hand side gives the Coulomb energy

$$\frac{e^2}{8\pi}\int d^3x\; d^3x' \;\frac{j^0(x)\; j^0(x')}{|\vec{x} - \vec{x}'|}$$

III - 1

$$D_\nu \mathscr{G}^{\mu\nu} + m^2 \phi^\mu = 0$$

where :

$$\mathscr{G}^{\mu\nu} = D^\nu \phi^\mu - D^\mu \phi^\nu,$$

$$D_\mu = \partial_\mu + i\, e\, A_\mu$$

From the above equation and

$$\left[D_\mu, D_\nu\right] = -\, i\, e\; F_{\mu\nu}$$

we obtain

$$D_\mu \phi^\mu = \frac{i\, e}{2\, m^2}\; F_{\mu\nu}\, \mathscr{G}^{\mu\nu}$$

and the second order equation is :

$$D^2 \phi^\mu + m^2 \phi^\mu + i\, e\; F^{\mu\nu} \phi_\nu - \frac{i\, e}{2\, m^2}\; F_{\alpha\beta}\, D^\mu \mathscr{G}^{\alpha\beta} -$$

$$- \frac{i\, e}{2\, m^2}\,(\partial_\mu\, F_{\alpha\beta})\, \mathscr{G}^{\alpha\beta} = 0$$

where :

$$D^2 \equiv D^\alpha\, D_\alpha$$

II - 2

a) Equations are

1) $\partial_\nu F^{\mu\nu} = e\, j^\mu,$

$$j^\mu = \bar{\psi}_\lambda \gamma^\mu \psi^\lambda - \frac{1}{3}(\bar{\psi}_\lambda \gamma^\lambda \psi^\mu + \bar{\psi}^\mu \gamma_\lambda \psi^\lambda) + \frac{1}{3}\bar{\psi}_\lambda \gamma^\lambda \gamma^\mu \gamma_\nu \psi^\nu \;;$$

2) $$(i\,\gamma^\alpha D_\alpha - m)\,\psi^\lambda - \frac{i}{3}\left[\gamma^\lambda D^\nu + \gamma^\nu D^\lambda\right]\psi_\nu +$$

$$+ \frac{1}{3}\gamma^\lambda (i\,\gamma^\alpha D_\alpha + m)\,\gamma^\nu \psi_\nu = 0$$

b) the subsidiary conditions, as derived from these equations, are :

$$\gamma^\alpha \psi_\alpha = - \frac{2i}{m}(D^\mu \psi_\mu)$$

$$D_\alpha \psi^\alpha = - \frac{e}{6m}\left\{ 3(\gamma^\mu \psi^\nu - \gamma^\nu \psi^\mu) + i\,\sigma^{\mu\nu}(\gamma_\alpha \psi^\alpha)\right\} F_{\mu\nu}$$

c) thus :

$$D_\alpha \psi^\alpha = \left[1 + \frac{e}{3\,m^2}\sigma^{\mu\nu} F_{\mu\nu}\right]^{-1}\left(- \frac{e}{2m}(\gamma^\mu \psi^\nu - \gamma^\nu \psi^\mu) F_{\mu\nu}\right)$$

The development of the inverse operator in $\gamma^\alpha \psi_\alpha$ gives for the current :

$$j^\mu = \bar{\psi}_\alpha \gamma^\mu \psi^\alpha - \frac{ie}{3\,m^2}\left\{(\bar{\psi}^\alpha \gamma^\beta - \bar{\psi}^\beta \gamma^\alpha)\psi^\mu + (\bar{\psi}^\mu \gamma^\alpha \psi^\beta - \bar{\psi}^\mu \gamma^\beta \psi^\alpha)\right\} \cdot$$

$$\cdot\; F_{\alpha\beta} - \frac{1}{3}\frac{e^2}{m^4}(\bar{\psi}^\alpha \gamma^\beta - \bar{\psi}^\beta \gamma^\alpha)\gamma^\mu(\gamma^\lambda \psi^\eta - \gamma^\eta \psi^\lambda) F_{\alpha\beta} F_{\lambda\eta} + \ldots$$

III - 3

a) Basis : $C, \vec{\sigma}\, C$

$$^t(\vec{\sigma}\, C) = (\vec{\sigma}\, C) \ ; \ ^tC = -\, C$$

b)

$$\psi_{\alpha\beta}(x) = \vec{F}(x) \cdot \left(\frac{\vec{\sigma}\, C}{2} \right)_{\alpha\beta}$$

c) Take the trace ; then multiply by $\vec{\sigma}$ and take the trace to find :

$$\vec{\nabla} \cdot \vec{F} = 0$$

$$\vec{\nabla} \times \vec{F} = i\, \partial_0\, \vec{F}$$

d) Choose :

$$\vec{F} = \vec{B} - i\, \vec{E}$$

and above equations give the free field Maxwell's equations.

e) $\vec{x} \to -\, \vec{x}$ induces the transformation $\vec{F} \to \vec{F}^*$, so that $\vec{B} \to \vec{B}$, $\vec{E} \to -\, \vec{E}$.
(see J. Leite Lopes and D. Spehler, Lett. Nuovo Cimento 25, 101 (1979).

IV - 1

a) Trace $(\tau_k) = 0$ follows from the commutation rule.

b) From $\tau^2_1 = I$ and $\tau_1\, \tau_2 = i\, \tau_3$ it follows that

$$(\det \tau_k)^2 = 1$$

$$\det \tau_1 \ \det \tau_2 = -\, \det \tau_3$$

hence

$$(\det \tau_k) = -\, 1$$

c) $$\tau_1 = \begin{pmatrix} 0 & e^{i\alpha} \\ e^{-i\alpha} & 0 \end{pmatrix} , \quad \tau_2 = \begin{pmatrix} 0 & -ie^{i\alpha} \\ ie^{-i\alpha} & 0 \end{pmatrix}$$

V - 2

From the commutation and anticommutation rules if follows :

$$Tr(\lambda_k \lambda_\ell) = 2 \delta_{k\ell} + (d_{k\ell n} + i f_{k\ell n}) Tr(\lambda_n)$$

$$Tr(\lambda_\ell \lambda_k) = 2 \delta_{k\ell} + (d_{k\ell n} - i f_{k\ell n}) Tr(\lambda_n)$$

hence :

a) $Tr(\lambda_n) = 0$

b) $Tr(\lambda_k \lambda_\ell) = 2 \delta_{k\ell}$

For the case $k = 1,2,3$ one obtains :

$$\lambda_1 \lambda_2 = i \lambda_3 ; \lambda_3 \lambda_1 = i \lambda_2 ; \lambda_2 \lambda_3 = i \lambda_1$$

and :

$$\lambda_k \lambda_\ell = - \lambda_\ell \lambda_k , \quad i, k = 1,2,3$$

which give :

c) $\det \lambda_1 = \det \lambda_2 = \det \lambda_3 = 0$

Let x_1, x_2, x_3 be the eigenvalues of a λ_k, $k = 1,2,3$. One has :

$$\det \lambda_k = x_1 x_2 x_3 = 0$$

$$Tr (\lambda_k) = x_1 + x_2 + x_3 = 0 , \quad k = 1,2,3.$$

So one x vanishes, let it be x_3. Then $x_1 = - x_2$. Thus $\lambda_1, \lambda_2, \lambda_3$ have each zero as an eingenvalue and the other two have opposite sign.

If λ'_k is a similarity transformed from λ_k

$$\lambda'_k = S^{-1} \lambda_k S$$

such that :

$$\lambda'_k = \begin{pmatrix} x_1 & 0 & 0 \\ 0 & -x_1 & 0 \\ 0 & 0 & 0 \end{pmatrix}$$

then :

$$\lambda'^2_k = S^{-1} \lambda^2_k S$$

$$\mathrm{Tr}\,(\lambda'^2_k) = \mathrm{Tr}\,(\lambda^2_k) = 2$$

whence :

$$x^2_1 = 1, \qquad x_1 = \pm 1.$$

d) 1, - 1, 0 are therefore the three eigenvalues of $\lambda_1, \lambda_2, \lambda_3$.

e) If one chooses :

$$\lambda_3 = \begin{pmatrix} 1 & 0 & 0 \\ 0 & -1 & 0 \\ 0 & 0 & 0 \end{pmatrix}$$

then

$$\lambda_1 = \begin{pmatrix} \tau_1 & 0 \\ 0 & 0 \end{pmatrix}, \quad \lambda_2 = \begin{pmatrix} \tau_2 & 0 \\ 0 & 0 \end{pmatrix}$$

From the commutation rules :

$$\lambda_1{}^2 = \lambda_2{}^2 = \lambda_3{}^2 = \frac{2}{3} I + \frac{1}{\sqrt{3}} \lambda_8$$

$$\lambda^2_8 = \frac{2}{3} I - \frac{1}{\sqrt{3}} \lambda_8$$

so that :

$$\lambda_1^{\,2} + \lambda_8^{\,2} = \frac{4}{3}\, I$$

therefore :

$$\lambda_8 = \frac{1}{\sqrt{3}} \begin{pmatrix} 1 & 0 & 0 \\ 0 & 1 & 0 \\ 0 & 0 & -2 \end{pmatrix}$$

One has also :

$$\lambda_4\, \lambda_8 - \lambda_8\, \lambda_4 = -\, i\, \sqrt{3}\;\; \lambda_5$$

$$\lambda_4\, \lambda_8 + \lambda_8\, \lambda_4 = -\, \frac{1}{\sqrt{3}}\;\; \lambda_4$$

$$\lambda_5\, \lambda_8 - \lambda_8\, \lambda_5 = i\, \sqrt{3}\;\; \lambda_4$$

$$\lambda_5\, \lambda_8 + \lambda_8\, \lambda_5 = -\, \frac{1}{\sqrt{3}}\;\; \lambda_5$$

$$\lambda_6\, \lambda_8 - \lambda_8\, \lambda_6 = -\, i\, \sqrt{3}\; \lambda_7$$

$$\lambda_6\, \lambda_8 + \lambda_8\, \lambda_6 = -\, \frac{1}{\sqrt{3}}\;\; \lambda_6$$

$$\lambda_7\, \lambda_8 - \lambda_8\, \lambda_7 = i\, \sqrt{3}\;\; \lambda_6$$

$$\lambda_7\, \lambda_8 + \lambda_8\, \lambda_7 = -\, \frac{1}{\sqrt{3}}\;\; \lambda_7$$

From

$$\lambda_k\, \lambda_8 + \lambda_8\, \lambda_k = -\, \frac{1}{\sqrt{3}}\;\; \lambda_k, \qquad k = 4,5,6,7$$

one deduces :

$$\lambda_k = \begin{pmatrix} 0 & 0 & a_{13} \\ 0 & 0 & a_{23} \\ a^*_{13} & a^*_{23} & 0 \end{pmatrix}$$

Let :

$$\lambda_4 = \begin{pmatrix} 0 & 0 & a_{13} \\ 0 & 0 & a_{23} \\ a^*_{13} & a^*_{23} & 0 \end{pmatrix} \quad , \quad \lambda_5 = \begin{pmatrix} 0 & 0 & b_{13} \\ 0 & 0 & b_{23} \\ b^*_{13} & b^*_{23} & 0 \end{pmatrix}$$

then :

$$a_{13} = i\, b_{13}$$

$$a_{23} = i\, b_{23}$$

From :

$$\lambda_4 \lambda_6 + \lambda_6 \lambda_4 = \lambda_1$$

$$\lambda_4 \lambda_6 - \lambda_6 \lambda_4 = i\, \lambda_2$$

we get :

$$\lambda_4 \lambda_6 = \frac{1}{2} (\lambda_1 + i\, \lambda_2)$$

so if

$$\lambda_6 = \begin{pmatrix} 0 & 0 & c_{13} \\ 0 & 0 & c_{23} \\ c^*_{13} & c^*_{23} & 0 \end{pmatrix}$$

then

$$a_{13}\, c^*_{13} = a_{23}\, c^*_{13} = a_{23}\, c^*_{23} = 0$$

$$a_{13}\, c^*_{23} = 1$$

so that :

$$a_{23} = c_{13} = 0$$

Similarly one obtains

$$\lambda_7 = \begin{pmatrix} 0 & 0 & 0 \\ 0 & 0 & -i\, a_{23} \\ 0 & i\, a^*_{23} & 0 \end{pmatrix}$$

From :

$$\lambda_1 \lambda_4 + \lambda_4 \lambda_1 = \lambda_6$$

$$\lambda_1 \lambda_4 - \lambda_4 \lambda_1 = i\, \lambda_7$$

there follow the relations :

$$a_{13} = c_{23} = a_{23}$$

so :

$$a_{13} = e^{i\alpha}$$

and

$$\lambda_4 = \begin{pmatrix} 0 & 0 & e^{i\alpha} \\ 0 & 0 & 0 \\ e^{-i\alpha} & 0 & 0 \end{pmatrix}, \quad \lambda_5 = \begin{pmatrix} 0 & 0 & -ie^{i\alpha} \\ 0 & 0 & 0 \\ ie^{-i\alpha} & 0 & 0 \end{pmatrix}$$

$$\lambda_6 = \begin{pmatrix} 0 & 0 & 0 \\ 0 & 0 & e^{i\alpha} \\ 0 & e^{-i\alpha} & 0 \end{pmatrix}, \quad \lambda_7 = \begin{pmatrix} 0 & 0 & 0 \\ 0 & 0 & -ie^{i\alpha} \\ 0 & ie^{i\alpha} & 0 \end{pmatrix}$$

and the choice $\alpha = 0$ gives the usual representation.

IV - 4

From equation (IV. 14) we obtain :

$$\vec{f}'(x) \cdot \left(\frac{\vec{\tau}\, C}{2} \right)_{\alpha\beta} = U_{\alpha a}\, U_{\beta b}\, \vec{f}\,(x) \cdot \left(\frac{\vec{\tau}\, C}{2} \right)_{ab}$$

a) Then the law (IV. 14) follows from the relationship :

$$U^{-1}(x) = C\ {}^{t}U(x)\ C^{-1}$$

so that :

$$f'_{k}(x) = a_{k\ell}\ f_{\ell}(x)$$

if :

$$a_{k\ell} = \frac{1}{2}\ \mathrm{Tr}\ (\tau_k\ U\ \tau_\ell\ U^{-1})$$

b) For infinitesimal transformations :

$$a_{k\ell}\ \delta_{k\ell} + g\ \varepsilon_{k\ell n}\ \Lambda_n(x)$$

IV. 5

a) As ψ_{abc} in the form given in the problem is symmetric in bc we assure its symmetry in ab by multiplying it by the antisymmetric matrix $C^{-1}{}_{ab}$ which then must vanish :

$$C^{-1}{}_{ab}\ \psi_{abc} = - \vec{f}_a \cdot \left(\frac{{}^{t}\vec{\tau}}{2} \right)_{ac} = 0$$

so that

$$\vec{\tau} \cdot \vec{f}(x) = 0$$

in the sense that :

$$\vec{\tau}_{ab} \cdot \vec{f}_b(x) = 0$$

b) Finite transformation law :

$$f'^{k}{}_{a}(x) = a^{k\ell}\, U_{ab}\, f^{\ell}{}_{b}(x)$$

where

$$a^{k\ell} = \frac{1}{2}\ \mathrm{Tr}(\tau^{k}\ U\ \tau^{\ell}\ U^{-1}).$$

Infinitesimal transformation law :

$$f'^{k}{}_{a}(x) = f^{k}{}_{a}(x) + i\ g\ \vec{\Lambda}\cdot\left(\frac{\vec{\tau}}{2}\right)_{ab} f^{k}{}_{b}$$

$$- g\ \varepsilon^{k\ell n}\ \Lambda^{\ell}\ f^{n}{}_{a}(x)$$

IV. 7

Apply the operator $i\ \gamma^{\alpha}\ D_{\alpha} + m$ to the left-hand side of Dirac's equation :

$$(i\ \gamma^{\mu}\ D_{\mu} - m)\ \psi(x) = 0$$

to obtain

$$((\gamma^{\alpha}\ D_{\alpha})^{2} + m^{2})\ \psi(x) = 0$$

so that, in view of the commutation rules for γ's and D's :

$$\left\{ D^{\alpha}\ D_{\alpha} + m^{2} - g\ \frac{\vec{\tau}}{2}\cdot\vec{F}_{\mu\nu}\ \frac{\sigma^{\mu\nu}}{2} \right\}\psi(x) = 0$$

Write this equation in terms of the Dalembertian operator and the field $A^{\mu}{}_{a}$ and discuss the significance of each term

V - 1

We find :

$$\Gamma^{\alpha}{}_{\lambda\alpha}(x) = -\frac{1}{2}\ g_{\alpha\beta}\ \partial_{\lambda}\ g^{\alpha\beta}$$

From

$$g = \det(g_{\mu\nu})$$

we have

$$g^{\mu\nu} = \frac{\Delta^{\mu\nu}}{g}$$

where $\Delta^{\mu\nu}$ is the cofactor of the element $g_{\mu\nu}$ so that :

$$g = \sum_{\beta} g_{\alpha o\beta} \Delta^{\alpha o\beta} , \qquad \frac{\partial g}{\partial g_{\alpha\beta}} = \Delta^{\alpha\beta}$$

$$g^{\mu\nu} = \frac{1}{g} \frac{\partial g}{\partial g_{\mu\nu}}$$

$$\partial_{\alpha} g = g\, g^{\mu\nu} \partial_{\alpha} g_{\mu\nu} = - g\, g_{\mu\nu} \partial_{\alpha} g^{\mu\nu}$$

from these relations the required formula is deduced.

V - 2

a) $R_{\mu\nu} = K (T_{\mu\nu} - \frac{1}{2} g_{\mu\nu} T)$

b) for a Yang-Mills field the trace of its energy momentum tensor vanishes as is the case of the energy-momentum tensor for an electromagnetic field $T^{YM} = 0$ so that :

$$R_{\mu\nu} = K\, T^{YM}_{\mu\nu}$$

V - 3

As $L(z^{\mu}, \frac{dz^{\mu}}{ds})$ is the lagrangean

$$\delta \int L\, ds = 0$$

then Lagrange's equations give :

$$\frac{d^2z^\mu}{ds^2} + \Gamma^\mu{}_{\alpha\beta}(z)\,\frac{dz^\alpha}{ds}\,\frac{dz^\beta}{ds} = 0$$

with

$$\Gamma^\mu{}_{\alpha\beta}(z) = \frac{1}{2}\,g^{\mu\lambda}\,(\partial_\beta\, g_{\alpha\lambda} + \partial_\alpha\, g_{\beta\lambda} - \partial_\lambda\, g_{\alpha\beta})$$

The force on the particle is such that :

$$m_i\, c^2\,\frac{d^2z^\mu}{ds^2} = -\,m_g\, c^2\,\Gamma^\mu{}_{\alpha\beta}\,\frac{dz^\alpha}{ds}\,\frac{dz^\beta}{ds}$$

with $m_i = m_g$, m_i is the inertial mass, m_g the gravitational mass. The gravitational force on the particle is $-\,m_g\, c^2\,\Gamma^\mu{}_{\alpha\beta}\,\frac{dz^\alpha}{ds}\,\frac{dz^\beta}{ds}$.

V - 4

In two-dimensional space there is only one independent component of the Riemann tensor $R_{\alpha\mu\beta\nu}(x)$; $R_{1010}(x)$, since it is antisymmetric in $\alpha\mu$ and in $\beta\nu$.

Now :

$$R_{\mu\nu} = g^{\alpha\beta}\,R_{\alpha\mu\beta\nu}$$

so that :

$$R_{00} = g^{11}\,R_{1010}\ ;\ R_{11} = g^{00}\,R_{1010}\ ;$$

$$R_{01} = -\,g^{10}\,R_{1010}$$

However if $g = \det(g_{\mu\nu}) = g_{00}\,g_{11} - (g_{01})^2$ we have :

$$g^{00} = \frac{g_{11}}{g}\ ;\ g^{11} = \frac{g_{00}}{g}\ ;\ g^{01} = -\,\frac{g_{01}}{g}$$

so :

$$R_{\mu\nu} = g_{\mu\nu}(x)\ F(x)$$

where :

$$F(x) = \frac{R_{1010}(x)}{g(x)}$$

Therefore :

$$R = g^{\mu\nu}\ R_{\mu\nu} = 2\ F$$

and

$$R_{\mu\nu} - \frac{1}{2}\ g_{\mu\nu}\ R \equiv 0$$

VI - 1

From :

$$\psi'(a + \ell\ x) = D(\ell)\ \psi(x),\ D(\ell) = e^{-\frac{i}{2}\varepsilon_{\mu\nu}\frac{\sigma^{\mu\nu}}{2}}$$

we have for infinitesimal a^μ and $\ell^\mu{}_\nu = \delta^\mu{}_\nu + \varepsilon^\mu{}_\nu$:

$$(I + \frac{i}{2}\ \varepsilon_{\mu\nu}\ J^{\mu\nu})(I - i\ a_\mu\ P^\mu)\ \psi(x')(I + i\ a_\mu\ P^\mu)\ \cdot$$

$$\cdot\ (I - \frac{i}{2}\ \varepsilon_{\alpha\beta}\ J^{\alpha\beta}) = (I - \frac{i}{2}\ \varepsilon_{\mu\nu}\ \frac{\sigma^{\mu\nu}}{2})\ \psi\ (x)$$

and as

$$\psi(x') = \psi(a + x + \varepsilon\ x) = \psi(x) + a^\mu\ \partial_\mu\ \psi + \varepsilon^\mu{}_\nu\ \partial^\nu\ \psi$$

the required equations follow.

VI - 2

As the charge operator is the generator of $U(\alpha)$, we have for infinitesimal α :

$$U(\alpha) = I + i\ e\ \alpha\ Q$$

hence :

$$(I - i\, e\, \alpha\, Q)\, \varphi(x)\, (I + i\, e\, Q) = (I + i\, e\, \alpha)\, \varphi(x)$$

therefore :

$$\left[\varphi(x), Q\right] = e\, \varphi(x)$$

VI - 3

a) $$\mathscr{L}_0 = \frac{1}{2} \sum_{\ell} \left\{ \bar{\ell}_L\, i\, \gamma^\alpha\, \partial_\alpha\, \ell_L + \bar{\ell}_L\, i\, \gamma^\alpha\, \partial_\alpha\, \ell_R - m_\ell\, (\bar{\ell}_L\, \ell_R + \bar{\ell}_R\, \ell_L) + \right.$$

$$\left. + (\bar{\nu}_\ell)_L\, i\, \gamma^\alpha\, \partial_\alpha\, (\nu_\ell)_L \right\} + \text{h.c.} \quad ;$$

$$j^\lambda_{(\gamma)} = \sum_\ell (\bar{\ell}_L\, \gamma^\lambda\, \ell_L + \bar{\ell}_R\, \gamma^\lambda\, \ell_R)$$

$$j^\lambda_{(W)} = 2 \sum_\ell ((\bar{\nu}_\ell)_L\, \gamma^\lambda\, \ell_L)$$

b) $$\nu_\ell \to - \gamma^5\, \nu_\ell \Rightarrow \bar{\nu}_\ell \to \bar{\nu}_\ell\, \gamma^5$$

hence

$$\frac{1}{2}\,(1 - \gamma^5)\, \nu_\ell \to \frac{1}{2}\,(1 - \gamma^5)\,(-\gamma^5\, \nu_\ell) = \frac{1}{2}\,(1 - \gamma^5)\, \nu_\ell$$

$$\bar{\nu}_\ell\, \frac{1}{2}\,(1 + \gamma^5) \to \bar{\nu}_\ell\, \frac{1}{2}\,(1 + \gamma^5)$$

c) Verify.

d)

$$U_1 = U_2 = \begin{pmatrix} 0 & 1 & 0 \\ 1 & 0 & 0 \\ 0 & 0 & 1 \end{pmatrix}$$

VI - 4

a)

$$U_1 = U_2 = \begin{pmatrix} e^{i\alpha 1} & 0 & 0 \\ 0 & e^{i\alpha 2} & 0 \\ 0 & 0 & e^{i\alpha 3} \end{pmatrix}$$

b)

$$N_\ell = \int j^0{}_{(\ell)} \, d^3x$$

where :

$$j^\lambda{}_{(\ell)} = \bar{\ell}_L \gamma^\lambda \ell_L + \bar{\ell}_R \gamma^\lambda \ell_R + \bar{\nu}_\ell \gamma^\lambda \nu_\ell \,, \qquad \ell = e,\mu,\tau$$

c) No

d)

$$\mathscr{L} = \sum_\ell \Big\{ \frac{1}{2} \Big[\bar{L}_\ell \, i \, \gamma^\alpha \partial_\alpha L_\ell + \bar{R}_\ell \, i \, \gamma^\alpha \partial_\alpha R_\ell \Big] + \text{h.c.} \; +$$

$$+ \, e \Big[\Big(\bar{L}_\ell \gamma^\lambda \Big(\frac{1 - \tau_3}{2} \Big) L_\ell \Big) + \bar{R}_\ell \gamma^\lambda R_\ell \Big] A_\lambda \; +$$

$$+ \, \frac{\mathscr{G}_F}{\sqrt{2}} \, (2 \, \bar{L}_\ell \gamma^\lambda \tau_+ L_\ell)^+ \, (2 \, \bar{L}_\ell \gamma_\lambda \tau_+ L_\ell) \Big\}$$

with

$$\tau_+ = \frac{1}{2} (\tau_1 + i \, \tau_2) = \begin{pmatrix} 0 & 1 \\ 0 & 0 \end{pmatrix}$$

$$\tau_- = (\tau_+)^+ = \frac{1}{2} (\tau_1 - i \, \tau_2) = \begin{pmatrix} 0 & 0 \\ 1 & 0 \end{pmatrix}$$

The currents are indicated in the terms in $e A_\lambda$ and $\frac{\mathscr{G}_F}{\sqrt{2}}$

e) Electromagnetic global gauge transformation :

$$L_\ell \doteq \begin{pmatrix} \nu_\ell \\ \ell \end{pmatrix}_L \rightarrow \begin{pmatrix} (\nu_\ell)_L \\ e^{ie\alpha}\,\ell_L \end{pmatrix} = \exp\left(ie\alpha\,\frac{1-\tau_3}{2}\right) L_\ell$$

$$R_\ell \rightarrow e^{ie\alpha} R_\ell$$

Noether Current :

$$j^\lambda_{(\gamma)} = \sum_\ell \left\{ \bar{L}_\ell \gamma^\lambda \frac{1-\tau_3}{2} L_\ell + \bar{R}_\ell \gamma^\lambda R_\ell \right\}$$

Leptonic global gauge transformation :

$$L_e \rightarrow e^{i\alpha 1} L_e$$

$$L_\mu \rightarrow e^{i\alpha 2} L_\mu$$

$$L_\tau \rightarrow e^{i\alpha 3} L_\tau$$

as given in a) and b).

f)

$$j^\lambda{}_a = \sum_\ell \bar{L}_\ell \gamma^\lambda \frac{\tau_a}{2} L_\ell$$

$$j^\lambda{}_{(w)} = 2\,(j^\lambda{}_1 + i\, j^\lambda{}_2)$$

I - 5

a) From

$$[AD, BC]_- = [A,B]_+ CD - B[A,C]_+ D + A[D,B]_+ C - AB[D,C]_+$$

we find :

$$\left[Q_{(w)}(t),\ Q^{+}_{(w)}(t)\right] = 2\int d^3x \sum_{\ell}\left[\bar{\nu}_{\ell}(1-\gamma^5)\,\nu_{\ell} - \bar{\ell}\,(1-\gamma^5)\,\ell\right]$$

b)

$$\left[K^{L}_{a},\ K^{L}_{b}\right] = i\,\varepsilon_{abc}\ K^{L}_{c}$$

$$K^{L}_{a} = \int d^3x \sum_{\ell} \left(L^{+}_{\ell}\ \frac{\tau_a}{2}\ L_{\ell}\right)$$

c)

$$K^{L}_{a} = \frac{1}{2}\left(K_a - K^{5}_{a}\right)$$

$$K_a = \int d^3X \sum_{\ell} \psi^{+}_{\ell}\ \frac{\tau_a}{2}\ \psi_{\ell}$$

$$K^{5}_{a} = \int d^3x \sum_{\ell} \psi^{+}_{\ell}\,\gamma^5\ \frac{\tau_a}{2}\ \psi_{\ell}$$

where

$$\psi_{\ell}(x) = \begin{pmatrix} \nu_{\ell}(x) \\ \ell(x) \end{pmatrix}$$

$$K^{R}_{a} = \frac{1}{2}\left(K_a + K^{5}_{a}\right)$$

then

$$\left[K_a,\ K_b\right] = i\,\varepsilon_{abc}\,K_c$$

$$\left[K_a,\ K^{5}_{b}\right] = i\,\varepsilon_{abc}\ K^{5}_{c}$$

hence

$$\left[K^{R}_{a},\ K^{R}_{b}\right] = i\,\varepsilon_{abc}\ K^{R}_{c}$$

$$\left[K^L_a, K^L_b \right] = i\, \varepsilon_{abc}\, K^L_c$$

$$\left[K^R_a, K^L_b \right] = 0 \; ; \qquad a, b = 1,2,3$$

which define the SU(2) ⊗ SU(2) algebra.

d) As the total lepton number is

$$N = \int d^3x \sum_\ell \psi^+_\ell(x)\, \psi_\ell(x) =$$

$$= \int d^3x \sum_\ell \left\{ \nu^+_\ell(x)\, \nu_\ell(x) + \ell^+(x)\, \ell(x) \right\}$$

and the charge is :

$$Q_{(\gamma)} = - \int d^3x \sum_\ell \psi^+_\ell(x) \frac{1 - \tau_3}{2} \psi_\ell(x)$$

$$= - \int d^3x \sum_\ell \ell^+(x)\, \ell(x)$$

then

$$Q_{(\gamma)} = K_3 - \frac{1}{2} N$$

As the total lepton number commutes with all the operators, then :

$$\left[Q_{(\gamma)}, K_+ \right] = K_+$$

$$\left[Q_{(\gamma)}, K_- \right] = - K_-$$

$$\left[Q_{(\gamma)}, K_3 \right] = 0$$

VI - 6

a) $$\vec{\nabla} \cdot \vec{\mathscr{E}}_{(v)} + m^2_v \phi^0 = j^0$$

$$\vec{\nabla} \times \vec{B}_{(v)} - \partial_0 \vec{\mathscr{E}}_{(v)} + m^2_v \vec{\phi} = \vec{j}$$

$$\vec{\nabla} \cdot \vec{B}_{(v)} = 0$$

$$\vec{\nabla} \times \vec{\mathscr{E}}_{(v)} + \partial_0 \vec{B}_{(v)} = 0$$

the coupling constant is incorporated in the current.

b) $$\vec{\nabla} \cdot \vec{B}_{(a)} + m^2_a a^0 = \rho^0$$

$$\vec{\nabla} \times \vec{\mathscr{E}}_{(a)} - \partial_0 \vec{B}_{(a)} + m^2_a \vec{a} = \vec{\rho}$$

$$\vec{\nabla} \cdot \vec{\mathscr{E}}_{(a)} = 0$$

$$\vec{\nabla} \times \vec{B}_{(a)} + \partial_0 \vec{\mathscr{E}}_{(a)} = 0$$

c) $$\vec{\nabla} \cdot \vec{\mathscr{E}}_{(w)} + m^2 W^0 = j^0 - \rho^0$$

$$\vec{\nabla} \times \vec{B}_{(w)} - \partial_0 \vec{\mathscr{E}}_{(w)} + m^2 \vec{W} = \vec{j} - \vec{\rho}$$

$$\vec{\nabla} \cdot \vec{B}_{(w)} = 0$$

$$\vec{\nabla} \times \vec{\mathscr{E}}_{(w)} + \partial_0 \vec{B}_{(w)} = 0$$

where :

$$W^{\mu} = \phi^{\mu} - a^{\mu}$$

$$\vec{\mathscr{E}}_{(w)} = \vec{\mathscr{E}}_{(v)} - \vec{B}_{(a)}, \quad \vec{B}_{(w)} = \vec{B}_{(v)} - \vec{\mathscr{E}}_{(a)}$$

VII - 1

$$L = \frac{1}{2} m (\dot{x})^2 - U(x)$$

$$U(x) = \frac{1}{2} m \omega^2 x^2 + \frac{\lambda}{4} x^4$$

$$\frac{\partial U}{\partial x} = x(m \omega^2 + \lambda n^2) = 0$$

$$x = \pm a \equiv \pm \sqrt{- \frac{m \omega^2}{\lambda}}, \quad m > 0, \lambda > 0, \omega^2 < 0$$

For $x' = x - a$

$$L = \frac{1}{2} m (\dot{x}')^2 - \frac{1}{2} m \alpha^2 x'^2 - \lambda a x'^3 - \frac{\lambda}{4} x'^4$$

where $\omega^2 = - \frac{1}{2} \alpha^2 < 0$

VIII - 1

For instance :

$$(1 - i \vec{\Lambda} \cdot \vec{T}) L (1 + i \vec{\Lambda} \cdot T) = (1 + i \vec{\Lambda} \cdot \frac{\vec{\tau}}{2}) L$$

leads to :

$$\left[L, \vec{T} \right] = \frac{\vec{\tau}}{2} L$$

VIII - 2

$$\mathscr{L} = -\frac{1}{4} F^{\mu\nu} F_{\mu\nu\,;\,a}$$

$$F^{\mu\nu}{}_a = \partial^\nu A^\mu{}_a - \partial^\mu A^\nu{}_a + g\,\varepsilon_{abc} A^\mu{}_b A^\nu{}_c$$

a) Equations of motion :

$$\partial_\nu F^{\mu\nu}{}_a + g\,\varepsilon_{abc} F^{\mu\nu}{}_b A_{\nu;\,c} = 0$$

or :

$$D_{\nu;ab} F^{\mu\nu}{}_b = 0$$

$$D_{\nu;ab} = \partial_\nu \delta_{ab} + g\,\varepsilon_{abc} A_{\nu;\,c}$$

Also :

$$D_{\alpha;ab} F_{\beta\lambda;b} + D_{\lambda;ab} F_{\alpha\beta;b} + D_{\beta;ab} F_{\lambda\alpha;b} = 0$$

The equations of motion are those which have time-derivative of fields :

$$\partial_o F^{jo}{}_a + \partial_k F^{jk}{}_a + g\,\varepsilon_{abc} F^{jn}{}_b A_{n;\,c} = 0$$

where

$$F^{oj}{}_a = -\partial_j A^o{}_a - \partial^o A^j{}_a + g\,\varepsilon_{abc} A^o{}_b A^j{}_c$$

The remaining equations are constraint relations on the field variables :

$$\partial_k F^{ok}{}_a + g\,\varepsilon_{abc} F^{ok}{}_b A_{k;\,c} = 0$$

and :

$$F^{jk}{}_a = \partial_j A^k - \partial_k A^j + g\, \varepsilon_{abc} A^j{}_b A^k{}_c$$

In terms of the electric-like and magnetic-like fields we have :

$$\vec{E}^k = - \partial_k \vec{A}^0 - \partial^0 \vec{A}^k + g\, (\vec{A}^0 \times \vec{A}^k)$$

$$\partial_k \vec{E}^k + g\, (\vec{A}^k \times \vec{E}^k) = 0$$

$$\vec{F}^{jk} = \partial_j \vec{A}^k - \partial_k \vec{A}^j + g\, (\vec{A}^j \times \vec{A}^k)$$

$$\partial_k \vec{F}^{jk} - \partial_0 \vec{E}^j + g(\vec{A}^k \times \vec{F}^{jk}) = 0$$

b) The canonical momenta are

$$\pi^k{}_a = \frac{\partial \mathscr{L}}{\partial\, \partial_0 A_{k;a}} F^{ok}{}_a \quad ; \quad \pi^0{}_a = 0$$

and the same result as in electrodynamics that $\pi^0{}_a = 0$ is found.

$$H = \int d^3x\, \frac{1}{2} \left\{ \vec{E}^{k\,2} + \vec{B}^{k\,2} \right\}$$

The commutators are :

$$\left[\pi^k{}_a(\vec{x}, \vec{t}),\, A^\ell{}_b(\vec{x}, \vec{t}) \right] = i\, \delta_{ab}\, \delta^{k\ell}\, \delta^3(x - x')$$

III - 3

The current in gauge theories

$$\partial_\nu F^{\mu\nu}{}_a = g\, j^\mu{}_a$$

is in the SU(2) case :

$$j^{\mu}{}_{a} = \bar{\psi}\, \gamma^{\mu}\, \frac{\tau_a}{2}\, \psi - \varepsilon_{abc}\, F^{\mu\nu}{}_{b}\, A_{\nu;\, c}$$

The corresponding charges are :

$$Q_a = \int d^3x\; j^{0}{}_{a}(x) = \int d^3x \left\{ \psi^{+}(x)\, \frac{\tau_a}{2}\, \psi(x) + \varepsilon_{abc}\, \pi^{k}{}_{b}(x)\, A^{k}{}_{c}(x) \right\}$$

From the commutation rules

$$\left[\psi_{\alpha k}(\vec{x}, t),\, \psi^{+}_{\beta \ell}(\vec{x}\,', t) \right]_{+} = \delta_{\alpha\beta}\, \delta_{k\ell}\, \delta^{3}(x - x')$$

$$\left[\pi^{\ell}{}_{\alpha}\, (\vec{x}, t),\, A^{k}{}_{\beta}(\vec{x}\,', t) \right]_{-} = i\, \delta_{\alpha\beta}\, \delta^{\ell k}\, \delta^{3}(x - x')$$

it follows that

$$\left[Q_a,\, Q_b \right]_{-} = i\, \varepsilon_{abc}\, Q_c$$

IX - 1

$$\left\{ \gamma^{\mu}\, (i\, \partial_{\mu} - e\, A_{\mu}) - m \right\} \psi = 0$$

gives rise to the equation for $\bar{\psi}$:

$$(i\, \partial_{\mu} + e\, A_{\mu})\, \bar{\psi}\, \gamma^{\mu} + m\, \bar{\psi} = 0$$

which, compared with :

$$\left\{ \gamma^{\mu}\, (i\, \partial_{\mu} + e\, A_{\mu}) - m \right\} \psi^{c} = 0$$

gives the relations.

IX - 2

As :

$$\mathscr{C}^{-1}\, Q\, \mathscr{C} = -\, Q$$

then

$$\mathscr{C}^{-1}\, e^{i\alpha Q}\, \mathscr{C}^{-1} = e^{-i\alpha Q}$$

IX - 3

a)

$$\frac{\vec{\Sigma} \cdot \vec{p}}{|\vec{p}|} u(p, s) = \gamma^5 u(p, s)$$

$$\frac{\vec{\Sigma} \cdot \vec{p}}{|\vec{p}|} v(p, s) = \gamma^5 v (p, s)$$

b) if

$$\frac{\vec{\Sigma} \cdot \vec{p}}{|\vec{p}|} u(p, s) = s\, u(p, s)$$

then :

$$\frac{\vec{\Sigma} \cdot \vec{p}}{|\vec{p}|} v(p, s) = - s\, v (p, s)$$

$$s = \pm 1.$$

Then :

$$\frac{1 - \gamma^5}{2} u (p, s) = \frac{1}{2} (I - \frac{\vec{\Sigma} \cdot \vec{p}}{|\vec{p}|}) u (p, s) =$$

$$= \frac{1}{2} (1 - s) u (p, s) = u_L(p)$$

where

$$u_L(p) = u (p, - 1) ;$$

$$\frac{1 - \gamma^5}{2} v (p, s) = \frac{1}{2} (1 + s) v (p, s) = v_R(p)$$

where

$$v_R(p) = v (p, 1)$$

c) From the Fourier development of ψ one obtains :

$$\psi_L(x) = \frac{1}{(2\pi)^{3/2}} \int \frac{d^3p}{2p^0} \Big\{ a\ (p_1 - 1)\ u_L(p)\ e^{-ipx} + $$

$$+ b^+\ (p, 1)\ v_R\ (p)\ e^{ipx} \Big\}$$

d) $$(\psi_L)^c = (\psi^c)_R \quad ; \quad (\psi_R)^c = (\psi^c)_L$$

IX - 4

Let us define the transformed spinor's ψ' and $\psi^{c\prime}$ by the equations

$$\psi' = \frac{1}{\sqrt{2}} (\psi - \gamma^5 \psi^c)$$

$$\psi^{c\prime} = \frac{1}{\sqrt{2}} (\psi^c + \gamma^5 \psi)$$

Now we can write the identities :

$$\psi = \frac{1}{\sqrt{2}} (M + N),$$

$$\psi^c = \frac{1}{\sqrt{2}} (M - N)$$

where :

$$M = \frac{1}{\sqrt{2}} (\psi + \psi^c) = M^c$$

$$N = \frac{1}{\sqrt{2}} (\psi - \psi^c) = - N^c$$

Therefore

$$\psi' = M_L + N_R$$

$$\psi^{c\,\prime} = M_R - N_L$$

For a left handed ψ'_L we have thus

$$\psi'_L = M_L$$

$$(\psi'_L)^c \equiv (\psi'^c)_R = M_R$$

As ψ' obeys the same equal time anticommutation rules as ψ there exists a unitary transformation between ψ and ψ' :

$$\psi' = U^{-1}\,\psi\,U$$

hence :

$$U^{-1}\,\psi_L\,U = M_L$$

One obtains also for the lagrangeans :

$$U^{-1}\,\mathscr{L}[\psi_L]\,U = \mathscr{L}(M_L)$$

A state with a left-handed Dirac neutrino is equivalent to a state with a left-handed Majorana neutrino. In the Majorana representation, a Majorana neutrino, self-charge conjugate spinor, can be taken as a real spinor. But a left-handed Majorana neutrino is not real and its charge conjugate is a right-handed Majorana neutrino.

X - 1

a) $[a_A, N] = a_A, \quad [a_B, N] = a_B$

b) $$a_A \mid n_A, n_B > \; = \; \sqrt{n_A} \mid n_A - 1, n_B >$$

$$a_B \mid n_A, n_B > \; = \; \sqrt{n_B} \mid n_A, n_B - 1 >$$

$$a^+_B \mid n_A, n_B > \; = \; \sqrt{n_A + 1} \mid n_A + 1, n_B >$$

$$a^+_B \mid n_A, n_B > \; = \; \sqrt{n_B + 1} \mid n_A, n_B + 1 >$$

c, d) $$\tau_+ \equiv a^+_A \, a_B = \begin{pmatrix} 0 & 1 \\ 0 & 0 \end{pmatrix}$$

$$\tau_- \equiv a^+_B \, a_A = \begin{pmatrix} 0 & 0 \\ 1 & 0 \end{pmatrix}$$

$$\tau_3 \equiv a^+_A \, a_A - a^+_B \, a_B = \begin{pmatrix} 1 & 0 \\ 0 & -1 \end{pmatrix}$$

whence

$$\tau_1 = \tau_+ + \tau_-$$

$$\tau_2 = i \, (\tau_- - \tau_+)$$

X - 2

b) For SU_4 let the four objects be demoted u, d, s, c. Find for the generators $\frac{\eta_k}{2}$:

$$\eta_1 = a^+_u \, a_d + a^+_d \, a_u = \left(\begin{array}{ccc|c} & & & 0 \\ & \lambda_1 & & 0 \\ & & & 0 \\ 0 & 0 & 0 & 0 \end{array}\right)$$

$$\eta_2 = - i \, (a^+_u \, a_d - a^+_d \, a_u) = \left(\begin{array}{ccc|c} & & & 0 \\ & \lambda_2 & & 0 \\ & & & 0 \\ 0 & 0 & 0 & 0 \end{array}\right)$$

$$\eta_3 = a^+_u a_u - a^+_d a_d = \begin{pmatrix} & \lambda_3 & & \begin{matrix}0\\0\\0\end{matrix} \\ 0 & 0 & 0 & 0 \end{pmatrix}$$

$$\eta_4 = a^+_u a_s + a^+_s a_u = \begin{pmatrix} & \lambda_4 & & \begin{matrix}0\\0\\0\end{matrix} \\ 0 & 0 & 0 & 0 \end{pmatrix}$$

$$\eta_5 = -i(a^+_u a_s - a^+_s a_u) = \begin{pmatrix} & \lambda_5 & & \begin{matrix}0\\0\\0\end{matrix} \\ 0 & 0 & 0 & 0 \end{pmatrix}$$

$$\eta_6 = a^+_d a_s + a^+_s a_d = \begin{pmatrix} & \lambda_6 & & \begin{matrix}0\\0\\0\end{matrix} \\ 0 & 0 & 0 & 0 \end{pmatrix}$$

$$\eta_7 = -i(a^+_d a_s - a^+_s a_d) = \begin{pmatrix} & \lambda_7 & & \begin{matrix}0\\0\\0\end{matrix} \\ 0 & 0 & 0 & 0 \end{pmatrix}$$

$$\eta_8 = \frac{1}{\sqrt{3}}(a^+_u a_u + a^+_d a_d - 2a^+_s a_s) = \begin{pmatrix} & \lambda_8 & & \begin{matrix}0\\0\\0\end{matrix} \\ 0 & 0 & 0 & 0 \end{pmatrix}$$

$$\eta_9 = a^+_u a_c + a^+_c a_u = \begin{pmatrix} 0 & 0 & 0 & 1 \\ 0 & 0 & 0 & 0 \\ 0 & 0 & 0 & 0 \\ 1 & 0 & 0 & 0 \end{pmatrix}$$

$$\eta_{10} = -i(a^+_u a_c - a^+_c a_u) = \begin{pmatrix} 0 & 0 & 0 & -i \\ 0 & 0 & 0 & 0 \\ 0 & 0 & 0 & 0 \\ i & 0 & 0 & 0 \end{pmatrix}$$

$$\eta_{11} = a^+_d a_c + a^+_c a_d = \begin{pmatrix} 0 & 0 & 0 & 0 \\ 0 & 0 & 0 & 1 \\ 0 & 0 & 0 & 0 \\ 0 & 1 & 0 & 0 \end{pmatrix}$$

$$\eta_{12} = -i(a^+_d a_c - a^+_c a_d) = \begin{pmatrix} 0 & 0 & 0 & 0 \\ 0 & 0 & 0 & -i \\ 0 & 0 & 0 & 0 \\ 0 & i & 0 & 0 \end{pmatrix}$$

$$\eta_{13} = a^+_s a_c + a^+_c a_s = \begin{pmatrix} 0 & 0 & 0 & 0 \\ 0 & 0 & 0 & 0 \\ 0 & 0 & 0 & 1 \\ 0 & 0 & 1 & 0 \end{pmatrix}$$

$$\eta_{14} = -i(a^+_s a_c - a^+_c a_s) = \begin{pmatrix} 0 & 0 & 0 & 0 \\ 0 & 0 & 0 & 0 \\ 0 & 0 & 0 & -i \\ 0 & 0 & i & 0 \end{pmatrix}$$

$$\eta_{15} = \frac{1}{\sqrt{6}} (a^+_u a_u + a^+_d a_d + a^+_s a_s - 3 a^+_c a_c) = \frac{1}{\sqrt{6}} \begin{pmatrix} 1 & 0 & 0 & 0 \\ 0 & 1 & 0 & 0 \\ 0 & 0 & 1 & 0 \\ 0 & 0 & 0 & -3 \end{pmatrix}$$

where the $\frac{\lambda_k}{2}$'s are the generators of the SU(3) group. Find the latter and those of SU(5).

The baryon number is :

$$B = \frac{1}{3} (a^+_u a_u + a^+_d a_d + a^+_s a_s + a^+_c a_c)$$

$$= \frac{1}{3} \begin{pmatrix} 1 & 0 & 0 & 0 \\ 0 & 1 & 0 & 0 \\ 0 & 0 & 1 & 0 \\ 0 & 0 & 0 & 1 \end{pmatrix}$$

and one has for the charm C

$$C = a^+_c a_c = \frac{3}{4} (B - \frac{\sqrt{6}}{3} \eta_{15})$$

for the hypercharge Y

$$Y = \frac{1}{\sqrt{3}} \eta_8 - \frac{2}{3} C$$

and for the charge Q

$$Q = \frac{1}{2} \eta_3 + \frac{1}{2} (Y + 2 C)$$

d) For SU(2) one diagonal matrix, τ_3 ; for SU(3), two diagonal matrices, λ_3, λ_8 ; for SU(4), three, η_3, η_8, η_{15} ; for SU(5), four, t_3, t_8, t_{15}, t_{24} ; study the general properties of SU(n).

CONCEPTUAL FOUNDATIONS OF THE UNIFIED THEORY OF WEAK AND ELECTROMAGNETIC INTERACTIONS

Nobel Lecture, December 8, 1979
by STEVEN WEINBERG
Lyman Laboratory of Physics Harvard University and Harvard-Smithsonian Center for Astrophysics Cambridge, Mass., USA.

Our job in physics is to see things simply, to understand a great many complicated phenomena in a unified way, in terms of a few simple principles. At times, our efforts are illuminated by a brilliant experiment, such as the 1973 discovery of neutral current neutrino reactions. But even in the dark times between experimental breakthroughs, there always continues a steady evolution of theoretical ideas, leading almost imperceptibly to changes in previous beliefs. In this talk, I want to discuss the development of two lines of thought in theoretical physics. One of them is the slow growth in our understanding of symmetry, and in particular, broken or hidden symmetry. The other is the old struggle to come to terms with the infinities in quantum field theories. To a remarkable degree, our present detailed theories of elementary particle interactions can be understood deductively, as consequences of symmetry principles and of a principle of renormalizability which is invoked to deal with the infinities. I will also briefly describe how the convergence of these lines of thought led to my own work on the unification of weak and electromagnetic interactions. For the most part, my talk will center on my own gradual education in these matters, because that is one subject on which I can speak with some confidence. With rather less confidence, I will also try to look ahead, and suggest what role these lines of thought may play in the physics of the future.

Symmetry principles made their appearance in twentieth century physics in 1905 with Einstein's identification of the invariance group of space and time. With this as a precedent, symmetries took on a character in physicists' minds as *a priori* principles of universal validity, expressions of the simplicity of nature at its deepest level. So it was painfully difficult in the 1930's to realize that there are internal symmetries, such as isospin conservation, [1] having nothing to do with space and time, symmetries which are far from self-evident, and that only govern what are now called the strong interactions. The 1950's saw the discovery of another internal symmetry – the conservation of strangeness [2] – which is not obeyed by the weak interactions, and even one of the supposedly sacred symmetries of space-time – parity – was also found to be violated by weak interactions. [3] Instead of moving toward unity, physicists were learning that different interactions are apparently governed by quite different symmetries. Matters became yet more confusing with the recognition in the early 1960's of a symmetry group – the "eightfold way" – which is not even an exact symmetry of the strong interactions. [4]

These are all "global" symmetries, for which the symmetry transformations do not depend on position in space and time. It had been recognized [5] in the 1920's that quantum electrodynamics has another symmetry of a far more powerful kind, a "local" symmetry under transformations in which the electron field suffers a phase change that can vary freely from point to point in space-time, and the electromagnetic vector potential undergoes a corresponding gauge transformation. Today this would be called a U(1) gauge symmetry, beacause a simple phase change can be thought of as multiplication by a 1×1 unitary matrix. The extension to more complicated groups was made by Yang and Mills [6] in 1954 in a seminal paper in which they showed how to construct an SU(2) gauge theory of strong interactions. (The name "SU(2)" means that the group of symmetry transformations consists of 2×2 unitary matrices that are "special," in that they have determinant unity). But here again it seemed that the symmetry if real at all would have to be approximate, because at least on a naive level gauge invariance requires that vector bosons like the photon would have to be massless, and it seemed obvious that the strong interactions are not mediated by massless particles. The old question remained: if symmetry principles are an expression of the simplicity of nature at its deepest level, then how can there be such a thing as an approximate symmetry? Is nature only approximately simple?

Some time in 1960 or early 1961, I learned of an idea which had originated earlier in solid state physics and had been brought into particle physics by those like Heisenberg, Nambu, and Goldstone, who had worked in both areas. It was the idea of "broken symmetry," that the Hamiltonian and commutation relations of a quantum theory could possess an exact symmetry, and that the physical states might nevertheless not provide neat representations of the symmetry. In particular, a symmetry of the Hamiltonian might turn out to be not a symmetry of the vacuum.

As theorists sometimes do, I fell in love with this idea. But as often happens with love affairs, at first I was rather confused about its implications. I thought (as turned out, wrongly) that the approximate symmetries – parity, isospin, strangeness, the eight-fold way – might really be exact *a priori* symmetry principles, and that the observed violations of these symmetries might somehow be brought about by spontaneous symmetry breaking. It was therefore rather disturbing for me to hear of a result of Goldstone, [7] that in at least one simple case the spontaneous breakdown of a continuous symmetry like isospin would necessarily entail the existence of a massless spin zero particle – what would today be called a "Goldstone boson." It seemed obvious that there could not exist any new type of massless particle of this sort which would not already have been discovered.

I had long discussions of this problems with Goldstone at Madison in the summer of 1961, and then with Salam while I was his guest at Imperial College in 1961–62. The three of us soon were able to show that Goldstone bosons must in fact occur whenever a symmetry like isospin or strangeness is spontaneously broken, and that their masses then remain zero to all orders of perturbation theory. I remember being so discouraged by these zero masses that when we wrote our joint paper on the subject, [8] I added an epigraph to the paper to underscore the futility of supposing that anything could be explained in terms of a non-invariant vacuum state: it was Lear's retort to Cordelia, "Nothing will come of nothing: speak again." Of course, The Physical Review protected the purity of the physics literature, and removed the quote. Considering the future of the non-invariant vacuum in theoretical physics, it was just as well.

There was actually an exception to this proof, pointed out soon afterwards by Higgs, Kibble, and others. [9] They showed that if the broken symmetry is a local, gauge symmetry, like electromagnetic gauge invariance, then although the Goldstone bosons exist formally, and are in some sense real, they can be eliminated by a gauge transformation, so that they do not appear as physical particles. The missing Goldstone bosons appear instead as helicity zero states of the vector particles, which thereby acquire a mass.

I think that at the time physicists who heard about this exception generally regarded it as a technicality. This may have been because of a new development in theoretical physics which suddenly seemed to change the role of Goldstone bosons from that of unwanted intruders to that of welcome friends.

In 1964 Adler and Weisberger [10] independently derived sum rules which gave the ratio g_A/g_V of axial-vector to vector coupling constants in beta decay in terms of pion-nucleon cross sections. One way of looking at their calculation, (perhaps the most common way at the time) was as an analogue to the old dipole sum rule in atomic physics: a complete set of hadronic states is inserted in the commutation relations of the axial vector currents. This is the approach memorialized in the name of "current algebra." [11] But there was another way of looking at the Adler-Weisberger sum rule. One could suppose that the strong interactions have an approximate symmetry, based on the group $SU(2) \times SU(2)$, and that this symmetry is spontaneously broken, giving rise among other things to the nucleon masses. The pion is then identified as (approximately) a Goldstone boson, with small non-zero mass, an idea that goes back to Nambu. [12] Although the $SU(2) \times SU(2)$ symmetry is spontaneously broken, it still has a great deal of predictive power, but its predictions take the form of approximate formulas, which give the matrix elements for low energy pionic reactions. In this approach, the Adler-Weisberger sum rule is obtained by using the predicted pion nucleon scattering lengths in conjunction with a well-known sum rule [13], which years earlier had been derived from the dispersion relations for pion-nucleon scattering.

In these calculations one is really using not only the fact that the strong interactions have a spontaneously broken approximate SU(2) × SU(2) symmetry, but also that the currents of this symmetry group are, up to an overall constant, to be identified with the vector and axial vector currents of beta decay. (With this assumption g_A/g_V gets into the picture through the Goldberger-Treiman relation, [14] which gives g_A/g_V in terms of the pion decay constant and the pion nucleon coupling.) Here, in this relation between the currents of the symmetries of the strong interactions and the physical currents of beta decay, there was a tantalizing hint of a deep connection between the weak interactions and the strong interactions. But this connection was not really understood for almost a decade.

I spent the years 1965–67 happily developing the implications of spontaneous symmetry breaking for the strong interactions. [15] It was this work that led to my 1967 paper on weak and electromagnetic unification. But before I come to that I have to go back in history and pick up one other line of though, having to do with the problem of infinities in quantum field theory.

I believe that it was Oppenheimer and Waller in 1930 [16] who independently first noted that quantum field theory when pushed beyond the lowest approximation yields ultraviolet divergent results for radiative self energies. Professor Waller told me last night that when he described this result to Pauli, Pauli did not believe it. It must have seemed that these infinities would be a disaster for the quantum field theory that had just been developed by Heisenberg and Pauli in 1929–30. And indeed, these infinites did lead to a sense of discouragement about quantum field theory, and many attempts were made in the 1930's and early 1940's to find alternatives. The problem was solved (at least for quantum electrodynamics) after the war, by Feynman, Schwinger, and Tomonaga [17] and Dyson [19]. It was found that all infinities disappear if one identifies the observed finite values of the electron mass and charge, not with the parameters m and e appearing in the Lagrangian, but with the electron mass and charge that are *calculated* from m and e, when one takes into account the fact that the electron and photon are always surrounded with clouds of virtual photons and electron-positron pairs [18]. Suddenly all sorts of calculations became possible, and gave results in spectacular agreement with experiment.

But even after this success, opinions differed as to the significance of the ultraviolet divergences in quantum field theory. Many thought – and some still do think – that what had been done was just to sweep the real problems under the rug. And it soon became clear that there was only a limited class of so-called "renormalizable" theories in which the infinities could be eliminated by absorbing them into a redefinition, or a "renormalization," of a finite number of physical parameters. (Roughly speaking, in renormalizable theories no coupling constants can have the dimensions of negative powers of mass. But every time we add a field or a space-time derivative to an interaction, we reduce the dimensionality of the associated coupling constant. So only a few simple types of interaction can be renormalizable.) In particular, the existing Fermi theory of weak interactions clearly was not renormalizable. (The Fermi coupling constant has the dimensions of $[\text{mass}]^{-2}$.) The sense of discouragement about quantum field theory persisted into the 1950's and 1960's.

I learned about renormalization theory as a graduate student, mostly by reading Dyson's papers. [19] From the beginning it seemed to me to be a wonderful thing that very few quantum field theories are renormalizable. Limitations of this sort are, after all, what we most *want,* not mathematical methods which can make sense of an infinite variety of physically irrelevant theories, but methods which carry constraints, because these constraints may point the way toward the one true theory. In particular, I was impressed by the fact that quantum electrodynamics could in a sense be *derived* from symmetry principles and the constraints of renormalizability; the only Lorentz invariant and gauge invariant renormalizable Lagrangian for photons and electrons is precisely the orginal Dirac Lagrangian of QED. Of course, that is not the way Dirac came to his theory. He had the benefit of the information gleaned in centuries of experimentation on electromagnetism, and in order to fix the final form of his theory he relied on ideas of simplicity (specifically, on what is sometimes called minimal electromagnetic coupling). But we have to look ahead, to try to make theories of phenomena which have not been so well studied experimentally, and we may not be able to trust purely formal ideas of simplicity. I thought that renormalizability might be the key criterion, which also in a more general context would impose a precise kind of simplicity on our theories and help us to pick out the one true physical theory out of the infinite variety of conceivable quantum field theories. As I will explain later, I would say this a bit differently today, but I am more convinced than ever that the use of renormalizability as a constraint on our theories of the observed interactions is a good strategy. Filled with enthusiasm for renormalization theory, I wrote my Ph.D. thesis under Sam Treiman in 1957 on the use of a limited version of renormalizability to set constraints on the weak interactions, [20] and a little later I worked out a rather tough little theorem [21] which completed the proof by Dyson [19] and Salam [22] that ultraviolet divergences really do cancel out to all orders in nominally renormalizable theories. But none of this seemed to help with the important problem, of how to make a renormalizable theory of weak interactions.

Now, back to 1967. I had been considering the implications of the broken $SU(2) \times SU(2)$ symmetry of the strong interactions, and I thought of trying out the idea that perhaps the $SU(2) \times SU(2)$ symmetry was a "local," not merely a "global," symmetry. That is, the strong interactions might be described by something like a Yang-Mills theory, but in addition to the vector ϱ mesons of the Yang-Mills theory, there would also be axial vector A1 mesons. To give the ϱ meson a mass, it was necessary to insert a common ϱ and A1 mass term in the Lagrangian, and the spontaneous breakdown of the SU(2) × SU(2) symmetry would then split the ϱ and A1 by something like the Higgs mechanism, but since the theory would not be gauge invariant the pions would remain as physical Goldstone bosons. This theory gave an intriguing result, that the A1/ϱ mass ratio should be $\sqrt{2}$, and in trying to understand this result without relying on perturbation theory, I discovered certain sum rules, the "spectral function sum rules," [23] which turned out to have variety of other uses. But the SU(2) × SU(2) theory was not gauge invariant, and hence it could not be renormalizable, [24] so I was not too enthusiastic about it. [25] Of course, if I did not insert the ϱ-A1 mass term in the Lagrangian, then the theory would be gauge invariant and renormalizable, and the A1 would be massive. But then there would be no pions and the ϱ mesons would be massless, in obvious contradiction (to say the least) with observation.

At some point in the fall of 1967, I think while driving to my office at M.I.T., it occurred to me that I had been applying the right ideas to the wrong problem. It is not the ϱ mesons that is massless: it is the photon. And its partner is not the A1, but the massive intermediate boson, which since the time of Yukawa had been suspected to be the mediator of the weak interactions. The weak and electromagnetic interactions could then be described [26] in a unified way in terms of an exact but spontaneously broken gauge symmetry. [Of course, not necessarily SU(2) × SU(2)]. And this theory would be renormalizable like quantum electrodynamics because it is gauge invariant like quantum electrodynamics.

It was not difficult to develop a concrete model which embodied these ideas. I had little confidence then in my understanding of strong interactions, so I decided to concentrate on leptons. There are two left-handed electron-type leptons, the ν_{eL} and e_L, and one right-handed electron-type lepton, the e_R, so I started with the group U(2) × U(1): all unitary 2 × 2 matrices acting on the left-handed e-type leptons, together with all unitary 1 × 1 matrices acting on the right-handed e-type lepton. Breaking up U(2) into unimodular transformations and phase transformations, one could say that the group was SU(2) × U(1) × U(1). But then one of the U(1)'s could be identified with ordinary lepton number, and since lepton number appears to be conserved and there is no massless vector particle coupled to it, I decided to exclude it from the group. This left the four-parameter group SU(2) × U(1). The spontaneous breakdown of SU(2) × U(1) to the U(1) of ordinary electromagnetic gauge invariance would give masses to three of the four vector gauge bosons: the charged bosons $W^{\pm}$, and a neutral boson that I called the Z^0. The fourth boson would automatically remain massless, and could be identified as the photon. Knowing the strength of the ordinary charged current weak interactions like beta decay which are mediated by $W^{\pm}$, the mass of the $W^{\pm}$ was then determined as about 40 GeV/sinθ, where θ is the γ-Z^0 mixing angle.

To go further, one had to make some hypothesis about the mechanism for the breakdown of SU (2) × U (1). The only kind of field in a renormalizable SU(2) × U(1) theory whose vacuum expectation values could give the electron a mass is a spin zero SU(2) doublet (Φ^+, Φ^0), so for simplicity I assumed that these were the only scalar fields in the theory. The mass of the Z^0 was then determined as about 80 GeV/sin 2Θ. This fixed the strength of the neutral current weak interactions. Indeed, just as in QED, once one decides on the menu of fields in the theory all details of the theory are completely determined by symmetry principles and renormalizability, with just a few free parameters: the lepton charge and masses, the Fermi coupling constant of beta decay, the mixing angle Θ, and the mass of the scalar particle. (It was of crucial importance to impose the constraint of renormalizability; otherwise weak interactions would receive contributions from SU(2)×U(I)−invariant four-fermion couplings as well as from vector boson exchange, and the theory would lose most of its predictive power.) The naturalness of the whole theory is well demonstrated by the fact that much the same theory was independently developed [27] by Salam in 1968.

The next question now was renormalizability. The Feynman rules for Yang–Mills theories with unbroken gauge symmetries had been worked out [28] by deWitt, Faddeev and Popov and others, and it was known that such theories are renormalizable. But in 1967 I did not know how to prove that this renormalizability was not spoiled by the spontaneous symmetry breaking. I worked on the problem on and off for several years, partly in collaboration with students, [29] but I made little progress. With hindsight, my main difficulty was that in quantizing the vector fields I adopted a gauge now known as the unitarity gauge [30]: this gauge has several wonderful advantages, it exhibits the true particle spectrum of the theory, but it has the disadvantage of making renormalizability totally obscure.

Finally, in 1971 't Hooft [31] showed in a beautiful paper how the problem could be solved. He invented a gauge, like the "Feynman gauge" in QED, in which the Feynman rules manifestly lead to only a finite number of types of ultraviolet divergence. It was also necessary to show that these infinities satisfied essentially the same constraints as the Lagrangian itself, so that they could be absorbed into a redefinition of the parameters of the theory. (This was plausible, but not easy to prove, because a gauge invariant theory can be quantized only after one has picked a specific gauge, so it is not obvious that the ultraviolet divergences satisfy the same gauge invariance constraints as the Lagrangian itself.) The proof was subsequently completed [32] by Lee and Zinn-Justin and by 't Hooft and Veltman. More recently, Becchi, Rouet and Stora [33] have invented an ingenious method for carrying out this sort of proof, by using a global supersymmetry of gauge theories which is preserved even when we choose a specific gauge.

I have to admit that when I first saw 't Hooft's paper in 1971, I was not convinced that he had found the way to prove renormalizability. The trouble was not with 't Hooft, but with me: I was simply not familiar enough with the path integral formalism on which 't Hooft's work was based, and I wanted to see a derivation of the Feynman rules in 't Hooft's gauge from canonical quantization. That was soon supplied (for a limited class of gauge theories) by a paper of Ben Lee, [34] and after Lee's paper I was ready to regard the renormalizability of the unified theory as essentially proved.

By this time, many theoretical physicists were becoming convinced of the general approach that Salam and I had adopted: that is, the weak and electromagnetic interactions are governed by some group of exact local gauge symmetries; this group is spontaneously broken to U(1), giving mass to all the vector bosons except the photon; and the theory is renormalizable. What was not so clear was that our specific simple model was the one chosen by nature. That, of course, was a matter for experiment to decide.

It was obvious even back in 1967 that the best way to test the theory would be by searching for neutral current weak interactions, mediated by the neutral intermediate vector boson, the Z^0. Of course, the possibility of neutral currents was nothing new. There had been speculations [35] about possible neutral currents as far back as 1937 by Gamow and Teller, Kemmer, and Wentzel, and again in 1958 by Bludman and Leite-Lopes. Attempts at a unified weak and electromagnetic theory had been made [36] by Glashow and Salam and Ward in the early 1960's, and these had neutral currents with many of the features that Salam and I encountered in developing the 1967–68 theory. But since one of the predictions of our theory was a value for the mass of the Z^0, it made a definite prediction of the strength of the neutral currents. More important, now we had a comprehensive quantum field theory of the weak and electromagnetic interactions that was physically and mathematically satisfactory in the same sense as was quantum electrodynamics—a theory that treated photons and intermediate vector bosons on the same footing, that was based on an exact symmetry principle, and that allowed one to carry calculations to any desired degree of accuracy. To test this theory, it had now become urgent to settle the question of the existence of the neutral currents.

Late in 1971, I carried out a study of the experimental possibilites. [37] The results were striking. Previous experiments had set upper bounds on the rates of neutral current processes which were rather low, and many people had received the impression that neutral currents were pretty well ruled out, but I found that in fact the 1967–68 theory *predicted* quite low rates, low enough in fact to have escaped clear detection up to that time. For instance, experiments [38] a few years earlier had found an upper bound of 0.12 ± 0.06 on the ratio of a neutral current process, the elastic scattering of muon neutrinos by protons, to the corresponding charged current process, in which a muon is produced. I found a predicted ratio of 0.15 to 0.25, depending on the value of the Z^0 -γ mixing angle θ. So there was every reason to look a little harder.

As everyone knows, neutral currents were finally discovered [39] in 1973. There followed years of careful experimental study on the detailed properties of the neutral currents. It would take me too far from my subject to survey these experiments, [40] so I will just say that they have confirmed the 1967–68 theory with steadily improving precision for neutrino-nucleon and neutrino-electron neutral current reactions, and since the remarkable SLAC-Yale experiment [41] last year, for the electron-nucleon neutral current as well.

This is all very nice. But I must say that I would not have been too disturbed if it had turned out that the correct theory was based on some other spontaneously broken gauge group, with very different neutral currents. One possibility was a clever SU(2) theory proposed in 1972 by Georgi and Glashow, [42] which has no neutral currents at all. The important thing to me was the idea of an exact spontaneously broken gauge symmetry, which connects the weak and electromagnetic interactions, and allows these interactions to be renormalizable. Of this I was convinced, if only because it fitted my conception of the way that nature ought to be.

There were two other relevant theoretical developments in the early 1970's, before the discovery of neutral currents, that I must mention here. One is the important work of Glashow, Iliopoulos, and Maiani on the charmed quark. [43] Their work provided a solution to what otherwise would have been a serious problem, that of neutral strangeness changing currents. I leave this topic for Professor Glashow's talk. The other theoretical development has to do specifically with the strong interactions, but it will take us back to one of the themes of my talk, the theme of symmetry.

In 1973, Politzer and Gross and Wilczek discovered [44] a remarkable property of Yang–Mills theories which they called "asymptotic freedom" – the effective coupling constant [45] decreases to zero as the characteristic energy of a process goes to infinity. It seemed that this might explain the experimental fact that the nucleon behaves in high energy deep inelastic electron scattering as if it consists of essentially free quarks. [46] But there was a problem. In order to give masses to the vector bosons in a gauge theory of strong interactions one would want to include strongly interacting scalar fields, and these would generally destroy asymptotic freedom. Another difficulty, one that particularly bothered me, was that in a unified theory of weak and electromagnetic interactions the fundamental weak coupling is of the same order as the electronic charge, e, so the effects of virtual intermediate vector bosons would introduce much too large violations of parity and strangeness conservation, of order 1/137, into the strong interactions of the scalars with each other and with the quarks. [47] At some point in the spring of 1973 it occurred to me (and independently to Gross and Wilczek) that one could do away with strongly interacting scalar fields altogether, allowing the strong interaction gauge symmetry to remain unbroken so that the vector bosons, or "gluons", are massless, and relying on the increase of the strong forces with increasing distance to explain why quarks as well as the massless gluons are not seen in the laboratory. [48] Assuming no strongly interacting scalars, three "colors" of quarks (as indicated by earlier work of several authors [49]), and an SU(3) gauge group, one then had a specific theory of strong interactions, the theory now generally known as quantum chromodynamics.

Experiments since then have increasingly confirmed QCD as the correct theory of strong interactions. What concerns me here, though, is its impact on our understanding of symmetry principles. Once again, the constraints of gauge invariance and renormalizability proved enormously powerful. These constraints force the Lagrangian to be so simple, that the strong interactions in QCD must conserve strangeness, charge conjugation, and (apart from problems [50] having to do with instantons) parity. One does not have to assume these symmetries as *a priori* principles; there is simply no way that the Lagrangian can be complicated enough to violate them. With one additional assumption, that the u and d quarks have relatively small masses, the strong interactions must also satisfy the approximate $SU(2) \times SU(2)$ symmetry of current algebra, which when spontaneously broken leaves us with isospin. If the s quark mass is also not too large, then one gets the whole eight-fold way as an approximate symmetry of the strong interactions. And the breaking of the $SU(3) \times SU(3)$ symmetry by quark masses has just the $(3,\bar{3})+(\bar{3},3)$ form required to account for the pion-pion scattering lengths [15] and Gell-Mann–Okubo mass formulas. Furthermore, with weak and electromagnetic interactions also described by a gauge theory, the weak currents are necessarily just the currents associated with these strong interaction symmetries. In other words, pretty much the whole pattern of approximate symmetries of strong, weak, and electromagnetic interactions that puzzled us so much in the 1950's and 1960's now stands explained as a simple consequence of strong, weak, and electromagnetic gauge invariance, plus renormalizability. Internal symmetry is now at the point where space-time symmetry was in Einstein's day. All the approximate internal symmetries are explained dynamically. On a fundamental level, there are no approximate or partial symmetries; there are only exact symmetries which govern all interactions.

I now want to look ahead a bit, and comment on the possible future development of the ideas of symmetry and renormalizability.

We are still confronted with the question whether the scalar particles that are responsible for the spontaneous breakdown of the electroweak gauge symmetry $SU(2) \times U(1)$ are really elementary. If they are, then spin zero semi-weakly decaying "Higgs bosons" should be found at energies comparable with those needed to produce the intermediate vector bosons. On the other hand, it may be that the scalars are composites. [51] The Higgs bosons would then be indistinct broad states at very high mass, analogous to the possible s-wave enhancement in π-π scattering. There would probably also exist lighter, more slowly decaying, scalar particles of a rather different type, known as pseudo-Goldstone bosons. [52] And there would have to exist a new class of "extra strong" interactions [53] to provide the binding force, extra strong in the sense that asymptotic freedom sets in not at a few hundred MeV, as in QCD, but at a few hundred GeV. This "extra strong" force would be felt by new families of fermions, and would give these fermions masses of the order of several hundred GeV. We shall see.

Of the four (now three) types of interactions, only gravity has resisted incorporation into a renormalizable quantum field theory. This may just mean that we are not being clever enough in our mathematical treatment of general relativity. But there is another possibility that seems to me quite plausible. The constant of gravity defines a unit of energy known as the Planck energy, about 10^{19} GeV. This is the energy at which gravitation becomes effectively a strong interaction, so that at this energy one can no longer ignore its ultraviolet divergences. It may be that there is a whole world of new physics with unsuspected degrees of freedom at these enormous energies, and that general relativity does not provide an adequate framework for understanding the physics of these superhigh energy degrees of freedom. When we explore gravitation or other ordinary phenomena, with particle masses and energies no greater than a TeV or so, we may be learning only about an "effective" field theory; that is, one in which superheavy degrees of freedom do not explicitly appear, but the coupling parameters implicitly represent sums over these hidden degrees of freedom.

To see if this makes sense, let us suppose it is true, and ask what kinds of interactions we would expect on this basis to find at ordinary energy. By "integrating out" the superhigh energy degrees of freedom in a fundamental theory, we generally encounter a very complicated effective field theory – so complicated, in fact, that it contains *all* interactions allowed by symmetry principles. But where dimensional analysis tells us that a coupling constant is a certain power of some mass, that mass is likely to be a typical superheavy mass, such as 10^{19} GeV. The infinite variety of non-renormalizable interactions in the effective theory have coupling constants with the dimensionality of negative powers of mass, so their effects are suppressed at ordinary energies by powers of energy divided by superheavy masses. Thus the only interactions that we can detect at ordinary energies are those that are renormalizable in the usual sense, plus any non-renormalizable interactions that produce effects which, although tiny, are somehow exotic enough to be seen.

One way that a very weak interaction could be detected is for it to be coherent and of long range, so that it can add up and have macroscopic effects. It has been shown [54] that the only particles whose exchange could produce such forces are massless particles of spin 0, 1, or 2. And furthermore, Lorentz's invariance alone is enough to show that the long-range interactions produced by any particle of mass zero and spin 2 must be governed by general relativity. [55] Thus from this point of view we should not be too surprised that gravitation is the only interaction discovered so far that does not seem to be described by a renormalizable field theory – it is almost the only superweak interaction that *could* have been detected. And we should not be surprised to find that gravity is well described by general relativity at macroscopic scales, even if we do not think that general relativity applies at 10^{19} GeV.

Non-renormalizable effective interactions may also be detected if they violate otherwise exact conservation laws. The leading candidates for violation are baryon and lepton conservation. It is a remarkable consequence of the SU(3) and SU(2) x U(1) gauge symmetries of strong, weak, and electromagnetic interactions, that all renormalizable interactions among known particles automatically conserve baryon and lepton number. Thus, the fact that ordinary matter seems pretty stable, that proton decay has not been seen, should not lead us to the conclusion that baryon and lepton conservation are fundamental conservation laws. To the accuracy with which they have been verified, baryon and lepton conservation can be explained as dynamical consequences of other symmetries, in the same way that strangeness conservation has been explained within QCD. But superheavy particles may exist, and these particles may have unusual SU(3) or SU(2) x SU(1) transformation properties, and in this case, there is no reason why their interactions should conserve baryon or lepton number. I doubt that they would. Indeed, the fact that the universe seems to contain an excess of baryons over antibaryons should lead us to suspect that baryon nonconserving processes have actually occurred. If effects of a tiny nonconservation of baryon or lepton number such as proton decay or neutrino masses are discovered experimentally, we will then be left with gauge symmetries as the only true internal symmetries of nature, a conclusion that I would regard as most satisfactory.

The idea of a new scale of superheavy masses has arisen in another way. [56] If any sort of "grand unification" of strong and electroweak gauge couplings is to be possible, then one would expect all of the SU(3) and SU(2) x U(1) gauge coupling constants to be of comparable magnitude. (In particular, if SU(3) and SU(2) x U(1) are subgroups of a larger simple group, then the ratios of the squared couplings are fixed as rational numbers of order unity.[57]) But this appears in contradiction with the obvious fact that the strong interactions are stronger than the weak and electromagnetic interactions. In 1974 Georgi, Quinn and I suggested that the grand unification scale, at which the couplings are comparable, is at an enormous energy, and that the reason that the strong coupling is so much larger than the electroweak couplings at ordinary energies is that QCD is asymptotically free, so that its effective coupling constant rises slowly as the energy drops from the grand unification scale to ordinary values. The change of the strong couplings is very slow (like $1/\sqrt{\ln E}$) so the grand unification scale must be enormous. We found that for a fairly large class of theories the grand unification scale comes out to be in the neighborhood of 10^{16} GeV, an energy not all that different from the Planck energy of 10^{19} GeV. The nucleon lifetime is very difficult to estimate accurately, but we gave a representative value of 10^{32} years, which may be accessible experimentally in a few years. (These estimates have been improved in more detailed calculations by several authors.) [58] We also calculated a value for the mixing parameter $\sin^2\theta$ of about 0.2, not far from the present experimental value[40] of 0.23 ± 0.01. It will be an important task for future experiments on neutral currents to improve the precision with which $\sin^2\Theta$ is known, to see if it really agrees with this prediction.

In a grand unified theory, in order for elementary scalar particles to be available to produce the spontaneous breakdown of the electroweak gauge symmetry at a few hundred GeV, it is necessary for such particles to escape getting superlarge masses from the spontaneous breakdown of the grand unified gauge group. There is nothing impossible in this, but I have not been able to think of any reason why it should happen. (The problem may be related to the old mystery of why quantum corrections do not produce an enormous cosmological constant; in both cases, one is concerned with an anomalously small "super-renormalizable" term in the effective Lagrangian which has to be adjusted to be zero. In the case of the cosmological constant, the adjustment must be precise to some fifty decimal places.) With elementary scalars of small or zero bare mass, enormous ratios of symmetry breaking scales can arise quite naturally [59]. On the other hand, if there are no elementary scalars which escape getting superlarge masses from the breakdown of the grand unified gauge group, then as I have already mentioned, there must be extra strong forces to bind the composite Goldstone and Higgs bosons that are associated with the spontaneous breakdown of SU(2) x U(1). Such forces can occur rather naturally in grand unified theories. To take one example, suppose that the grand gauge group breaks, not into SU(3) x SU(2) x U(1), but into SU(4) x SU(3) x SU(2) x U(1). Since SU(4) is a bigger group than SU(3), its coupling constant rises with decreasing energy more rapidly than the QCD coupling, so the SU(4) force becomes strong at a much higher energy than the few hundred MeV at which the QCD force becomes strong. Ordinary quarks and leptons would be neutral under SU(4), so they would not feel this force, but other fermions might carry SU(4) quantum numbers, and so get rather large masses. One can even imagine a sequence of increasingly large subgroups of the grand gauge group, which would fill in the vast energy range up to 10^{15} or 10^{19} GeV with particle masses that are produced by these successively stronger interactions.

If there are elementary scalars whose vacuum expectation values are responsible for the masses of ordinary quarks and leptons, then these masses can be affected in order α by radiative corrections involving the superheavy vector bosons of the grand gauge group, and it will probably be impossible to explain the value of quantities like m_e/m_μ without a complete grand unified theory. On the other hand, if there are no such elementary scalars, then almost all the details of the grand unified theory are forgotten by the effective field theory that describes physics at ordinary energies, and it ought to be possible to calculate quark and lepton masses purely in terms of processes at accessible energies. Unfortunately, no one so far has been able to see how in this way anything resembling the observed pattern of masses could arise. [60]

Putting aside all these uncertainties, suppose that there is a truly fundamental theory, characterized by an energy scale of order 10^{16} to 10^{19} GeV, at which strong, electroweak, and gravitational interactions are all united. It might be a conventional renormalizable quantum field theory, but at the moment, if we include gravity, we do not see how this is possible. (I leave the topic of supersymmetry and supergravity for Professor Salam's talk.) But if it is not renormalizable, what then determines the infinite set of coupling constants that are needed to absorb all the ultraviolet divergences of the theory?

I think the answer must lie in the fact that the quantum field theory, which was born just fifty years ago from the marriage of quantum mechanics with relativity, is a beautiful but not very robust child. As Landau and Källén recognized long ago, quantum field theory at superhigh energies is susceptible to all sorts of diseases–tachyons, ghosts, etc.–and it needs special medicine to survive. One way that a quantum field theory can avoid these diseases is to be renormalizable and asymptotically free, but there are other possibilities. For instance, even an infinite set of coupling constants may approach a non-zero fixed point as the energy at which they are measured goes to infinity. However, to require this behavior generally imposes so many constraints on the couplings that there are only a finite number of free parameters left[61]–just as for theories that are renormalizable in the usual sense. Thus, one way or another, I think that quantum field theory is going to go on being very stubborn, refusing to allow us to describe all but a small number of possible worlds, among which, we hope, is ours.

I suppose that I tend to be optimistic about the future of physics. And nothing makes me more optimistic than the discovery of broken symmetries. In the seventh book of the *Republic,* Plato describes prisoners who are chained in a cave and can see only shadows that things outside cast on the cave wall. When released from the cave at first their eyes hurt, and for a while they think that the shadows they saw in the cave are more real than the objects they now see. But eventually their vision clears, and they can understand how beautiful the real world is. We are in such a cave, imprisoned by the limitations on the sorts of experiments we can do. In particular, we can study matter only at relatively low temperatures, where symmetries are likely to be spontaneously broken, so that nature does not appear very simple or unified. We have not been able to get out of this cave, but by looking long and hard at the shadows on the cave wall, we can at least make out the shapes of symmetries, which though broken, are exact principles governing all phenomena, expressions of the beauty of the world outside.

***

It has only been possible here to give references to a very small part of the literature on the subjects discussed in this talk. Additional references can be found in the following reviews:

Abers, E.S. and Lee, B.W., *Gauge Theories* (Physics Reports *9C,* No. 1, 1973).

Marciano, W. and Pagels, H., *Quantum Chromodynamics* (Physics Reports *36C,* No. 3, 1978).

Taylor, J.C., *Gauge Theories of Weak Interactions* (Cambridge Univ. Press, 1976).

REFERENCES

1. Tuve, M. A., Heydenberg, N. and Hafstad, L. R. Phys. Rev. *50,* 806 (1936); Breit, G., Condon, E. V. and Present, R. D. Phys. Rev. *50,* 825 (1936); Breit, G. and Feenberg, E. Phys. Rev. *50,* 850 (1936).
2. Gell-Mann, M. Phys. Rev. *92,* 833 (1953); Nakano. T. and Nishijima, K. Prog. Theor. Phys. *10,* 581 (1955).
3. Lee, T. D. and Yang, C. N. Phys. Rev. *104,* 254 (1956); Wu. C. S. et.al. Phys. Rev. *105,* 1413 (1957); Garwin, R., Lederman, L. and Weinrich, M. Phys. Rev. *105,* 1415 (1957); Friedman, J. I. and Telegdi V. L. Phys. Rev. *105,* 1681 (1957).
4. Gell-Mann, M. Cal. Tech. Synchotron Laboratory Report CTSL-20 (1961), unpublished; Ne'eman, Y. Nucl. Phys. *26,* 222 (1961).
5. Fock, V. Z. f. Physik *39,* 226 (1927); Weyl, H. Z. f. Physik *56,* 330 (1929). The name "gauge invariance" is based on an analogy with the earlier speculations of Weyl, H. in *Raum, Zeit, Materie,* 3rd edn, (Springer, 1920). Also see London, F. Z. f. Physik *42,* 375 (1927). (This history has been reviewed by Yang, C. N. in a talk at City College, (1977).)
6. Yang, C. N. and Mills, R. L. Phys. Rev. *96,* 191 (1954).
7. Goldstone, J. Nuovo Cimento *19,* 154 (1961).
8. Goldstone, J., Salam, A. and Weinberg, S. Phys. Rev. *127,* 965 (1962).
9. Higgs, P. W. Phys. Lett. *12,* 132 (1964); *13,* 508 (1964); Phys. Rev. *145,* 1156 (1966); Kibble, T. W. B. Phys. Rev. *155,* 1554 (1967); Guralnik, G. S., Hagen, C. R. and Kibble, T. W. B. Phys. Rev. Lett. *13,* 585 (1964); Englert, F. and Brout, R. Phys. Rev. Lett. *13,* 321 (1964); Also see Anderson, P. W. Phys. Rev. *130,* 439 (1963).
10. Adler, S. L. Phys. Rev. Lett. *14,* 1051 (1965); Phys Rev. *140,* B736 (1965); Weisberger, W. I. Phys. Rev. Lett. *14,* 1047 (1965); Phys Rev. *143,* 1302 (1966).
11. Gell-Mann, M. Physics *1,* 63 (1964).
12. Nambu, Y. and Jona-Lasinio, G. Phys. Rev. *122,* 345 (1961); *124,* 246 (1961); Nambu, Y, and Lurie, D. Phys. Rev. *125,* 1429 (1962); Nambu. Y. and Shrauner, E. Phys. Rev. *128,* 862 (1962); Also see Gell-Mann, M. and Lévy, M., Nuovo Cimento *16,* 705 (1960).
13. Goldberger, M. L., Miyazawa, H. and Oehme, R. Phys Rev. *99,* 986 (1955).
14. Goldberger, M. L., and Treiman, S. B. Phys. Rev. *111,* 354 (1958).
15 .Weinberg, S. Phys. Rev. Lett. *16,* 879 (1966); *17,* 336 (1966); *17,* 616 (1966); *18,* 188 (1967); Phys Rev. *166,* 1568 (1967).
16. Oppenheimer, J. R. Phys. Rev. *35,* 461 (1930); Waller, I. Z. Phys. *59,* 168 (1930); ibid., *62,* 673 (1930).
17. Feynman, R. P. Rev. Mod. Phys. *20,* 367 (1948); Phys. Rev. *74,* 939, 1430 (1948); *76,* 749, 769 (1949); *80,* 440 (1950); Schwinger, J. Phys. Rev. *73,* 146 (1948); *74,* 1439 (1948); *75,* 651 (1949); *76,* 790 (1949); *82,* 664, 914 (1951); *91,* 713 (1953); Proc. Nat. Acad. Sci. *37,* 452 (1951); Tomonaga, S. Progr. Theor. Phys. (Japan) *1,* 27 (1946); Koba, Z., Tati, T. and Tomonaga, S. ibid. *2,* 101 (1947); Kanazawa, S. and Tomonaga, S. ibid. *3,* 276 (1948); Koba, Z. and Tomonaga, S. ibid *3,* 290 (1948).
18. There had been earlier suggestions that infinities could be eliminated from quantum field theories in this way, by Weisskopf, V. F. Kong. Dansk. Vid. Sel. Mat.-Fys. Medd. *15* (6) 1936, especially p. 34 and pp. 5–6; Kramers, H. (unpublished).
19. Dyson, F. J. Phys. Rev. *75,* 486, 1736 (1949).
20. Weinberg, S. Phys. Rev. *106,* 1301 (1957).
21. Weinberg, S. Phys. Rev. *118,* 838 (1960).
22. Salam, A. Phys. Rev. *82,* 217 (1951); *84,* 426 (1951).

23. Weinberg, S. Phys. Rev. Lett. *18,* 507 (1967).
24. For the non-renormalizability of theories with intrinsically broken gauge symmetries, see Komar, A. and Salam, A. Nucl. Phys. *21,* 624 (1960); Umezawa, H. and Kamefuchi, S. Nucl. Phys. *23,* 399 (1961); Kamefuchi, S., O'Raifeartaigh, L. and Salam, A. Nucl. Phys. *28,* 529 (1961); Salam, A. Phys. Rev. *127,* 331 (1962); Veltman, M. Nucl. Phys. B7, 637 (1968); B*21,* 288 (1970); Boulware, D. Ann. Phys. (N, Y,) *56,* 140 (1970).
25. This work was briefly reported in reference 23, footnote 7.
26. Weinberg, S. Phys. Rev. Lett. *19,* 1264 (1967).
27. Salam, A. In *Elementary Particle Physics* (Nobel Symposium No. 8), ed. by Svartholm, N. (Almqvist and Wiksell, Stockholm, 1968), p. 367.
28. deWitt, B. Phys. Rev. Lett. *12,* 742 (1964); Phys. Rev. *162,* 1195 (1967); Faddeev L. D., and Popov, V. N. Phys. Lett. *B25,* 29 (1967); Also see Feynman, R. P. Acta. Phys. Pol. *24,* 697 (1963); Mandelstam, S. Phys. Rev. *175,* 1580, 1604 (1968).
29. See Stuller, L. M. I. T., Thesis, Ph. D. (1971), unpublished.
30. My work with the unitarity gauge was reported in Weinberg, S. Phys. Rev. Lett. *27,* 1688 (1971), and described in more detail in Weinberg, S. Phys. Rev. *D7,* 1068 (1973).
31. 't Hooft, G Nucl. Phys. *B35,* 167 (1971).
32. Lee, B. W. and Zinn-Justin, J. Phys. Rev. *D5,* 3121, 3137, 3155 (1972); 't Hooft, G. and Veltman, M. Nucl. Phys. *B44,* 189 (1972), *B50,* 318 (1972). There still remained the problem of possible Adler–Bell–Jackiw anomalies, but these nicely cancelled; see D. J. Gross and R. Jackiw, Phys. Rev. *D6,* 477 (1972) and C. Bouchiat, J. Iliopoulos, and Ph. Meyer, Phys. Lett. *38B,* 519 (1972).
33. Beechi, C., Rouet, A. and Stora R. Comm. Math. Phys. *42,* 127 (1975).
34. Lee, B. W. Phys. Rev. *D5,* 823 (1972).
35. Gamow, G. and Teller, E. Phys. Rev. *51,* 288 (1937); Kemmer, N. Phys. Rev. *52,* 906 (1937); Wentzel, G. Helv. Phys. Acta. *10,* 108 (1937); Bludman, S. Nuovo Cimento *9,* 433 (1958); Leite-Lopes, J. Nucl. Fhys. *8,* 234 (1958).
36. Glashow, S. L. Nucl. Phys. *22,* 519 (1961); Salam, A. and Ward, J. C. Phys. Lett. *13,* 168 (1964).
37. Weinberg, S. Phys. Rev. *5,* 1412 (1972).
38. Cundy, D. C. et.al., Phys. Lett. *31B,* 478 (1970).
39. The first published discovery of neutral currents was at the Gargamelle Bubble Chamber at CERN: Hasert, F. J. et.al., Phys. Lett. *46B,* 121, 138 (1973). Also see Musset, P. Jour. de Physique 11/12 T34 (1973). Muonless events were seen at about the same time by the HPWF group at Fermilab, but when publication of their paper was delayed, they took the opportunity to rebuild their detector, and then did not at first find the same neutral current signal. The HPWF group published evidence for neutral currents in Benvenuti, A. et.al., Phys. Rev. Lett. *32,* 800 (1974).
40. For a survey of the data see Baltay, C. *Proceedings of the 19th International Conference on High Energy Physics,* Tokyo, 1978. For theoretical analyses, see Abbott, L. F. and Barnett, R. M. Phys. Rev. *D19,* 3230 (1979); Langacker, P., Kim, J. E., Levine, M., Williams, H. H. and Sidhu, D. P. Neutrino Conference '79; and earlier references cited therein.
41. Prescott, C. Y. et.al., Phys. Lett. *77B,* 347 (1978).
42. Glashow, S. L. and Georgi, H. L. Phys. Rev. Lett. *28,* 1494 (1972). Also see Schwinger, J. Annals of Physics (N. Y.) *2,* 407 (1957).
43. Glashow, S. L., Iliopoulos, J. and Maiani, L. Phys. Rev. *D2,* 1285 (1970). This paper was cited in ref. 37 as providing a possible solution to the problem of strangeness changing neutral currents. However, at that time I was skeptical about the quark model, so in the calculations of ref. 37 baryons were incorporated in the theory by taking the protons and neutrons to form an SU(2) doublet, with strange particles simply ignored.
44. Politzer, H. D. Phys. Rev. Lett. *30,* 1346 (1973); Gross, D. J. and Wilczek, F. Phys. Rev. Lett. *30,* 1343 (1973).
45. Energy dependent effective couping constants were introduced by Gell-Mann, M. and Low, F. E. Phys. Rev. *95,* 1300 (1954).
46. Bloom, E. D. et.al., Phys. Rev. Lett. *23,* 930 (1969); Breidenbach, M. et.al., Phys. Rev. Lett. *23,* 935 (1969).

47. Weinberg, S. Phys. Rev. *D8,* 605 (1973).

48. Gross, D. J. and Wilczek, F. Phys. Rev. *D8,* 3633 (1973); Weinberg, S. Phys. Rev. Lett. *31,* 494 (1973). A similar idea had been proposed before the discovery of asymptotic freedom by Fritzsch, H., Gell-Mann, M. and Leutwyler, H. Phys. Lett. *47B,* 365 (1973).

49. Greenberg, O. W. Phys. Rev. Lett. *13,* 598 (1964); Han, M. Y. and Nambu, Y. Phys. Rev. *139,* B1006 (1965); Bardeen, W. A., Fritzsch, H. and Gell-Mann, M. in *Scale and Conformal Symmetry in Hadron Physics,* ed. by Gatto, R. (Wiley, 1973), p. 139; etc.

50. 't Hooft, G. Phys. Rev. Lett. *37,* 8 (1976).

51. Such "dynamical" mechanisms for spontaneous symmetry breaking were first discussed by Nambu, Y. and Jona-Lasinio, G. Phys. Rev. *122,* 345 (1961); Schwinger, J. Phys. Rev. *125,* 397 (1962); *128,* 2425 (1962); and in the context of modern gauge theories by Jackiw, R. and Johnson, K. Phys. Rev. *D8,* 2386 (1973); Cornwall, J. M. and Norton, R. E. Phys. Rev. *D8,* 3338 (1973). The implications of dynamical symmetry breaking have been considered by Weinberg, S. Phys. Rev. *D13,* 974 (1976); *D19,* 1277 (1979); Susskind, L. Phys. Rev. *D20,* 2619 (1979).

52. Weinberg, S. ref 51. The possibility of pseudo-Goldstone bosons was originally noted in a different context by Weinberg, S. Phys. Rev. Lett. *29,* 1698 (1972).

53. Weinberg, S. ref. 51. Models involving such interactions have also been discussed by Susskind, L. ref. 51.

54. Weinberg, S. Phys. Rev. *135,* B1049 (1964).

55. Weinberg. S. Phys. Lett. *9,* 357 (1964); Phys. Rev. *B138,* 988 (1965); *Lectures in Particles and Field Theory,* ed. by Deser, S. and Ford, K. (Prentice-Hall, 1965), p. 988; and ref. 54. The program of deriving general relativity from quantum mechanics and special relativity was completed by Boulware, D. and Deser, S. Ann. Phys. *89,* 173 (1975). I understand that similar ideas were developed by Feynman, R. in unpublished lectures at Cal. Tech.

56. Georgi, H., Quinn, H. and Weinberg, S. Phys. Rev. Lett. *33,* 451 (1974).

57. An example of a simple gauge group for weak and electromagnetic interactions (forwhich $\sin^2\theta=\frac{1}{4}$) was given by S. Weinberg, Phys. Rev. *D5,* 1962 (1972). There are a number of specific models of weak, electromagnetic, and strong interactions based on simple gauge groups, including those of Pati, J. C. and Salam, A. Phys. Rev. *D10,* 275 (1974); Georgi, H. and Glashow, S. L. Phys. Rev. Lett. *32,* 438 (1974); Georgi, H. in *Particles and Fields* (American Institute of Physics, 1975); Fritzsch, H. and Minkowski, P. Ann. Phys. *93,* 193 (1975); Georgi, H. and Nanopoulos, D. V. Phys. Lett. *82B,* 392 (1979); Gürsey, F. Ramond, P. and Sikivie, P. Phys. Lett. *B60,* 177 (1975); Gürsey, F. and Sikivie, P. Phys. Rev. Lett. *36,* 775 (1976); Ramond, P. Nucl. Phys, *B110,* 214 (1976); etc; all these violate baryon and lepton conservation, because they have quarks and leptons in the same multiplet; see Pati, J. C. and Salam, A. Phys. Rev. Lett. *31,* 661 (1973); Phys. Rev. D *8,* 1240 (1973).

58. Buras, A., Ellis, J., Gaillard, M. K. and Nanopoulos, D. V. Nucl. Phys. *B135,* 66 (1978); Ross, D. Nucl. Phys. *B140,* 1 (1978); Marciano, W. J. Phys. Rev. *D20,* 274 (1979); Goldman, T. and Ross, D. CALT 68-704, to be published; Jarlskog, C. and Yndurain, F. J. CERN preprint, to be published. Machacek, M. Harvard preprint HUTP-79/AO21, to be published in Nuclear Physics; Weinberg, S. paper in preparation. The phenomenonology of nucleon decay has been discussed in general terms by Weinberg, S. Phys. Rev. Lett. *43,* 1566 (1979); Wilczek, F. and Zee, A. Phys. Rev. Lett. *43,* 1571 (1979).

59. Gildener, E. and Weinberg, S. Phys. Rev. *D13,* 3333 (1976); Weinberg, S. Phys. Letters *82B,* 387 (1979). In general there should exist at least one scalar particle with physical mass of order 10 GeV. The spontaneous symmetry breaking in models with zero bare scalar mass was first considered by Coleman, S. and Weinberg, E., Phys. Rev. D *7,* 1888 (1973).

60. This problem has been studied recently by Dimopoulos, S. and Susskind, L. Nucl. Phys. *B155,* 237 (1979); Eichten, E. and Lane, K. Physics Letters, to be published; Weinberg, S. unpublished.

61. Weinberg, S. in *General Relativity – An Einstein Centenary Survey,* ed. by Hawking, S. W. and Israel, W. (Cambridge Univ. Press, 1979), Chapter 16.

GAUGE UNIFICATION OF FUNDAMENTAL FORCES

Nobel lecture, 8 December, 1979

by

ABDUS SALAM
Imperial College of Science and Technology, London, England
and International Centre for Theoretical Physics, Trieste, Italy

Introduction: In June 1938, Sir George Thomson, then Professor of Physics at Imperial College, London, delivered his 1937 Nobel Lecture. Speaking of Alfred Nobel, he said: "The idealism which permeated his character led him to ... (being) as much concerned with helping science as a whole, as individual scientists. ... The Swedish people under the leadership of the Royal Family and through the medium of the Royal Academy of Sciences have made Nobel Prizes one of the chief causes of the growth of the prestige of science in the eyes of the world ... As a recipient of Nobel's generosity, I owe sincerest thanks to them as well as to him."

I am sure I am echoing my colleagues' feelings as well as my own, in reinforcing what Sir George Thomson said–in respect of Nobel's generosity and its influence on the growth of the prestige of science. Nowhere is this more true than in the developing world. And it is in this context that I have been encouraged by the Permanent Secretary of the Academy – Professor Carl Gustaf Bernhard–to say a few words before I turn to the scientific part of my lecture.

Scientific thought and its creation is the common and shared heritage of mankind. In this respect, the history of science, like the history of all civilization, has gone through cycles. Perhaps I can illustrate this with an actual example.

Seven hundred and sixty years ago, a young Scotsman left his native glens to travel south to Toledo in Spain. His name was Michael, his goal to live and work at the Arab Universities of Toledo and Cordova, where the greatest of Jewish scholars, Moses bin Maimoun, had taught a generation before.

Michael reached Toledo in 1217 AD. Once in Toledo, Michael formed the ambitious project of introducing Aristotle to Latin Europe, translating not from the original Greek, which he did not know, but from the Arabic translation then taught in Spain. From Toledo, Michael travelled to Sicily, to the Court of Emperor Frederick II.

Visiting the medical school at Salerno, chartered by Frederick in 1231, Michael met the Danish physician, Henrik Harpestraeng – later to become Court Physician of King Erik Plovpenning. Henrik had come to Salerno to compose his treatise on blood-letting and surgery. Henrik's sources were the medical canons of the great clinicians of Islam, Al-Razi and Avicenna, which only Michael the Scot could translate for him.

Toledo's and Salerno's schools, representing as they did the finest synthesis of Arabic, Greek, Latin and Hebrew scholarship, were some of the most memorable of international assays in scientific collaboration. To Toledo and Salerno came scholars not only from the rich countries of the East and the South, like Syria, Egypt, Iran and Afghanistan, but also from developing lands of the West and the North like Scotland and Scandinavia. Then, as now, there were obstacles to this international scientific concourse, with an economic and intellectual disparity between different parts of the world. Men like Michael the Scot or Henrik Harpestraeng were singularities. They did not represent any flourishing schools of research in their own countries. With all the best will in the world their teachers at Toledo and Salerno doubted the wisdom and value of training them for advanced scientific research. At least one of his masters counselled young Michael the Scot to go back to clipping sheep and to the weaving of woollen cloth.

In respect of this cycle of scientific disparity, perhaps I can be more quantitative. George Sarton, in his monumental five-volume History of Science chose to divide his story of achievement in sciences into ages, each age lasting half a century. With each half century he associated one central figure. Thus 450 BC–400 BC Sarton calls the Age of Plato; this is followed by half centuries of Aristotle, of Euclid, of Archimedes and so on. From 600 AD to 650 AD is the Chinese half century of Hsiian Tsang, from 650 to 700 AD that of I-Ching, and then from 750 AD to 1100 AD–350 years continuously–it is the unbroken succession of the Ages of Jabir, Khwarizmi, Razi, Masudi, Wafa, Biruni and Avicenna, and then Omar Khayam–Arabs, Turks, Afghans and Persians–men belonging to the culture of Islam. After 1100 appear the first Western names; Gerard of Cremona, Roger Bacon–but the honours are still shared with the names of Ibn-Rushd (Averroes), Moses Bin Maimoun, Tusi and Ibn-Nafis–the man who anticipated Harvey's theory of circulation of blood. No Sarton has yet chronicled the history of scientific creativity among the pre-Spanish Mayas and Aztecs, with their invention of the zero, of the calendars of the moon and Venus and of their diverse pharmacological discoveries, including quinine, but the outline of the story is the same–one of undoubted superiority to the Western contemporary correlates.

After 1350, however, the developing world loses out except for the occasional flash of scientific work, like that of Ulugh Beg–the grandson of Timurlane, in Samarkand in 1400 AD; or of Maharaja Jai Singh of Jaipur in 1720–who corrected the serious errors of the then Western tables of eclipses of the sun and the moon by as much as six minutes of arc. As it was, Jai Singh's techniques were surpassed soon after with the development of the telescope in Europe. As a contemporary Indian chronicler wrote: "With him on the funeral pyre, expired also all science in the East." And this brings us to this century when the cycle begun by Michael the Scot turns full circle, and it is we in the developing world who turn to the Westwards for science. As Al-Kindi wrote 1100 years ago: "It is fitting then for us not to be ashamed to acknowledge and to assimilate it from whatever source it comes to us. For him who scales the truth there is nothing of higher value than truth itself; it never cheapens nor abases him."

Ladies and Gentlemen,

It is in the spirit of Al-Kindi that I start my lecture with a sincere expression of gratitude to the modern equivalents of the Universities of Toledo and Cordova, which I have been privileged to be associated with – Cambridge, Imperial College, and the Centre at Trieste.

I. FUNDAMENTAL PARTICLES, FUNDAMENTAL FORCES AND GAUGE UNIFICATION

The Nobel lectures this year are concerned with a set of ideas relevant to the gauge unification of the electromagnetic force with the weak nuclear force. These lectures coincide nearly with the 100th death-anniversary of Maxwell, with whom the first unification of forces (electric with the magnetic) matured and with whom gauge theories originated. They also nearly coincide with the 100th anniversary of the birth of Einstein—the man who gave us the vision of an ultimate unification of *all* forces.

The ideas of today started more than twenty years ago, as gleams in several theoretical eyes. They were brought to predictive maturity over a decade back. And they started to receive experimental confirmation some six years ago.

In some senses then, our story has a fairly long background in the past. In this lecture I wish to examine some of the theoretical gleams of today and ask the question if these may be the ideas to watch for maturity twenty years from now.

From time immemorial, man has desired to comprehend the complexity of nature in terms of as few elementary concepts as possible. Among his quests—in Feynman's words—has been the one for "wheels within wheels"—the task of natural philosophy being to discover the innermost wheels if any such exist. A second quest has concerned itself with the fundamental forces which make the wheels go round and enmesh with one another. The greatness of gauge ideas—of gauge field theories—is that they reduce these two quests to just one; elementary particles (described by relativistic quantum fields) are representations of certain charge operators, corresponding to gravitational mass, spin, flavour, colour, electric charge and the like, while the fundamental forces are the forces of attraction or repulsion between these same *charges*. A third quest seeks for a *unification* between the charges (and thus of the forces) by searching for a single entity, of which the various charges are components in the sense that they can be transformed one into the other.

But are all fundamental forces gauge forces? Can they be understood as such, in terms of charges—and their corresponding currents—only? And if they are, how many charges? What unified entity are the charges components of?

What is the nature of charge? Just as Einstein comprehended the nature of gravitational charge in terms of space-time curvature, can we comprehend the nature of the other charges—the nature of the entire unified set, *as a set*, in terms of something equally profound? This briefly is the dream, much reinforced by the verification of gauge theory predictions. But before I examine the new theoretical ideas on offer for the future in this particular context, I would like your indulgence to range over a one-man, purely subjective, perspective in respect of the developments of the last twenty years themselves. The point I wish to emphasize during this part of my talk was well made by G. P. Thomson in his 1937 Nobel Lecture. G. P. said ". . . The goddess of learning is fabled to have sprung full grown from the brain of Zeus, but it is seldom that a scientific conception is born in its final form, or owns a single parent. More often it is the product of a series of minds, each in turn modifying the ideas of those that came before, and providing material for those that come after."

II. THE EMERGENCE OF SPONTANEOUSLY BROKEN SU(2)×U(1) GAUGE THEORY

I started physics research thirty years ago as an experimental physicist in the Cavendish, experimenting with tritium-deuterium scattering. Soon I knew the craft of experimental physics was beyond me—it was the sublime quality of patience—patience in accumulating data, patience with recalcitrant equipment—which I sadly lacked. Reluctantly I turned my papers in, and started instead on quantum field theory with Nicholas Kemmer in the exciting department of P. A. M. Dirac.

The year 1949 was the culminating year of the Tomonaga-Schwinger-Feynman-Dyson reformulation of renormalized Maxwell-Dirac gauge theory, and its triumphant experimental vindication. A field theory must be renormalizable and be capable of being made free of infinities—first discussed by Waller—if perturbative calculations with it are to make any sense. More—a renormalizable theory, with no dimensional parameter in its interaction term, connotes *somehow* that the fields represent "structureless" elementary entities. With Paul Matthews, we started on an exploration of renormalizability of meson theories. Finding that renormalizability held only for spin-zero mesons and that these were the only mesons that empirically existed then, (pseudoscalar pions, invented by Kemmer, following Yukawa) one felt thrillingly euphoric that with the triplet of pions (considered as the carriers of the strong nuclear force between the proton-neutron doublet) one might resolve the dilemma of the origin of this particular force which is responsible for fusion and fission. By the same token, the so-called weak nuclear force—the force responsible for β-radioactivity (and described then by Fermi's non-renormalizable theory) had to be mediated by some unknown spin-zero mesons if it was to be renormalizable, If massive charged spin-one mesons were to mediate this interaction, the theory would be non-renormalizable, according to the ideas then.

Now this agreeably renormalizable spin-zero theory for the pion was a field theory, but not a gauge field theory. There was no conserved charge which determined the pionic interaction. As is well known, shortly after the theory was elaborated, it was found wanting. The $(\frac{3}{2}, \frac{3}{2})$ resonance Δ effectively killed it off as a fundamental theory; we were dealing with a complex dynamical system, not "structureless" in the field-theoretic sense.

For me, personally, the trek to gauge theories as candidates for fundamental physical theories started in earnest in September 1956—the year I heard at the Seattle Conference Professor Yang expound his and Professor Lee's ideas[1] on the possibility of the hitherto sacred principle of left-right symmetry, being violated in the realm of the *weak nuclear force*. Lee and Yang had been led to consider abandoning left-right symmetry for weak nuclear interactions as a possible resolution of the (τ, θ) puzzle. I remember travelling back to London on an American Air Force (MATS) transport flight. Although I had been granted, for that night, the status of a Brigadier or a Field Marshal—I don't quite remember which—the plane was very uncomfortable, full of crying service-men's children—that is, the children were crying, not the servicemen. I could not sleep. I kept reflecting on why Nature should violate left-right symmetry in weak interactions. Now the hallmark of most weak interactions was the involvement in radioactivity phenomena of Pauli's neutrino. While crossing over the Atlantic, came back to me a deeply perceptive question about the neutrino which Professor Rudolf Peierls had asked when he was examining me for a Ph. D. a few years before. Peierls' question was: "The photon mass is zero because of Maxwell's principle of a gauge symmetry for electromagnetism; tell me, why is the neutrino mass zero?" I had then felt somewhat uncomfortable at Peierls, asking for a Ph. D. viva, a question of which he himself said he did not know the answer. But during that comfortless night the answer came. The analogue for the neutrino, of the gauge symmetry for the photon existed; it had to do with the masslessness of the neutrino, with symmetry under the γ_5 transformation[2] (later christened "chiral symmetry"). The existence of this symmetry for the massless neutrino must imply a combination $(1+\gamma_5)$ or $(1-\gamma_5)$ for the neutrino interactions. Nature had the choice of an aesthetically satisfying but a left-right symmetry violating theory, with a neutrino which travels exactly with the velocity of light; or alternatively a theory where left-right symmetry is preserved, but the neutrino has a tiny mass—some ten thousand times smaller than the mass of the electron.

It appeared at that time clear to me what choice Nature must have made. Surely, left-right symmetry must be sacrificed in all neutrino interactions. I got off the plane the next morning, naturally very elated. I rushed to the Cavendish, worked out the Michel parameter and a few other consequences of γ_5 symmetry, rushed out again, got into a train to Birmingham where Peierls lived. To Peierls I presented my idea; he had asked the original question; could he approve of the answer? Peierls' reply was kind but firm. He said "I do not believe left-right symmetry is violated in weak nuclear forces at all. I would not touch such ideas with a pair of tongs." Thus rebuffed in Birmingham, like Zuleika Dobson, I wondered where I could go next and the obvious place was CERN in Geneva, with Pauli—the father of the neutrino—nearby in Zurich.

At that time CERN lived in a wooden hut just outside Geneva airport. Besides my friends, Prentki and d'Espagnat, the hut contained a gas ring on which was cooked the staple diet of CERN—Entrecôte à la creme. The hut also contained Professor Villars of MIT, who was visiting Pauli the same day in Zurich. I gave him my paper. He returned the next day with a message from the Oracle; "Give my regards to my friend Salam and tell him to think of something better". This was discouraging, but I was compensated by Pauli's excessive kindness a few months later, when Mrs. Wu's[3], Lederman's[4] and Telegdi's[5] experiments were announced showing that left-right symmetry was indeed violated and ideas similar to mine about chiral symmetry were expressed independently by Landau[6] and Lee and Yang[7]. I received Pauli's first somewhat apologetic leatter on 24 January 1957. Thinking that Pauli's spirit should by now be suitably crushed, I sent him two short notes[8] I had written in the meantime. These contained suggestions to extend chiral symmetry to electrons and muons, assuming that their masses were a consequence of what has come to be known as dynamical spontaneous symmetry breaking. With chiral symmetry for electrons, muons and neutrinos, the only mesons that could mediate weak decays of the muons would have to carry spin one. Reviving thus the notion of charged intermediate *spin-one* bosons, one could then postulate for these a type of gauge invariance which I called the "neutrino gauge". Pauli's reaction was swift and terrible. He wrote on 30th January 1957, then on 18 February and later on 11, 12 and 13 March: "I am reading (along the shores of Lake Zurich) in bright sunshine quietly your paper . . ." "I am very much startled on the title of your paper 'Universal Fermi interaction' . . . For quite a while I have for myself the rule if a theoretician says *universal* it just means pure nonsense. This holds particularly in connection with the Fermi interaction, but otherwise too, and now you too, Brutus, my son, come with this word. . . ." Earlier, on 30 January, he had written "There is a similarity between this type of gauge invariance and that which was published by Yang and Mills . . . In the latter, of course, no γ_5 was used in the exponent." and he gave me the full reference of Yang and Mills' paper; (Phys. Rev. *96*, 191 (1954)). I quote from his letter: "However, there are dark points in your paper regarding the vector field B_μ. If the rest mass is infinite (or very large), how can this be compatible with the gauge transformation $B_\mu \rightarrow B_\mu - \partial_\mu \Lambda$?" and he concludes his letter with the remark: "Every reader will realize that you deliberately conceal here something and will ask you the same questions". Although he signed himself "With friendly regards", Pauli had forgotten his earlier penitence. He was clearly and rightly on the warpath.

Now the fact that I was using gauge ideas similar to the Yang—Mills (non-Abelian SU(2)-invariant) gauge theory was no news to me. This was because the Yang—Mills theory[9] (which married gauge ideas of Maxwell with the internal symmetry SU(2) of which the proton-neutron system constituted a doublet) had been independently invented by a Ph. D. pupil of mine, Ronald Shaw,[10] at Cambridge at the same time as Yang and Mills had written. Shaw's work is relatively unknown; it remains buried in his Cambridge thesis. I must admit I was taken aback by Pauli's fierce prejudice against universalism—against what we would today call unification of basic forces—but I did not take this too seriously. I felt this was a legacy of the exasperation which Pauli had always felt at Einstein's somewhat formalistic attempts at unifying gravity with electromagnetism—forces which in Pauli's phrase "cannot be joined—for God hath rent them asunder". But Pauli was absolutely right in accusing me of darkness about the problem of the masses of the Yang—Mills fields; one could not obtain a mass without wantonly destroying the gauge symmetry one had started with. And this was particularly serious in this context, because Yang and Mills had conjectured the desirable renormalizability of their theory with a proof which relied heavily and exceptionally on the masslessness of their spin-one intermediate mesons. The problem was to be solved only seven years later with the understanding of what is now known as the Higgs mechanism, but I will come back to this later.

Be that as it may, the point I wish to make from this exchange with Pauli is that already in early 1957, just after the first set of parity experiments, many ideas coming to fruition now, had started to become clear. These are:

1. First was the idea of chiral symmetry leading to a V-A theory. In those early days my humble suggestion[2], [8] of this was limited to neutrinos, electrons and muons only, while shortly after, that year, Sudarshan and Marshak,[11] Feynman and Gell-Mann,[12] and Sakurai[13] had the courage to postulate γ_5 symmetry for baryons as well as leptons, making this into a universal principle of physics.[1]

Concomitant with the (V-A) theory was the result that if weak interactions are mediated by intermediate mesons, these must carry spin one.

2. Second, was the idea of spontaneous breaking of chiral symmetry to generate electron and muon masses: though the price which those latter-day Shylocks, Nambu and Jona-Lasinio[14] and Goldstone[15] exacted for this (i.e. the appearance of massless scalars), was not yet appreciated.
3. And finally, though the use of a Yang—Mills—Shaw (non-Abelian) gauge theory for describing spin-one intermediate charged mesons was suggested already in 1957, the giving of masses to the intermediate bosons through spontaneous symmetry breaking, in a manner to preserve the renormalizability of the theory, was to be accomplished only during a long period of theoretical development between 1963 and 1971.

Once the Yang—Mills—Shaw ideas were accepted as relevant to the charged weak currents—to which the charged intermediate mesons were coupled in this theory—during 1957 and 1958 was raised the question of what was the third component of the SU(2) triplet, of which the charged weak currents were the two members. There were the two alternatives: the electroweak unification suggestion, where the electromagnetic current was assumed to be this third component; and the rival suggestion that the third component was a neutral current unconnected with electroweak unification. With hindsight, I shall

[1] Today we believe protons and neutrons are composites of quarks, so that γ_5 symmetry is now postulated for the elementary entities of today—the quarks.

call these the Klein[16] (1938) and the Kemmer[17] (1937) alternatives. The Klein suggestion, made in the context of a Kaluza—Klein five-dimensional space-time, is a real tour-de-force; it combined two hypothetical spin-one charged mesons with the photon in one multiplet, deducing from the compactification of the fifth dimension, a theory which looks like Yang—Mills—Shaw's. Klein intended his charged mesons for *strong* interactions, but if we read charged *weak* mesons for Klein's *strong* ones, one obtains the theory independently suggested by Schwinger[18] (1957), though Schwinger, unlike Klein, did not build in any non-Abelian gauge aspects. With just these non-Abelian Yang—Mills gauge aspects very much to the fore, the idea of uniting weak interactions with electromagnetism was developed by Glashow[19] and Ward and myself[20] in late 1958. The rival Kemmer suggestion of a global SU(2)-invariant triplet of weak charged and neutral currents was independently suggested by Bludman[21] (1958) in a gauge context and this is how matters stood till 1960.

To give you the flavour of, for example, the year 1960, there is a paper written that year of Ward and myself[22] with the statement: "Our basic postulate is that it should be possible to generate strong, weak and electromagnetic interaction terms with all their correct symmetry properties (as well as with clues regarding their relative strengths) by making local gauge transformations on the kinetic energy terms in the free Lagrangian for all particles. This is the statement of an ideal which, in this paper at least, is only very partially realized". I am not laying a claim that we were the only ones who were saying this, but I just wish to convey to you the temper of the physics of twenty years ago—qualitatively no different today from then. But what a quantitative difference the next twenty years made, first with new and far-reaching developments in theory—and then, thanks to CERN, Fermilab, Brookhaven, Argonne, Serpukhov and SLAC in testing it!

So far as theory itself is concerned, it was the next seven years between 1961—67 which were the crucial years of quantitative comprehension of the phenomenon of spontaneous symmetry breaking and the emergence of the $SU(2)\times U(1)$ theory in a form capable of being tested. The story is well known and Steve Weinberg has already spoken about it. So I will give the barest outline. First there was the realization that the two alternatives mentioned above a pure electromagnetic current versus a pure neutral current—Klein—Schwinger versus Kemmer—Bludman—were not alternatives; they were complementary. As was noted by Glashow[23] and independently by Ward and myself[24], both types of currents and the corresponding gauge particles ($W^{\pm}$, Z^0 and γ) were needed in order to build a theory that could simultaneously accommodate parity violation for weak and parity conservation for the electromagnetic phenomena. Second, there was the influential paper of Goldstone[25] in 1961 which, utilizing a non-gauge self-interaction between scalar particles, showed that the price of spontaneous breaking of a continuous internal symmetry was the appearance of zero mass scalars—a result foreshadowed earlier by Nambu. In giving a proof of this theorem[26] with Goldstone I collaborated with Steve Weinberg, who spent a year at Imperial College in London.

I would like to pay here a most sincerely felt tribute to him and to Sheldon Glashow for their warm and personal friendship.

I shall not dwell on the now well-known contributions of Anderson[27], Higgs[28], Brout & Englert[29], Guralnik, Hagen and Kibble[30] starting from 1963, which showed the way how spontaneous symmetry breaking using spin-zero fields could generate vector-meson masses, defeating Goldstone at the same time. This is the so-called Higgs mechanism.

The final steps towards the electroweak theory were taken by Weinberg[31] and myself[32] (with Kibble at Imperial College tutoring me about the Higgs phenomena). We were able to complete the present formulation of the spontaneously broken SU(2)×U(1) theory so far as leptonic weak interactions were concerned—with one parameter $\sin^2\theta$ describing all weak and electromagnetic phenomena and with one isodoublet Higgs multiplet. An account of this development was given during the contribution[32] to the Nobel Symposium (organized by Nils Svartholm and chaired by Lamek Hulthén held at Gothenburg after some postponements, in early 1968). As is well known, we did not then, and still do not, have a prediction for the scalar Higgs mass.

Both Weinberg and I suspected that this theory was likely to be renormalizable.[2] Regarding spontaneously broken Yang—Mills—Shaw theories in general this had earlier been suggested by Englert, Brout and Thiry[29]. But this subject was not pursued seriously except at Veltman's school at Utrecht, where the proof of renormalizability was given by 't Hooft[33] in 1971. This was elaborated further by that remarkable physicist the late Benjamin Lee[34], working with Zinn Justin, and by 't Hooft and Veltman[35]. This followed on the earlier basic advances in Yang—Mills calculational technology by Feynman[36], DeWitt[37], Faddeev and Popov[38], Mandelstam[39], Fradkin and Tyutin[40], Boulware[41], Taylor[42], Slavnov[43], Strathdee[44] and Salam. In Coleman's eloquent phrase "'t Hooft's work turned the Weinberg—Salam frog into an enchanted prince". Just before had come the GIM (Glashow, Iliopoulos and Maiani) mechanism[45], emphasising that the existence of the fourth charmed quark (postulated earlier by several authors) was essential to the natural resolution of the dilemma posed by the absence of strangeness--violating currents. This tied in naturally with the understanding of the Steinberger—Schwinger—Rosenberg—Bell—Jackiw—Adler anomaly[46] and its removal for SU(2)×U(1) by the parallelism of four quarks and four leptons, pointed out by Bouchiat, Iliopoulos and Meyer and independently by Gross and Jackiw.[47]

[2] When I was discussing the final version of the SU(2)×U(1) theory and its possible renormalizability in Autumn 1967 during a post-doctoral course of lectures at Imperial College, Nino Zichichi from CERN happened to be present. I was delighted because Zichichi had been badgering me since 1958 with persistent questioning of what theoretical avail his precise measurements on (g-2) for the muon as well as those of the muon lifetime were, when not only the magnitude of the electromagnetic corrections to weak decays was uncertain, but also conversely the effect of non-renormalizable weak interactions on "renormalized" electromagnetism was so unclear.

If one has kept a count, I have so far mentioned around fifty theoreticians. As a failed experimenter, I have always felt envious of the ambience of large experimental teams and it gives me the greatest pleasure to acknowledge the direct or the indirect contributions of the "series of minds" to the spontaneously broken SU(2)×U(1) gauge theory. My profoundest personal appreciation goes to my collaborators at Imperial College, Cambridge, and the Trieste Centre, John Ward, Paul Matthews, Jogesh Pati, John Strathdee, Tom Kibble and to Nicholas Kemmer.

In retrospect, what strikes me most about the early part of this story is how uninformed all of us were, not only of each other's work, but also of work done earlier. For example, only in 1972 did I learn of Kemmer's paper written at Imperial College in 1937.

Kemmer's argument essentially was that Fermi's weak theory was not globally SU(2) invariant and should be made so—though not for its own sake but as a prototype for strong interactions. Then this year I learnt that earlier, in 1936, Kemmer's Ph. D. supervisor, Gregor Wentzel[48], had introduced (the yet undiscovered) analogues of lepto-quarks, whose mediation could give rise to neutral currents after a Fierz reshuffle. And only this summer, Cecilia Jarlskog at Bergen rescued Oscar Klein's paper from the anonymity of the Proceedings of the International Institute of Intellectual Cooperation of Paris, and we learnt of his anticipation of a theory similar to Yang—Mills—Shaw's long before these authors. As I indicated before, the interesting point is that Klein was using his triplet, of two charged mesons plus the photon, not to describe weak interaction but for strong nuclear force unification with the electromagnetic—something our generation started on only in 1972—and not yet experimentally verified. Even in this recitation I am sure I have inadvertantly left off some names of those who have in some way contributed to SU(2)×U(1). Perhaps the moral is that not unless there is the prospect of quantitative verification, does a qualitative idea make its impress in physics.

And this brings me to experiment, and the year of the Gargamelle[49]. I still remember Paul Matthews and I getting off the train at Aix-en-Provence for the 1973 European Conference and foolishly deciding to walk with our rather heavy luggage to the student hostel where we were billeted. A car drove from behind us, stopped, and the driver leaned out. This was Musset whom I did not know well personally then. He peered out of the window and said: "Are you Salam?" I said "Yes". He said: "Get into the car. I have news for you. We have found neutral currents." I will not say whether I was more relieved for being given a lift because of our heavy luggage or for the discovery of neutral currents. At the Aix-en-Provence meeting that great and modest man, Lagarrigue, was also present and the atmosphere was that of a carnival—at least this is how it appeared to me. Steve Weinberg gave the rapporteur's talk with T. D. Lee as the chairman. T. D. was kind enough to ask me to comment after Weinberg finished. That summer Jogesh Pati and I had predicted proton decay within the context of what is now called grand unification and in the flush of this excitement I am afraid I ignored weak neutral currents as a subject which had already come to a successful conclusion, and concentrated on speaking of the possible decays of the proton. I understand now that proton decay experiments are being planned in the United States by the Brookhaven, Irvine and Michigan and the Wisconsin—Harvard groups and also by a European collaboration to be mounted in the Mont Blanc Tunnel Garage No. 17. The later quantitative work on neutral currents at CERN, Fermilab., Brookhaven, Argonne and Serpukhov is, of course, history, but a special tribute is warranted to the beautiful SLAC-Yale-CERN experiment[50] of 1978 which exhibited the effective Z^0-photon interference in accordance with the predictions of the theory. This was foreshadowed by Barkov *et al*'s experiments[51] at Novosibirsk in the USSR in their exploration of parity violation in the atomic potential for bismuth. There is the apocryphal story about Einstein, who was asked what he would have thought if experiment had not confirmed the light deflection predicted by him. Einstein is supposed to have said, "Madam, I would have thought the Lord has missed a most marvellous opportunity." I believe, however, that the following quote from Einstein's Herbert Spencer lecture of 1933 expresses his, my colleagues' and my own views more accurately. "Pure logical thinking cannot yield us any knowledge of the empirical world; all knowledge of reality starts from experience and ends in it." This is exactly how I feel about the Gargamelle-SLAC experience.

III. THE PRESENT AND ITS PROBLEMS

Thus far we have reviewed the last twenty years and the emergence of $SU(2)\times U(1)$, with the twin developments of a gauge theory of basic interactions, linked with internal symmetries, and of the spontaneous breaking of these symmetries. I shall first summarize the situation as we believe it to exist now and the immediate problems. Then we turn to the future.

1. To the level of energies explored, we believe that the following sets of particles are "structureless" (in a field-theoretic sense) and, at least to the level of energies explored hitherto, constitute the elementary entities of which all other objects are made.

		$SU_c(3)$ triplets		
Family I	quarks	$\begin{Bmatrix} u_R, u_Y, u_B \\ d_R, d_Y, d_B \end{Bmatrix}$	leptons $\begin{pmatrix} \nu_e \\ e \end{pmatrix}$	SU(2) doublets
Family II	quarks	$\begin{Bmatrix} c_R, c_Y, c_B \\ s_R, s_Y, s_B \end{Bmatrix}$	leptons $\begin{pmatrix} \nu_\mu \\ \mu \end{pmatrix}$	,,
Family III	quarks	$\begin{Bmatrix} t_R, t_Y, t_B \\ b_R, b_Y, b_B \end{Bmatrix}$	leptons $\begin{pmatrix} \nu_\tau \\ \tau \end{pmatrix}$	,,

Together with their antiparticles each family consists of 15 or 16 two-component fermions (15 or 16 depending on whether the neutrino is massless or not). The third family is still conjectural, since the top quark (t_R, t_Y, t_B) has not yet been discovered. Does this family really follow the pattern of the other two? Are there more families? Does the fact that the families are replicas of each other imply that Nature has discovered a dynamical stability about a system of 15 (or 16) objects, and that by this token there is a more basic layer of structure underneath?[52]

2. Note that quarks come in three colours; Red (R), Yellow (Y) and Blue (B). Parallel with the electroweak $SU(2)\times U(1)$, a *gauge* field[3] theory ($SU_c(3)$) of strong (quark) interactions (quantum chromodynamics, QCD)[53] has emerged which gauges the three colours. The indirect discovery of the (eight) gauge bosons associated with QCD (gluons), has already been surmised by the groups at DESY.[54]
3. All known baryons and mesons are singlets of colour $SU_c(3)$. This has led to a hypothesis that colour is always confined. One of the major unsolved problems of field theory is to determine if QCD—treated non-perturbatively—is capable of confining quarks and gluons.
4. In respect of the electroweak $SU(2)\times U(1)$, all known experiments on weak and electromagnetic phenomena below 100 GeV carried out to date agree with the theory which contains one theoretically undetermined parameter $\sin^2\theta = 0.230\pm0.009$.[55] The predicted values of the associated gauge boson ($W^{\pm}$ and Z^0) masses are: $m_W \approx 77$—84 GeV, $m_Z \approx 89$—95 GeV, for $0.25 \geqslant \sin^2\theta \geqslant 0.21$.
5. Perhaps the most remarkable measurement in electroweak physics is that of the parameter $\rho = \left(\frac{m_W}{m_Z \cos\theta}\right)^2$. Currently this has been determined from the ratio of neutral to charged current cross-sections. The predicted value $\rho = 1$ for weak *iso-doublet Higgs* is to be compared with the experimental[4] $\rho = 1.00\pm0.02$.
6. Why does Nature favour the simplest suggestion in $SU(2)\times U(1)$ theory of the Higgs scalars being iso-doublet?[5] Is there just one physical Higgs?

[3] "To my mind the most striking feature of theoretical physics in the last thirty-six years is the fact that not a single new theoretical idea of a fundamental nature has been successful. The notions of relativistic quantum theory . . . have in every instance proved stronger than the revolutionary ideas . . . of a great number of talented physicists. We live in a dilapidated house and we seem to be unable to move out. The difference between this house and a prison is hardly noticeable"—Res Jost (1963) in Praise of Quantum Field Theory (Siena European Conference).

[4] The one-loop radiative corrections to ρ suggest that the maximum mass of leptons contributing to ρ is less than 100 GeV.[56]

[5] To reduce the arbitrariness of the Higgs couplings and to motivate their iso-doublet character, one suggestion is to use supersymmetry. Supersymmetry is a Fermi-Bose symmetry, so that iso-doublet leptons like (ν_e, e) or (ν_μ, μ) in a supersymmetric theory must be accompanied in the same multiplet by iso-doublet Higgs.

Alternatively, one may identify the Higgs as composite fields associated with bound states of a yet new level of elementary particles and new (so-called techni-colour) forces (Dimopoulos & Susskind[57], Weinberg[58] and 't Hooft) of which, at present low energies, we have no cognisance and which may manifest themselves in the 1—100 TeV range. Unfortunately, both these ideas at first sight appear to introduce complexities, though in the context of a wider theory, which spans energy scales upto much higher masses, a satisfactory theory of the Higgs phenomena, incorporating these, may well emerge.

Of what mass? At present the Higgs interactions with leptons, quarks as well as their self-interactions are non-gauge interactions. For a three-family (6-quark) model, 21 out of the 26 parameters needed, are attributable to the Higgs interactions. Is there a basic principle, as compelling and as economical as the gauge principle, which embraces the Higgs sector? Alternatively, could the Higgs phenomenon itself be a manifestation of a dynamical breakdown of the gauge symmetry.[5]

7. Finally there is the problem of the families; is there a distinct SU(2) for the first, another for the second as well as a third SU(2), with spontaneous symmetry breaking such that the SU(2) apprehended by present experiment is a diagonal sum of these "family" SU(2)'s? To state this in another way, how far in energy does the e-μ universality (for example) extend? Are there more[59] Z^0's than just one, effectively differentially coupled to the e and the μ systems? (If there are, this will constitute mini-modifications of the theory, but not a drastic revolution of its basic ideas.)

In the next section I turn to a direct extrapolation of the ideas which went into the electroweak unification, so as to include strong interactions as well. Later I shall consider the more drastic alternatives which may be needed for the unification of all forces (including gravity)—ideas which have the promise of providing a deeper understanding of the charge concept. Regretfully, by the same token, I must also become more technical and obscure for the non-specialist. I apologize for this. The non-specialist may sample the flavour of the arguments in the next section (Sec. IV), ignoring the Appendices and then go on to Sec. V which is perhaps less technical.

IV. DIRECT EXTRAPOLATION FROM THE ELECTROWEAK TO THE ELECTRONUCLEAR

4.1 *The three ideas*

The three main ideas which have gone into the electronuclear—also called grand—unification of the electroweak with the *strong* nuclear force (and which date back to the period 1972—1974), are the following:

1. First: the psychological break (for us) of grouping quarks and leptons in the *same* multiplet of a unifying group G, suggested by Pati and myself in 1972[60]. The group G must contain $SU(2) \times U(1) \times SU_c(3)$; must be simple, if all quantum numbers (flavour, colour, lepton, quark and family numbers) are to be automatically quantized and the resulting gauge theory asymptotically free.
2. Second: an extension, proposed by Georgi and Glashow (1974)[61] which places not only (left-handed) quarks and leptons but also their antiparticles in the same multiplet of the unifying group.

Appendix I displays some examples of the unifying groups presently considered.

Now a gauge theory based on a "simple" (or with discrete symmetrics, a "semi-simple") group G contains one basic gauge constant. This constant would manifest itself physically above the "grand unification mass" M, exceeding all particle masses in the theory—these themselves being generated (if possible) hierarchially through a suitable spontaneous symmetry-breaking mechanism.

3. The third crucial development was by Georgi, Quinn and Weinberg (1974)[62] who showed how, using renormalization group ideas, one could relate the observed low-energy couplings $\alpha(\mu)$, and $\alpha_s(\mu)$ ($\mu \sim 100$ GeV) to the magnitude of the grand unifying mass M and the observed value of $\sin^2\theta(\mu)$; ($\tan\theta$ is the ratio of the U(1) to the SU(2) couplings).
4. If one extrapolates with Jowett[6], that nothing essentially new can possibly be discovered—i.e. one assumes that there are no new features, no new forces, or no new "types" of particles to be discovered, till we go beyond the grand unifying energy M—then the Georgi, Quinn, Weinberg method leads to a startling result: this featureless "plateau" with no "new physics" heights to be scaled stretches to fantastically high energies. More precisely, if $\sin^2\theta(\mu)$ is as large as 0.23, then the grand unifying mass M cannot be smaller than 1.3×10^{13} GeV.[63] (Compare with Planck mass $m_P \approx 1.2 \times 10^{19}$ GeV related to Newton's constant where gravity must come in.)[7] The result follows from the formula[63], [64]

$$\frac{11\alpha}{3\pi} \ell n \frac{M}{\mu} = \frac{\sin^2\theta(M) - \sin^2\theta(\mu)}{\cos^2\theta(M)}, \qquad (I)$$

if it is assumed that $\sin^2\theta(M)$—the magnitude of $\sin^2\theta$ for energies of the order of the unifying mass M—equals 3/8 (see Appendix II).

This startling result will be examined more closely in Appendix II. I show there that it is very much a consequence of the assumption that the $SU(2) \times U(1)$ symmetry survives intact from the low regime energies μ right upto the grand unifying mass M. I will also show that there already is some experimental indication that this assumption is too strong, and that there may be likely peaks of new physics at energies of 10 TeV upwards (Appendix II).

[6] The universal urge to extrapolate from what we know to-day and to believe that nothing new can possibly be discovered, is well expressed in the following:

"I come first, My name is Jowett
I am the Master of this College,
Everything that is, I know it
If I don't, it isn't knowledge"

—The Balliol Masque.

[7] On account of the relative proximity of $M \approx 10^{13}$ GeV to m_P (and the hope of eventual unification with gravity), Planck mass m_P is now the accepted "natural" mass scale in Particle Physics. With this large mass as the input, the great unsolved problem of Grand Unification is the "natural" emergence of mass hierarchies (m_P, αm_P, $\alpha^2 m_P$, . . .) or $m_P \exp(-c_n/\alpha)$, where c_n's are constants.

$$\left(\frac{m_e}{m_P} \sim 10^{-22}.\right)$$

4.2 *Tests of electronuclear grand unification*

The most characteristic prediction from the existence of the ELECTRONUCLEAR force is proton decay, first discussed in the context of grand unification at the Aix-en-Provence Conference (1973)[65]. For "semi-simple" unifying groups with multiplets containing quarks and leptons only, (but no antiquarks nor antileptons) the lepto-quark composites have masses (determined by renormalization group arguments), of the order of $\approx 10^5-10^6$ GeV[66]. For such theories the characteristic proton decays (proceeding through exchanges of *three* lepto-quarks) conserve quark number+lepton number, i.e. $P = qqq \rightarrow \ell\ell\ell$, $\tau_P \sim 10^{29}-10^{34}$ years. On the contrary, for the "simple" unifying family groups like SU(5)[61] or SO(10)[67] (with multiplets containing antiquarks and antileptons) proton decay proceeds through an exchange of *one* lepto-quark into an antilepton (plus pions etc.) $(P \rightarrow \bar{\ell})$.

An intriguing possibility in this context is that investigated recently for the maximal unifying group SU(16)—the largest group to contain a 16-fold fermionic $(q, \ell, \bar{q}, \bar{\ell})$. This can permit four types of decay modes: $P \rightarrow 3\bar{\ell}$ *as well as* $P \rightarrow \bar{\ell}$, $P \rightarrow \ell$ (e.g. $P \rightarrow \ell^- + \pi^+ + \pi^+$) and $P \rightarrow 3\ell$ (e.g. $N \rightarrow 3\nu + \pi^0$, $P \rightarrow 2\nu + e^+ + \pi^0$), the relative magnitudes of these alternative decays being model-dependent on how precisely SU(16) breaks down to $SU(3) \times SU(2) \times U(1)$. Quite clearly, it is the central fact of the existence of the proton decay for which the present generation of experiments must be designed, rather than for any specific type of decay modes.

Finally, grand unifying theories predict mass relations like:[68]

$$\frac{m_d}{m_e} = \frac{m_s}{m_\mu} = \frac{m_b}{m_\tau} \approx 2.8$$

for 6 (or at most 8) flavours *below the unification mass*. The important remark for proton decay and for mass relations of the above type as well as for an understanding of baryon excess[69] in the Universe[8], is that for the present *these are essentially characteristic of the fact of grand unification—rather than of specific models*.

"Yet each man kills the thing he loves" sang Oscar Wilde in his famous Ballad of Reading Gaol. Like generations of physicists before us, some in our generation also (through a direct extrapolation of the electroweak gauge methodology to the electronuclear)—and with faith in the assumption of

[8] The calculation of baryon excess in the Universe—arising from a combination of CP and baryon number violations—has recently been claimed to provide teleological arguments for grand unification. For example, Nanopoulos[70] has suggested that the "existence of human beings to measure the ratio n_B/n_γ (where n_B is the numbers of baryons and n_γ the numbers of photons in the Universe) necessarily imposes severe bounds on this quantity: i.e. $10^{-11} \approx (m_e/m_p)^{1/2} \lesssim n_B/n_\gamma \lesssim 10^{-4}$ $(\approx 0(\alpha^2))$". Of importance in deriving these constraints are the upper (and lower) bound on the numbers of flavours (≈ 6) deduced (1) from mass relations above, (2) from cosmological arguments which seek to limit the numbers of massless neutrinos, (3) from asymptotic freedom and (4) from numerous (one-loop) radiative calculations. It is clear that lack of accelerators as we move up in energy scale will force particle physics to reliance on teleology and cosmology (which in Landau's famous phrase is "often wrong, but never in doubt").

no "new physics", which leads to a grand unifying mass ~ 10^{13} GeV—are beginning to believe that the end of the problems of elementarity as well as of fundamental forces is nigh. They may be right, but before we are carried away by this prospect, it is worth stressing that even for the simplest grand unifying model (Georgi and Glashow's SU(5) with just two Higgs (a 5 and a 24)), the number of presently *ad hoc* parameters needed by the model is still unwholesomely large—22, to compare with 26 of the six-quark model based on the humble $SU(2) \times U(1) \times SU_c(3)$. We cannot feel proud.

V. ELEMENTARITY: UNIFICATION WITH GRAVITY AND NATURE OF CHARGE

In some of the remaining parts of this lecture I shall be questioning two of the notions which have gone into the direct extrapolation of Sec. IV—first, do quarks and leptons represent the correct elementary[9] fields, which should appear in the matter Lagrangian, and which are structureless for renormalizaibility; second, could some of the presently considered gauge fields themselves be composite? This part of the lecture relies heavily on an address I was privileged to give at the European Physical Society meeting in Geneva in July this year.[64]

5.1 *The quest for elementarity, prequarks (preons and pre-preons)*

If quarks and leptons are elementary, we are dealing with $3 \times 15 = 45$ elementary entities. The "natural" group of which these constitute the fundamental representation is SU(45) with 2024 elementary gauge bosons. It is possible to reduce the size of this group to SU(11) for example (see Appendix I), with only 120 gauge bosons, but then the number of elementary fermions increases to 561, (of which presumably $3 \times 15 = 45$ objects are of low and the rest of Planckian mass). Is there any basic reason for one's instinctive revulsion when faced with these vast numbers of elementary fields.

The numbers by themselves would perhaps not matter so much. After all, Einstein in his description of gravity,[71] chose to work with *10* fields ($g_{\mu\nu}(x)$) rather than with just one (scalar field) as Nordström[72] had done before him. Einstein was not perturbed by the multiplicity he chose to introduce, since he relied on the sheet-anchor of a fundamental principle—(the equivalence principle)—which permitted him to relate the 10 fields for gravity $g_{\mu\nu}$ with the 10 components of the physically relevant quantity, the tensor $T_{\mu\nu}$ of energy and momentum. *Einstein knew that nature was not economical of structures:* only of principles of fundamental applicability. The question we must ask ourselves is this: have we yet discovered such principles in our quest for elementarity, to justify having fields with such large numbers of components as elementary.

[9] I would like to quote Feynman in a recent interview to the "Omni" magazine: "As long as it looks like the way things are built with wheels within wheels, then you are looking for the innermost wheel—but it might not be that way, in which case you are looking for whatever the hell it is you find!". In the same interview he remarks "a few years ago I was very sceptical about the gauge theories . . . I was expecting mist, and now it looks like ridges and valleys after all."

Recall that quarks carry at least three charges (colour, flavour and a family number). Should one not, by now, entertain the notions of quarks (and possibly of leptons) as being composites of some more basic entities[10] (PRE-QUARKS or PREONS), which each carry but *one* basic charge[52]. These ideas have been expressed before but they have become more compulsive now, with the growing multiplicity of quarks and leptons. Recall that it was similar ideas which led from the eight-fold of baryons to a triplet of (Sakatons and) quarks in the first place.

The preon notion is not new. In 1975, among others, Pati, Salam and Strathdee[52] introduced 4 chromons (the fourth colour corresponding to the lepton number) and 4 flavons, the basic group being SU(8)—of which the family group $SU_F(4)\times SU_C(4)$ was but a subgroup. As an extension of these ideas, we now believe these preons carry magnetic charges and are bound together by very strong short-range forces, with quarks and leptons as their magnetically neutral composites[73]. The important remark in this context is that in a theory containing *both* electric and magnetic generalized charges, the analogues of the well-known Dirac quantization condition[74] gives relations like $\frac{eg}{4\pi} = \frac{n}{2}$ for the strength of the two types of charges. Clearly, magnetic monopoles[11] of strength $\pm g$ and mass $\approx m_w/d \approx 10^4$ GeV, are likely to bind much more tightly than electric charges, yielding composites whose non-elementary nature will reveal itself only for very high energies. This appears to be the situation at least for leptons if they are composites.

In another form the preon idea has been revived this year by Curtwright and Freund[52], who motivated by ideas of extended supergravity (to be discussed in the next subsection), reintroduce an SU(8) of 3 chromons (R, Y, B), 2 flavons and 3 familons (horrible names). The family group SU(5) could be a subgroup of this SU(8). In the Curtwright-Freund scheme, the $3\times 15 = 45$ fermions of SU(5)[61] can be found among the 8+28+56 of SU(8) (or alternatively the $3\times 16 = 48$ of SO(10) among the vectorial 56 fermions of SU(8)). (The next succession after the preon level may be the pre-preon level. It was suggested at the Geneva Conference[64] that with certain developments in field theory of composite fields it could be that just two-preons may suffice. But at this stage this is pure speculation.)

Before I conclude this section, I would like to make a prediction regarding the course of physics in the next decade, extrapolating from our past experience of the decades gone by:

[10] One must emphasise however that zero mass neutrinos are the hardest objects to conceive of as composites.

[11] According to 't Hooft's theorem, a monopole corresponding to the $SU_L(2)$ gauge symmetry is expected to possess a mass with the lower limit $\frac{m_W}{\alpha}$.[75], [76] Even if such monopoles are confined, their indirect effects must manifest themselves, if they exist. (Note that $\frac{m_W}{\alpha}$ is very much a lower limit. For a grand unified theory like SU(5) for which the monopole mass is α^{-1} times the heavy lepto-quark mass.) The monopole force may be the techni-colour force of Footnote 5.

DECADE	1950—1960	1960—1970	1970—1980	1980→
Discovery in early part of the decade	The strange particles	The 8-fold way, Ω^-	Confirmation of neutral currents	W, Z, Proton decay
Expectation for the rest of the decade		SU(3) resonances		Grand Unification, Tribal Groups
Actual discovery		Hit the next level of elementarity with quarks		May hit the preon level, and composite structure of quarks

5.2 *Post-Planck physics, supergravity and Einstein's dreams*

I now turn to the problem of a deeper comprehension of the charge concept (the basis of gauging)—*which, in my humble view, is the real quest of particle physics*. Einstein, in the last thirty-five years of his life lived with two dreams: one was to unite gravity with matter (the photon)—he wished to see the "base wood" (as he put it) which makes up the stress tensor $T_{\mu\nu}$ on the right-hand side of his equation $R_{\mu\nu}-\frac{1}{2}g_{\mu\nu}R = -T_{\mu\nu}$ transmuted through this union, into the "marble" of gravity on the left-hand side. The second (and the complementary) dream was to use this unification to comprehend the nature of electric charge in terms of space-time geometry in the same manner as he had successfully comprehended the nature of gravitational charge in terms of space-time curvature.

In case some one imagines[12] that such deeper comprehension is irrelevant to quantitative physics, let me adduce the tests of Einstein's theory versus the proposed modifications to it (Brans-Dicke[77] for example). Recently (1976), the *strong* equivalence principle (i.e. the proposition that gravitational forces contribute equally to the inertial and the gravitational masses) was tested[13] to one part in 10^{12} (i.e. to the same accuracy as achieved in particle physics for $(g\text{-}2)_e$) through lunar-laser ranging measurements[78]. These measurements determined departures from Kepler equilibrium distances, of the moon, the earth and the sun to better than ± 30 cms. and triumphantly vindicated Einstein.

There have been four major developments in realizing Einstein's dreams:

1. The Kaluza-Klein[79] miracle: An Einstein Lagrangian (scalar curvature) in five-dimensional space-time (where the fifth dimension is compactified in

[12] The following quotation from Einstein is relevant here. "We now realize, with special clarity, how much in error are those theorists who believe theory comes inductively from experience. Even the great Newton could not free himself from this error (Hypotheses non fingo)." This quote is complementary to the quotation from Einstein at the end of Sec. II.

[13] The *weak* equivalence principle (the proposition that all but the gravitational force contribute equally to the inertial and the gravitational masses) was verified by Eötvös to $1 : 10^8$ and by Dicke and Braginsky and Panov to $1 : 10^{12}$.

the sense of all fields being explicitly independent of the fifth co-ordinate) precisely reproduces the *Einstein-Maxwell* theory in four dimensions, the $g_{\mu 5}$ ($\mu = 0, 1, 2, 3$) components of the metric in five dimensions being identified with the Maxwell field A_μ. From this point of view, Maxwell's field is associated with the extra components of curvature implied by the (conceptual) existence of the fifth dimension.

2. The second development is the recent realization by Cremmer, Scherk, Englert, Brout, Minkowski and others that the compactification of the extra dimensions[80]—(their curling up to sizes perhaps smaller than Planck length $\lesssim 10^{-33}$ cms. and the very high curvature associated with them)—might arise through a spontaneous symmetry breaking (in the first 10^{-43} seconds) which reduced the higher dimensional space-time effectively to the four-dimensional that we apprehend directly.
3. So far we have considered Einstein's second dream, i.e. the unification of of electromagnetism (and presumably of other gauge forces) with gravity, giving a space-time significance to gauge charges as corresponding to extended curvature in extra bosonic dimensions. A full realization of the first dream (unification of spinor matter with gravity and with other gauge fields) had to await the development of supergravity[81], [82]—and an extension to extra fermionic dimensions of superspace[83] (with extended torsion being brought into play in addition to curvature). I discuss this development later.
4. And finally there was the alternative suggestion by Wheeler[84] and Schemberg that electric charge may be associated with space-time topology—with worm-holes, with space-time Gruyère-cheesiness. This idea has recently been developed by Hawking[14] and his collaborators[85].

5.3 *Extended supergravity, SU(8) preons and composite gauge fields*

Thus far I have reviewed the developments in respect of Einstein's dreams as reported at the Stockholm Conference held in 1978 in this hall and organized by the Swedish Academy of Sciences.

A remarkable new development was reported during 1979 by Julia and Cremmer[87] which started with an attempt to use the ideas of Kaluza and Klein to formulate extended supergravity theory in a higher (compactified) spacetime—more precisely in eleven dimensions. This development links up, as we shall see, with preons and composite Fermi fields—and even more important—possibly with the notion of composite gauge fields.

Recall that simple supergravity[81] is the gauge theory of supersymmetry[88]—the gauge particles being the (helicity ±2) gravitons and

[14] The Einstein Langrangian allows large fluctuations of metric and topology on Planck-length scale. Hawking has surmised that the dominant contributions to the path integral of quantum gravity come from metrics which carry one unit of topology per Planck volume. On account of the intimate connection (de Rham, Atiyah-Singer)[86] of curvature with the measures of space-time topology (Euler number, Pontryagin number) the extended Kaluza-Klein and Wheeler-Hawking points of view may find consonance after all.

(helicity $\pm\frac{3}{2}$) gravitinos[15]. *Extended supergravity* gauges supersymmetry combined with SO(N) internal symmetry. For N = 8, the (tribal) supergravity multiplet consists of the following SO(8) families:[81], [87]

Helicity ± 2	1
$\pm\frac{3}{2}$	8
± 1	28
$\pm\frac{1}{2}$	56
0	70

As is well known, SO(8) is too small to contain $SU(2)\times U(1)\times SU_c(3)$. Thus this tribe has no place for $W^\pm$ (though Z^0 and γ are contained) and no places for μ or τ or the t quark.

This was the situation last year. This year, Cremmer and Julia[87] attempted to write down the N = 8 supergravity Langrangian explicitly, using an extension of the Kaluza-Klein ansatz which states that *extended supergravity* (with SO(8) internal symmetry) has the same Lagrangian in four space-time dimensions as *simple supergravity* in (compactified) eleven dimensions. This formal—and rather formidable ansatz—when carried through yielded a most agreeable bonus. *The supergravity Lagrangian possesses an unsuspected SU(8) "local" internal symmetry* although one started with an internal SO(8) only.

The tantalizing questions which now arise are the following.

1. Could this internal SU(8) be the symmetry group of the 8 preons (3 chromons, 2 flavons, 3 familons) introduced earlier?
2. When SU(8) is gauged, there should be 63 spin-one fields. The supergravity tribe contains only 28 spin-one fundamental objects which are not minimally coupled. Are the 63 fields of SU(8) to be identified with composite gauge fields made up of the 70 spin-zero objects of the form $V^{-1}\partial_\mu V$; Do these composites propagate, in analogy with the well-known recent result in CP^{n-1} theories[89], where a composite gauge field of this form propagates as a consequence of quantum effects (quantum completion)?

The entire development I have described—the unsuspected extension of SO(8) to SU(8) when extra compactified space-time dimensions are used—and the possible existence and quantum propagation of composite gauge fields—is of such crucial importance for the future prospects of gauge theories that one begins to wonder how much of the extrapolation which took $SU(2)\times U(1)\times$

[15] Supersymmetry algebra extends Poincaré group algebra by adjoining to it supersymmetric charges Q_α which transform bosons to fermions, $\{Q_\alpha, Q_\beta\} = (\gamma_\mu P_\mu)_{\alpha\beta}$. The currents which correspond to these charges (Q_α and P_μ) are $J_{\mu\alpha}$ and $T_{\mu\nu}$—these are essentially the currents which in gauged supersymmetry (i.e. supergravity) couple to the gravitino and the graviton respectively.

$SU_c(3)$ into the electronuclear grand unified theories is likely to remain unaffected by these new ideas now unfolding.

But where in all this is the possibility to appeal directly to experiment? For grand unified theories, it was the proton decay. What is the analogue for supergravity? Perhaps the spin $\frac{3}{2}$ massive gravitino, picking its mass from a super-Higgs effect[90] provides the answer. Fayet[91] has shown that for a spontaneously broken globally supersymmetric weak theory the introduction of a local gravitational interaction leads to a super-Higgs effect. Assuming that supersymmetry breakdown is at mass scale m_W, the gravitino acquires a mass and an effective interaction, but of conventional weak rather than of the gravitational strength—an enhancement by a factor of 10^{34}. One may thus search for the gravitino among the neutral decay modes of J/ψ—the predicted rate being 10^{-3}—10^{-5} times smaller than the observed rate for $J/\psi \rightarrow e^+e^-$. This will surely tax all the ingenuity of the great men (and women) at SLAC and DESY. Another effect suggested by Scherk[92] is antigravity—a cancellation of the attractive gravitational force with the force produced by spin-one gravi-photons which exist in all extended supergravity theories, Scherk shows that the Compton wave length of the gravi-photon is either smaller than 5 cms. or comprised between 10 and 850 metres in order that there is no conflict with what is presently known about the strenght of the gravitational force.

Let me summarize: it is conceivable of course, that there is indeed a grand plateau—extending even to Planck energies. If so, the only eventual laboratory for particle physics will be the Early Universe, where we shall have to seek for the answers to the questions on the nature of charge. There may, however, be indications of a next level of structure around 10 TeV; there are also beautiful ideas (like, for example, of electric and magnetic monopole duality) which may manifest at energies of the order of α^{-1} m_W (= 10 TeV). Whether even this level of structure will give us the final clues to the nature of charge, one cannot predict. All I can say is that I am for ever and continually being amazed at the depth revealed at each successive level we explore. I would like to conclude, as I did at the 1978 Stockholm Conference, with a prediction which J. R. Oppenheimer made more than twenty-five years ago and which has been fulfilled to-day in a manner he did not live to see. More than anything else, it expresses the faith for the future with which this greatest of decades in particle physics ends: "Physics will change even more . . . If it is radical and unfamiliar . . . we think that the future will be only more radical and not less, only more strange and not more familiar, and that it will have its own new insights for the inquiring human spirit."

J. R. Oppenheimer
Reith Lectures BBC 1953

APPENDIX I. EXAMPLES OF GRAND UNIFYING GROUPS

*Semi-simple groups**	*Multiplet*	*Exotic gauge particles*	*Proton decay*
(with left-right symmetry)	$G_L \to \begin{pmatrix} q \\ \ell \end{pmatrix}_L, \; G_R \to \begin{pmatrix} q \\ \ell \end{pmatrix}_R$	Lepto-quarks→($\bar{q}\ell$)	Lepto-quarks→W +(Higgs) or
Example $[SU(6)_F \times SU(6)_c]_{L\to R}$	$G = G_L \times G_R$	Unifying mass $\approx 10^6$ GeV	Proton = qqq→$\ell\ell\ell$
Simple groups		Diquarks→(qq)	qq→$\bar{q}\ell$ i.e.
Examples		Dileptons→($\ell\ell$)	Proton P = qqq→$\bar{\ell}$
Family groups→ {SU(5) ↓ SU(11)} ←Tribal groups	or {SO(10) ↓ SO(22)} $G \to \begin{pmatrix} q \\ \ell \\ \bar{q} \\ \bar{\ell} \end{pmatrix}_L$	Lepto-quarks→($\bar{q}\ell$), ($q\ell$)	Also possible, P→ℓ, P→$3\bar{\ell}$, P→3ℓ
		Unifying mass $\approx 10^{13}-10^{15}$ GeV	

APPENDIX II

The following assumptions went into the derivation of the formula (I) in the text.

a) $SU_L(2)\times U_{L,R}(1)$ survives intact as the electroweak symmetry group from energies $\approx \mu$ right upto M. This intact survival implies that one eschews, for example, all suggestions that i) low-energy $SU_L(2)$ may be the diagonal sum of $SU_L^I(2)$, $SU_L^{II}(2)$, $SU_L^{III}(2)$, where I, II, III refer to the (three?) known families; ii) or that the $U_{L,R}(1)$ is a sum of pieces, where $U_R(1)$ may have differentially descended from a (V+A)-symmetric $SU_R(2)$ contained in G, or iii) that U(1) contains a piece from a four-colour symmetry $SU_c(4)$ (with lepton number as the fourth colour) and with $SU_c(4)$ breaking at an intermediate mass scale to $SU_c(3)\times U_c(1)$.

b) The second assumption which goes into the derivation of the formula above is that there are no unexpected heavy fundamental fermions, which might make $\sin^2\theta(M)$ differ from $\frac{3}{8}$—its value for the low mass fermions presently known to exist.$^+$

* Grouping quarks (q) and leptons (ℓ) together, implies treating lepton number as the fourth colour, i.e. $SU_c(3)$ extends to $SU_c(4)$ (Pati and Salam)[93]. A Tribal group, by definition, contains all known families in its basic representation. Favoured representations of Tribal SU(11) (Georgi)[94] and Tribal SO(22) (Gell-Mann[95] *et al.*) contain 561 and 2048 fermions!

$^+$ If one does not know G, one way to infer the parameter $\sin^2\theta(M)$ is from the formula:

$$\sin^2\theta(M) = \frac{\Sigma T_{3L}^2}{\Sigma Q^2}\left(= \frac{9\,N_q + 3\,N_\ell}{20\,N_q + 12\,N_\ell}\right).$$

Here N_q and N_ℓ are the numbers of the fundamental quark and lepton SU(2) doublets (assuming these are the only multiplets that exist). If we make the further *assumption* that $N_q = N_\ell$ (from the requirement of anomaly cancellation between quarks and leptons) we obtain $\sin^2\theta(M) = \frac{3}{8}$. This assumption however is not compulsive; for example anomalies cancel also if (heavy) mirror fermions exist[98]. This is the case for $[SU(6)]^4$ for which $\sin^2\theta(M) = \frac{9}{28}$.

c) If these assumptions are relaxed, for example, for the three family group $G = [SU_F(6) \times SU_c(6)]_{L \to R}$, where $\sin^2\theta(M) = \frac{9}{28}$, we find the grand unifying mass M tumbles down to 10^6 GeV.

d) The introduction of intermediate mass scales (for example, those connoting the breakdown of family universality, or of left-right symmetry, or of a breakdown of 4-colour $SU_c(4)$ down to $SU_c(3) \times U_c(1)$) will as a rule push the magnitude of the grand unifying mass M upwards[96]. In order to secure a proton decay life, consonant with present empirical lower limits ($\sim 10^{30}$ years)[97] this is desirable anyway. (τ_{proton} for $M \sim 10^{13}$ GeV is unacceptably low $\sim 6 \times 10^{23}$ years unless there are 15 Higgs.) There is, from this point of view, an indication of there being in Particle Physics one or several intermediate mass scales which can be shown to start from around 10^4 GeV upwards. *This is the end result which I wished this Appendix to lead up to.*

REFERENCES

1. Lee, T. D. and Yang, C. N., Phys. Rev. *104,* 254 (1956).
2. Abdus Salam, Nuovo Cimento *5,* 299 (1957).
3. Wu, C. S., *et al.,* Phys. Rev. *105,* 1413 (1957).
4. Garwin, R., Lederman, L. and Weinrich, M., Phys. Rev. *105,* 1415 (1957).
5. Friedman, J. I. and Telegdi, V. L., Phys. Rev. *105,* 1681 (1957).
6. Landau, L., Nucl. Phys. *3,* 127 (1957).
7. Lee, T. D. and Yang, C. N., Phys. Rev. *105,* 1671 (1957).
8. Abdus Salam, Imperial College, London, preprint (1957). For reference, see Footnote 7, p. 89, of *Theory of Weak Interactions in Particle Physics,* by Marshak, R. E., Riazuddin and Ryan, C. P., (Wiley-Interscience, New York 1969), and W. Pauli's letters (CERN Archives).
9. Yang, C. N. and Mills, R. L., Phys. Rev. *96,* 191 (1954).
10. Shaw, R., "The problem of particle types and other contributions to the theory of elementary particles", Cambridge Ph. D. Thesis (1955), unpublished.
11. Marshak, R. E. and Sudarshan, E. C. G., Proc. Padua-Venice Conference on Mesons and Recently Discovered Particles (1957), and Phys. Rev. *109,* 1860 (1958). The idea of a universal Fermi interaction for (P,N), (υ_e,e) and (Υ_μ,μ) doublets goes back to Tiomno, J. and Wheeler, J. A., Rev. Mod. Phys. *21,* 144 (1949); *21,* 153 (1949) and by Yang, C. N. and Tiomno, J., Phys. Rev. *75,* 495 (1950). Tiomno, J., considered γ_5 transformations of Fermi fields linked with mass reversal in Il Nuovo Cimento *1,* 226 (1956).
12. Feynman, R. P. and Gell-Mann, M., Phys. Rev. *109,* 193 (1958).
13. Sakurai, J. J., Nuovo Cimento *7,* 1306 (1958).
14. Nambu, Y. and Jona-Lasinio, G., Phys. Rev. *122,* 345 (1961).
15. Nambu, Y., Phys. Rev. Letters *4,* 380 (1960); Goldstone, J., Nuovo Cimento *19,* 154 (1961).
16. Klein, O., "On the theory of charged fields", Proceedings of the Conference organized by International Institute of Intellectual Cooperation, Paris (1939).
17. Kemmer, N., Phys. Rev. *52,* 906 (1937).
18. Schwinger, J., Ann. Phys. (N. Y.) *2,* 407 (1957).
19. Glashow, S. L., Nucl. Phys. *10,* 107 (1959).
20. Abdus Salam and Ward, J. C., Nuovo Cimento *11,* 568 (1959).
21. Bludman, S., Nuovo Cimento *9,* 433 (1958).

22. Abdus Salam and Ward, J. C., Nuovo Cimento *19,* 165 (1961).
23. Glashow, S. L., Nucl. Phys. *22,* 579 (1961).
24. Abdus Salam and Ward, J. C., Phys. Letters *13,* 168 (1964).
25. Goldstone, J., see Ref. 15.
26. Goldstone, J., Abdus Salam and Weinberg, S., Phys. Rev. *127,* 965 (1962).
27. Anderson, P. W., Phys. Rev. *130,* 439 (1963).
28. Higgs, P. W., Phys. Letters *12,* 132 (1964); Phys. Rev. Letters *13,* 508 (1964); Phys. Rev. *145,* 1156 (1966).
29. Englert, F. and Brout, R., Phys. Rev. Letters *13,* 321 (1964);
Englert, F., Brout, R. and Thiry, M. F., Nuovo Cimento *48,* 244 (1966).
30. Guralnik, G. S., Hagen, C. R. and Kibble, T. W. B., Phys. Rev. Letters *13,* 585 (1964); Kibble, T. W. B., Phys. Rev. *155,* 1554 (1967).
31. Weinberg, S., Phys. Rev. Letters *27,* 1264 (1967).
32. Abdus Salam, Proceedings of the 8th Nobel Symposium, Ed. Svartholm, N., (Almqvist and Wiksell, Stockholm 1968).
33. 't Hooft, G., Nucl. Phys. *B33,* 173 (1971); *ibid. B35,* 167 (1971).
34. Lee, B. W., Phys. Rev. *D5,* 823 (1972); Lee, B. W. and Zinn-Justin, J., Phys. Rev. *D5,* 3137 (1972); *ibid. D7,* 1049 (1973).
35. 't Hooft, G. and Veltman, M., Nucl. Phys. *B44,* 189 (1972); *ibid. B50,* 318 (1972). An important development in this context was the invention of the dimensional regularization technique by Bollini, C. and Giambiagi, J., Nuovo Cimento *B12,* 20 (1972);
Ashmore, J., Nuovo Cimento Letters *4,* 289 (1972) and by 't Hooft, G. and Veltman, M.
36. Feynman, R. P., Acta Phys. Polonica *24,* 297 (1963).
37. DeWitt, B. S., Phys. Rev. *162,* 1195 and 1239 (1967).
38. Faddeev, L. D. and Popov, V. N., Phys. Letters *25B,* 29 (1967).
39. Mandelstam, S., Phys. Rev. *175,* 1588 and 1604 (1968).
40. Fradkin, E. S. and Tyutin, I. V., Phys. Rev. *D2,* 2841 (1970).
41. Boulware, D. G., Ann. Phys. (N.Y.) *56,* 140 (1970).
42. Taylor, J. C., Nucl. Phys. *B33,* 436 (1971).
43. Slavnov, A., Theor. Math. Phys. *10,* 99 (1972).
44. Abdus Salam and Strathdee, J., Phys. Rev. *D2,* 2869 (1970).
45. Glashow, S., Iliopoulos, J. and Maiani, L., Phys. Rev. *D2,* 1285 (1970).
46. For a review, see Jackiw, R., in *Lectures on Current Algebra and its Applications,* by Treiman, S. B., Jackiw, R. and Gross, D. J., (Princeton Univ. Press, 1972).
47. Bouchiat, C., Iliopoulos, J. and Meyer, Ph., Phys. Letters *38B,* 519 (1972); Gross, D. J. and Jackiw, R., Phys. Rev. *D6,* 477 (1972).
48. Wentzel, G., Helv. Phys. Acta *10,* 108 (1937).
49. Hasert, F. J., *et al.,* Phys. Letters *46B,* 138 (1973).
50. Taylor, R. E., Proceedings of the 19th International Conference on High Energy Physics, Tokyo, Physical Society of Japan, 1979, p. 422.
51. Barkov, L. M., Proceedings of the 19th International Conference on High Energy Physics, Tokyo, Physical Society of Japan, 1979, p. 425.
52. Pati, J. C. and Abdus Salam, ICTP, Trieste, IC/75/106, Palermo Conference, June 1975; Pati, J. C., Abdus Salam and Strathdee, J., Phys. Letters *59B,* 265 (1975); Harari, H., Phys. Letters *86B,* 83 (1979); Schupe, M., *ibid. 86B,* 87 (1979); Curtwright, T. L. and Freund, P. G.O., Enrico Fermi Inst. preprint EFI 79/25, University of Chicago, April 1979.
53. Pati, J. C. and Abdus Salam, see the review by Bjorken, J. D., Proceedings of the 16th International Conference on High Energy Physics, Chicago-Batavia, 1972, Vol. 2, p. 304; Fritsch, H. and Gell-Mann, M., *ibid.* p. 135; Fritzsch, H., Gell-Mann, M. and Leutwyler, H., Phys. Letters *47B,* 365 (1973);
Weinberg, S., Phys. Rev. Letters *31,* 494 (1973); Phys. Rev. *D8,* 4482 (1973); Gross, D. J. and Wilczek, F., Phys. Rev. *D8,* 3633 (1973); For a review see Marciano, W. and Pagels, H., Phys. Rep. *36C,* 137 (1978).
54. Tasso Collaboration, Brandelik *et al.,* Phys. Letters *86B,* 243 (1979); Mark-J. Collaboration, Barber *et al.,* Phys. Rev. Letters *43,* 830 (1979); See also reports of the Jade, Mark-J,

Pluto and Tasso Collaborations to the International Symposium on Lepton and Photon Interactions at High Energies, Fermilab, August 1979.

55. Winter, K., Proceedings of the International Symposium on Lepton and Photon Interactions at High Energies, Fermilab, August 1979.
56. Ellis, J., Proceedings of the "Neutrino-79" International Conference on Neutrinos, Weak Interactions and Cosmology, Bergen, June 1979.
57. Dimopoulos, S. and Susskind, L., Nucl. Phys. *B155,* 237 (1979).
58. Weinberg, S., Phys. Rev. *D19,* 1277 (1979).
59. Pati, J. C. and Abdus Salam, Phys. Rev. *D10,* 275 (1974);
    Mohapatra, R. N. and Pati, J. C., Phys. Rev. *D11,* 566, 2558 (1975);
    Elias, V., Pati, J. C. and Abdus Salam, Phys. Letters *73B,* 450 (1978);
    Pati, J. C. and Rajpoot, S., Phys. Letters *79B,* 65 (1978).
60. See Pati, J. C. and Abdus Salam, Ref. 53 above and Pati, J. C. and Abdus Salam, Phys. Rev. *D8,* 1240 (1973).
61. Georgi, H. and Glashow, S. L., Phys. Rev. Letters *32,* 438 (1974).
62. Georgi, H., Quinn, H. R. and Weinberg, S., Phys. Rev. Letters *33,* 451 (1974).
63. Marciano, W. J., Phys. Rev. *D20,* 274 (1979).
64. See Abdus Salam, Proceedings of the European Physical Society Conference, Geneva, August 1979, ICTP, Trieste, preprint IC/79/142, with references to H. Harari's work.
65. Pati, J. C. and Abdus Salam, Phys, Rev. Letters *31,* 661 (1973).
66. Elias, V., Pati, J. C. and Abdus Salam, Phys. Rev. Letters *40,* 920 (1978);
    Rajpoot, S. and Elias, V., ICTP, Trieste, preprint IC/78/159.
67. Fritzsch, H. and Minkowski, P., Ann. Phys. (N. Y.) *93,* 193 (1975); Nucl. Phys. *B103,* 61 (1976);
    Georgi, H., Particles and Fields (APS/OPF Williamsburg), Ed. Carlson, C. E., AIP, New York, 1975, p. 575;
    Georgi, H. and Nanopoulos, D. V., Phys. Letters *82B,* 392 (1979).
68. Buras, A., Ellis, J., Gaillard, M. K. and Nanopoulos, D. V., Nucl. Phys. *B135,* 66 (1978).
69. Yoshimura, M., Phys. Rev. Letters *41,* 381 (1978);
    Dimopoulos, S. and Susskind, L., Phys. Rev. *D18,* 4500 (1978);
    Toussaint, B., Treiman, S. B., Wilczek, F. and Zee, A., Phys. Rev. *D19,* 1036 (1979);
    Ellis J., Gaillard, M. K. and Nanopoulos, D. V., Phys. Letters *80B,* 360 (1979);
    Erratum *82B,* 464 (1979);
    Weinberg, S., Phys. Rev. Letters *42,* 850 (1979); Nanopoulos, D. V. and Weinberg, S., Harvard University preprint HUTP-79/A023.
70. Nanopoulos, D. V., CERN preprint TH2737, September 1979.
71. Einstein, A., Annalen der Phys. *49,* 769 (1916). For an English translation, see *The Principle of Relativity* (Methuen, 1923, reprinted by Dover Publications), p. 35.
72. Nordström, G., Phys. Z. *13,* 1126 (1912); Ann. Phys. (Leipzig) *40,* 856 (1913); *ibid. 42,* 533 (1913); *ibid. 43,* 1101 (1914); Phys. Z. *15,* 375 (1914); Ann. Acad. Sci. Fenn. *57* (1914, 1915);
    See also Einstein, A., Ann. Phys. Leipzig *38,* 355, 433 (1912).
73. Pati, J. C. and Abdus Salam, in preparation.
74. Dirac, P. A. M., Proc. Roy. Soc. (London) *A133,* 60 (1931).
75. 't Hooft, G., Nucl. Phys. *B79,* 276 (1974).
76. Polyakov, A. M., JETP Letters *20,* 194 (1974).
77. Brans, C. H. and Dicke, R. H., Phys. Rev. *124,* 925 (1961).
78. Williams, J. G., *et al.,* Phys. Rev. Letters *36,* 551 (1976);
    Shapiro, I. I., *et al.,* Phys. Rev. Letters *36,* 555 (1976);
    For a discussion, see Abdus Salam, in *Physics and Contemporary Needs* Ed. Riazuddin (Plenum Publishing Corp., 1977), p. 301.
79. Kaluza, Th., Sitzungsber. Preuss. Akad. Wiss. p. 966 (1921);
    Klein, O., Z. Phys. *37,* 895 (1926).
80. Cremmer, E. and Scherk, J., Nucl. Phys. *B103,* 399 (1976); *ibid. B108,* 409 (1976); *ibid. B118,* 61 (1976);
    Minkowski, P., Univ. of Berne preprint, October 1977.

81. Freedman, D. Z., van Nieuwenhuizen, P. and Ferrara, S., Phys. Rev. *D13,* 3214 (1976); Deser, S. and Zumino, B., Phys. Letters *62B,* 335 (1976);
For a review and comprehensive list of references, see D. Z. Freedman's presentation to the 19th International Conference on High Energy Physics, Tokyo, Physical Society of Japan, 1979.
82. Arnowitt, R., Nath, P. and Zumino, B., Phys. Letters *56B,* 81 (1975); Zumino, B., in Proceedings of the Conference on Gauge Theories and Modern Field Theory, Northeastern University, September 1975, Eds. Arnowitt, R. and Nath, P., (MIT Press);
Wess, J. and Zumino, B., Phys. Letters *66B,* 361 (1977);
Akulov, V. P., Volkov, D. V. and Soroka, V. A., JETP Letters *22,* 187 (1975);
Brink, L., Gell-Mann, M., Ramond, P. and Schwarz, J. H., Phys. Letters *74B,* 336 (1978);
Taylor, J. G., King's College, London, preprint, 1977 (unpublished);
Siegel, W., Harvard University preprint HUTP-77/AO68, 1977 (unpublished);
Ogievetsky, V. and Sokatchev, E., Phys. Letters *79B,* 222 (1978);
Chamseddine, A. H. and West, P. C., Nucl. Phys. *B129,* 39 (1977);
MacDowell, S. W. and Mansouri, F., Phys. Rev. Letters *38,* 739 (1977).
83. Abdus Salam and Strathdee, J., Nucl. Phys. *B79,* 477 (1974).
84. Fuller, R. W. and Wheeler, J. A., Phys. Rev. *128,* 919 (1962);
Wheeler, J. A., in *Relativity Groups and Topology,* Proceedings of the Les Houches Summer School, 1963, Eds. DeWitt, B. S. and DeWitt, C. M., (Gordon and Breach, New York 1964).
85. Hawking, S. W., in *General Relativity: An Einstein Centenary Survey* (Cambridge University Press, 1979);
See also "Euclidean quantum gravity", DAMTP, Univ. of Cambridge preprint, 1979;
Gibbons, G. W., Hawking, S. W. and Perry, M. J., Nucl. Phys. *B138,* 141 (1978);
Hawking, S. W., Phys. Rev. *D18,* 1747 (1978).
86. Atiyah, M. F. and Singer, I. M., Bull. Am. Math. Soc. *69,* 422 (1963).
87. Cremmer, E., Julia, B. and Scherk, J., Phys. Letters *76B,* 409 (1978); Cremmer, E. and Julia, B., Phys. Letters *80B,* 48 (1978); Ecole Normale Supérieure preprint, LPTENS 79/6, March 1979;
See also Julia, B., in Proceedings of the Second Marcel Grossmann Meeting, Trieste, July 1979 (in preparation).
88. Gol'fand, Yu. A. and Likhtman, E. P., JETP Letters *13,* 323 (1971);
Volkov, D. V. and Akulov, V. P., JETP Letters *16,* 438 (1972);
Wess, J. and Zumino, B., Nucl. Phys. *B70,* 39 (1974);
Abdus Salam and Strathdee, J., Nucl. Phys. *B79,* 477 (1974); *ibid. B80,* 499 (1974); Phys. Letters *51B,* 353 (1974);
For a review, see Abdus Salam and Strathdee, J., Fortschr. Phys. *26,* 57 (1978).
89. D'Adda, A., Lüscher, M. and Di Vecchia, P., Nucl. Phys. *B146,* 63 (1978).
90. Cremmer, E., *et al.,* Nucl. Phys. *B147,* 105 (1979);
See also Ferrara, S., in Proceedings of the Second Marcel Grossmann Meeting, Trieste, July 1979 (in preparation), and references therein.
91. Fayet, P., Phys. Letters *70B,* 461 (1977); *ibid. 84B,* 421 (1979).
92. Scherk, J., Ecole Normale Supérieure preprint, LPTENS 79/17, September 1979.
93. Pati, J. C. and Abdus Salam, Phys. Rev. *D10,* 275 (1974).
94. Georgi, H., Harvard University Report No. HUTP-29/AO13 (1979).
95. Gell-Mann, M., (unpublished).
96. See Ref. 64 above and also Shafi, Q. and Wetterich, C., Phys. Letters *85B,* 52 (1979).
97. Learned, J., Reines, F. and Soni, A., Phys. Letters *43,* 907 (1979).
98. Pati, J. C., Abdus Salam and Strathdee, J., Nuovo Cimento *26A,* 72 (1975);
Pati, J. C. and Abdus Salam, Phys. Rev. *D11,* 1137, 1149 (1975);
Pati, J. C., invited talk, Proceedings Second Orbis Scientiae, Coral Gables, Florida, 1975, Eds. Perlmutter, A. and Widmayer, S., p. 253.

TOWARDS A UNIFIED THEORY – THREADS IN A TAPESTRY

Nobel Lecture, 8 December, 1979

by

SHELDON LEE GLASHOW
Lyman Laboratory of Physics Harvard University Cambridge, Mass., USA

INTRODUCTION

In 1956, when I began doing theoretical physics, the study of elementary particles was like a patchwork quilt. Electrodynamics, weak interactions, and strong interactions were clearly separate disciplines, separately taught and separately studied. There was no coherent theory that described them all. Developments such as the observation of parity violation, the successes of quantum electrodynamics, the discovery of hadron resonances and the appearance of strangeness were well-defined parts of the picture, but they could not be easily fitted together.

Things have changed. Today we have what has been called a "standard theory" of elementary particle physics in which strong, weak, and electromagnetic interactions all arise from a local symmetry principle. It is, in a sense, a complete and apparently correct theory, offering a qualitative description of all particle phenomena and precise quantitative predictions in many instances. There is no experimental data that contradicts the theory. In principle, if not yet in practice, all experimental data can be expressed in terms of a small number of "fundamental" masses and coupling constants. The theory we now have is an integral work of art: the patchwork quilt has become a tapestry.

Tapestries are made by many artisans working together. The contributions of separate workers cannot be discerned in the completed work, and the loose and false threads have been covered over. So it is in our picture of particle physics. Part of the picture is the unification of weak and electromagnetic interactions and the prediction of neutral currents, now being celebrated by the award of the Nobel Prize. Another part concerns the reasoned evolution of the quark hypothesis from mere whimsy to established dogma. Yet another is the development of quantum chromodynamics into a plausible, powerful, and predictive theory of strong interactions. All is woven together in the tapestry; one part makes little sense without the other. Even the development of the electroweak theory was not as simple and straightforward as it might have been. It did not arise full blown in the mind of one physicist, nor even of three. It, too, is the result of the collective endeavor of many scientists, both experimenters and theorists.

Let me stress that I do not believe that the standard theory will long survive as a correct and complete picture of physics. All interactions may be gauge interactions, but surely they must lie within a unifying group. This would imply the existence of a new and very weak interaction which mediates the decay of protons. All matter is thus inherently unstable, and can be observed to decay. Such a synthesis of weak, strong, and electromagnetic interactions has been called a "grand unified theory", but a theory is neither grand nor unified unless it includes a description of gravitational phenomena. We are still far from Einstein's truly grand design.

Physics of the past century has been characterized by frequent great but unanticipated experimental discoveries. If the standard theory is correct, this age has come to an end. Only a few important particles remain to be discovered, and many of their properties are alleged to be known in advance. Surely this is not the way things will be, for Nature must still have some surprises in store for us.

Nevertheless, the standard theory will prove useful for years to come. The confusion of the past is now replaced by a simple and elegant synthesis. The standard theory may survive as a part of the ultimate theory, or it may turn out to be fundamentally wrong. In either case, it will have been an important way-station, and the next theory will have to be better.

In this talk, I shall not attempt to describe the tapestry as a whole, nor even that portion which is the electroweak synthesis and its empirical triumph. Rather, I shall describe several old threads, mostly overwoven, which are closely related to my own researches. My purpose is not so much to explain who did what when, but to approach the more difficult question of why things went as they did. I shall also follow several new threads which may suggest the future development of the tapestry.

EARLY MODELS

In the 1920's, it was still believed that there were only two fundamental forces: gravity and electromagnetism. In attempting to unify them, Einstein might have hoped to formulate a universal theory of physics. However, the study of the atomic nucleus soon revealed the need for two additional forces: the strong force to hold the nucleus together and the weak force to enable it to decay. Yukawa asked whether there might be a deep analogy between these new forces and electromagnetism. All forces, he said, were to result from the exchange of mesons. His conjectured mesons were originally intended to mediate both the strong and the weak interactions: they were strongly coupled to nucleons and weakly coupled to leptons. This first attempt to unify strong and weak interactions was fully forty years premature. Not only this, but Yukawa could have predicted the existence of neutral currents. His neutral meson, essential to provide the charge independence of nuclear forces, was also weakly coupled to pairs of leptons.

Not only is electromagnetism mediated by photons, but it arises from the requirement of local gauge invariance. This concept was generalized in 1954 to apply to non-Abelian local symmetry groups. [1] It soon became clear that a more far-reaching analogy might exist between electromagnetism and the other forces. They, too, might emerge from a gauge principle.

A bit of a problem arises at this point. All gauge mesons must be massless, yet the photon is the only massless meson. How do the other gauge bosons get their masses? There was no good answer to this question until the work of Weinberg and Salam [2] as proven by 't Hooft [3] (for spontaneously broken gauge theories) and of Gross, Wilczek, and Politzer [4] (for unbroken gauge theories). Until this work was done, gauge meson masses had simply to be put in *ad hoc.*

Sakurai suggested in 1960 that strong interactions should arise from a gauge principle. [5] Applying the Yang–Mills construct to the isospin-hypercharge symmetry group, he predicted the existence of the vector mesons ϱ and ω. This was the first phenomenological SU(2) × U(1) gauge theory. It was extended to local SU(3) by Gell-Mann and Ne'eman in 1961. [6] Yet, these early attempts to formulate a gauge theory of strong interactions were doomed to fail. In today's jargon, they used "flavor" as the relevant dynamical variable, rather than the hidden and then unknown variable "color". Nevertheless, this work prepared the way for the emergence of quantum chromodynamics a decade later.

Early work in nuclear beta decay seemed to show that the relevant interaction was a mixture of S, T, and P. Only after the discovery of parity violation, and the undoing of several wrong experiments, did it become clear that the weak interactions were in reality V-A. The synthesis of Feynman and Gell-Mann and of Marshak and Sudarshan was a necessary precursor to the notion of a gauge theory of weak interactions. [7] Bludman formulated the first SU(2) gauge theory of weak interactions in 1958. [8] No attempt was made to include electromagnetism. The model included the conventional charged-current interactions, and in addition, a set of neutral current couplings. These are of the same strength and form as those of today's theory in the limit in which the weak mixing angle vanishes. Of course, a gauge theory of weak interactions alone cannot be made renormalizable. For this, the weak and electromagnetic interactions must be unified.

Schwinger, as early as 1956, believed that the weak and electromagnetic interactions should be combined together into a gauge theory. [9] The charged massive vector intermediary and the massless photon were to be the gauge mesons. As his student, I accepted this faith. In my 1958 Harvard thesis, I wrote: "It is of little value to have a potentially renormalizable theory of beta processes without the possibility of a renormalizable electrodynamics. We should care to suggest that a fully acceptable theory of these interactions may only be achieved if they are treated together. . ." [10] We used the original SU(2) gauge interaction of Yang and Mills. Things had to be arranged so that the charged current, but not the neutral (electromagnetic) current, would violate parity and strangeness. Such a theory is technically possible to construct, but it is both ugly and experimentally false. [11] We know now that neutral currents do exist and that the electroweak gauge group must be larger than SU(2).

Another electroweak synthesis without neutral currents was put forward by Salam and Ward in 1959. [12] Again, they failed to see how to incorporate the experimental fact of parity violation. Incidentally, in a continuation of their work in 1961, they suggested a gauge theory of strong, weak, and electromagnetic interactions based on the local symmetry group SU(2) × SU(2). [13] This was a remarkable portent of the SU(3) × SU(2) × U(1) model which is accepted today.

We come to my own work [14] done in Copenhagen in 1960, and done independently by Salam and Ward. [15] We finally saw that a gauge group larger than SU(2) was necessary to describe the electroweak interactions. Salam and Ward were motivated by the compelling beauty of gauge theory. I thought I saw a way to a renormalizable scheme. I was led to the group SU(2) × U(1) by analogy with the approximate isospin-hypercharge group which characterizes strong interactions. In this model there were two electrically neutral intermediaries: the massless photon and a massive neutral vector meson which I called B but which is now known as Z. The weak mixing angle determined to what linear combination of SU(2) × U(1) generators B would correspond. The precise form of the predicted neutral-current interaction has been verified by recent experimental data. However, the strength of the neutral current was not prescribed, and the model was not in fact renormalizable. These glaring omissions were to be rectified by the work of Salam and Weinberg and the subsequent proof of renormalizability. Furthermore, the model was a model of leptons—it could not evidently be extended to deal with hadrons.

RENORMALIZABILITY

In the late 50's, quantum electrodynamics and pseudoscalar meson theory were known to be renormalizable, thanks in part to work of Salam. Neither of the customary models of weak interactions – charged intermediate vector bosons or direct four-fermion couplings – satisfied this essential criterion. My thesis at Harvard, under the direction of Julian Schwinger, was to pursue my teacher's belief in a unified electroweak gauge theory. I had found some reason to believe that such a theory was less singular than its alternatives. Feinberg, working with charged intermediate vector mesons discovered that a certain type of divergence would cancel for a special value of the meson anomalous magnetic moment.[16] It did not correspond to a "minimal electromagnetic coupling", but to the magnetic properties demanded by a gauge theory. Tzou Kuo-Hsien examined the zero-mass limit of charged vector meson electrodynamics.[17] Again, a sensible result is obtained only for a very special choice of the magnetic dipole moment and electric quadrupole moment, just the values assumed in a gauge theory. Was it just coincidence that the electromagnetism of a charged vector meson was least pathological in a gauge theory?

Inspired by these special properties, I wrote a notorious paper.[18] I alleged that a softly-broken gauge theory, with symmetry breaking provided by explicit mass terms, was renormalizable. It was quickly shown that this is false.

Again, in 1970, Iliopoulos and I showed that a wide class of divergences that might be expected would cancel in such a gauge theory.[19] We showed that the naive divergences of order $(\alpha\Lambda^4)^n$ were reduced to "merely" $(\alpha\Lambda^2)^n$, where Λ is a cut-off momentum. This is probably the most difficult theorem that Iliopoulos or I had even proven. Yet, our labors were in vain. In the spring of 1971, Veltman informed us that his student Gerhart 't Hooft had established the renormalizability of spontaneously broken gauge theory.

In pursuit of renormalizability, I had worked diligently but I completely missed the boat. The gauge symmetry is an exact symmetry, but it is hidden. One must not put in mass terms by hand. The key to the problem is the idea of spontaneous symmetry breakdown: the work of Goldstone as extended to gauge theories by Higgs and Kibble in 1964.[20] These workers never thought to apply their work on formal field theory to a phenomenologically relevant model. I had had many conversations with Goldstone and Higgs in 1960. Did I neglect to tell them about my SU (2)×U (1) model, or did they simply forget?

Both Salam and Weinberg had had considerable experience in formal field theory, and they had both collaborated with Goldstone on spontaneous symmetry breaking. In retrospect, it is not so surprising that it was they who first used the key. Their SU (2)×U (1) gauge symmetry was spontaneously broken. The masses of the W and Z and the nature of neutral current effects depend on a single measurable parameter, not two as in my unrenormalizable model. The strength of the neutral currents was correctly predicted. The daring Weinberg-Salam conjecture of renormalizability was proven in 1971. Neutral currents were discovered in 1973[21], but not until 1978 was it clear that they had just the predicted properties.[22]

THE STRANGENESS-CHANGING NEUTRAL CURRENT

I had more or less abandoned the idea of an electroweak gauge theory during the period 1961–1970. Of the several reasons for this, one was the failure of my naive foray into renormalizability. Another was the emergence of an empirically successful description of strong interactions – the SU(3) unitary symmetry scheme of Gell-Mann and Ne'eman. This theory was originally phrased as a gauge theory, with ϱ, ω, and K* as gauge mesons. It was completely impossible to imagine how both strong and weak interactions could be gauge theories: there simply wasn't room enough for commuting structures of weak and strong currents. Who could foresee the success of the quark model, and the displacement of SU(3) from the arena of flavor to that of color? The predictions of unitary symmetry were being borne out – the predicted Ω^- was discovered in 1964. Current algebra was being successfully exploited. Strong interactions dominated the scene.

When I came upon the SU(2)×U(1) model in 1960, I had speculated on a possible extension to include hadrons. To construct a model of leptons alone seemed senseless: nuclear beta decay, after all, was the first and foremost problem. One thing seemed clear. The fact that the charged current violated strangeness would force the neutral current to violate strangeness as well. It was already well known that strangeness-changing neutral currents were either strongly suppressed or absent. I concluded that the Z^0 had to be made very much heavier than the W. This was an arbitrary but permissible act in those days: the symmetry breaking mechanism was unknown. I had "solved" the problem of strangeness-changing neutral currents by suppressing all neutral currents: the baby was lost with the bath water.

I returned briefly to the question of gauge theories of weak interactions in a collaboration with Gell-Mann in 1961.[23] From the recently developing ideas of current algebra we showed that a gauge theory of weak interactions would inevitably run into the problem of strangeness-changing neutral currents. We concluded that something essential was missing. Indeed it was. Only after quarks were invented could the idea of the fourth quark and the GIM mechanism arise.

From 1961 to 1964, Sidney Coleman and I devoted ourselves to the exploitation of the unitary symmetry scheme. In the spring of 1964, I spent a short leave of absence in Copenhagen. There, Bjorken and I suggested that the Gell-Mann–Zweig-system of three quarks should be extended to four.[24] (Other workers had the same idea at the same time).[25] We called the fourth quark the charmed quark. Part of our motivation for introducing a fourth quark was based on our mistaken notions of hadron spectroscopy. But we also wished to enforce an analogy between the weak leptonic current and the weak hadronic current. Because there were two weak doublets of leptons, we believed there had to be two weak doublets of quarks as well. The basic idea was correct, but today there seem to be three doublets of quarks and three doublets of leptons.

The weak current Bjorken and I introduced in 1964 was precisely the GIM current. The associated neutral current, as we noted, conserved strangeness. Had we inserted these currents into the earlier electroweak theory, we would have solved the problem of strangeness-changing neutral currents. We did not. I had apparently quite forgotten my earlier ideas of electroweak synthesis. The problem which was explicitly posed in 1961 was solved, in principle, in 1964. No one, least of all me, knew it. Perhaps we were all befuddled by the chimera of relativistic SU(6), which arose at about this time to cloud the minds of theorists.

Five years later, John Iliopoulos, Luciano Maiani and I returned to the question of strangeness-changing neutral currents.[26] It seems incredible that the problem was totally ignorned for so long. We argued that unobserved effects (a large K_1K_2 mass difference; decays like $K \rightarrow \pi\nu\bar{\nu}$; etc.) would be expected to arise in any of the known weak interaction models: four fermion couplings; charged vector meson models; or the electroweak gauge theory. We worked in terms of cut-offs, since no renormalizable theory was known at the time. We showed how the unwanted effects would be eliminated with the conjectured existence of a fourth quark. After languishing for a decade, the problem of the selection rules of the neutral current was finally solved. Of course, not everyone believed in the predicted existence of charmed hadrons.

This work was done fully three years after the epochal work of Weinberg and Salam, and was presented in seminars at Harvard and at M. I. T. Neither I, nor my coworkers, nor Weinberg, sensed the connection between the two endeavors. We did not refer, nor were we asked to refer, to the Weinberg–Salam work in our paper.

The relevance became evident only a year later. Due to the work of 't Hooft, Veltman, Benjamin Lee, and Zinn-Justin, it became clear that the Weinberg-Salam *ansatz* was in fact a renormalizable theory. With GIM, it was trivially extended from a model of leptons to a theory of weak interactions. The ball was now squarely in the hands of the experimenters. Within a few years, charmed hadrons and neutral currents were discovered, and both had just the properties they were predicted to have.

FROM ACCELERATORS TO MINES

Pions and strange particles were discovered by passive experiments which made use of the natural flux of cosmic rays. However, in the last three decades, most discoveries in particle physics were made in the active mode, with the artificial aid of particle accelerators. Passive experimentation stagnates from a lack of funding and lack of interest. Recent developments in theoretical particle physics and in astrophysics may mark an imminent rebirth of passive experimentation. The concentration of virtually all high-energy physics endeavors at a small number of major accelerator laboratories may be a thing of the past.

This is not to say that the large accelerator is becoming extinct; it will remain an essential if not exclusive tool of high-energy physics. Do not forget that the existence of Z^0 at ~ 100 GeV is an essential but quite untested prediction of the electroweak theory. There will be additional dramatic discoveries at accelerators, and these will not always have been predicted in advance by theorists. The construction of new machines like LEP and ISABELLE is mandatory.

Consider the successes of the electroweak synthesis, and the fact that the only plausible theory of strong interactions is also a gauge theory. We must believe in the ultimate synthesis of strong, weak, and electromagnetic interactions. It has been shown how the strong and electroweak gauge groups may be put into a larger but simple gauge group.[27] Grand unification – perhaps along the lines of the original SU (5) theory of Georgi and me – must be essentially correct. This implies that the proton, and indeed all nuclear matter, must be inherently unstable. Sensitive searches for proton decay are now being launched. If the proton lifetime is shorter than 10^{32} years, as theoretical estimates indicate, it will not be long before it is seen to decay.

Once the effect is discovered (and I am sure it will be), further experiments will have to be done to establish the precise modes of decay of nucleons. The selection rules, mixing angles, and space-time structure of a new class of effective four-fermion couplings must be established. The heroic days of the discovery of the nature of beta decay will be repeated.

The first generation of proton decay experiments is cheap, but subsequent generations will not be. Active and passive experiments will compete for the same dwindling resources.

Other new physics may show up in elaborate passive experiments. Today's theories suggest modes of proton decay which violate both baryon number and lepton number by unity. Perhaps this $\Delta B = \Delta L = 1$ law will be satisfied. Perhaps $\Delta B = -\Delta L$ transitions will be seen. Perhaps, as Pati and Salam suggest, the proton will decay into three leptons. Perhaps two nucleons will annihilate in $\Delta B = 2$ transitions. The effects of neutrino oscillations resulting from neutrino masses of a fraction of an election volt may be detectable. "Superheavy isotopes" which may be present in the Earth's crust in small concentrations could reveal themselves through their multi-GeV decays. Neutrino bursts arising from distant astronomical catastrophes may be seen. The list may be endless or empty. Large passive experiments of the sort now envisioned have never been done before. Who can say what results they may yield?

PREMATURE ORTHODOXY

The discovery of the J/Ψ in 1974 made it possible to believe in a system involving just four quarks and four leptons. Very quickly after this a third charged lepton (the tau) was discovered, and evidence appeared for a third $Q = -1/3$ quark (the b quark). Both discoveries were classic surprises. It became immediately fashionable to put the known fermions into families or generations:

$$\begin{pmatrix} u & \nu_e \\ d & e \end{pmatrix} \quad \begin{pmatrix} c & \nu_\mu \\ s & \mu \end{pmatrix} \quad \begin{pmatrix} t & \nu_\tau \\ b & \tau \end{pmatrix}.$$

The existence of a third $Q = 2/3$ quark (the t quark) is predicted. The Cabibbo-GIM scheme is extended to a system of six quarks. The three family system is the basis to a vast and daring theoretical endeavor. For example, a variety of papers have been written putting experimental constraints on the four parameters which replace the Cabibbo angle in a six quark system. The detailed manner of decay of particles containing a single b quark has been worked out. All that is wanting is experimental confirmation. A new orthodoxy has emerged, one for which there is little evidence, and one in which I have little faith.

The predicted t quark has not been found. While the upsilon mass is less than 10 GeV, the analogous $t\bar{t}$ particle, if it exists at all, must be heavier than 30 GeV. Perhaps it doesn't exist.

Howard Georgi and I, and other before us, have been working on models with no t quark.[28] We believe this unorthodox view is as attractive as its alternative. And, it suggests a number of exciting experimental possibilities.

We assume that b and τ share a quantum number, like baryon number, that is essentially exactly conserved. (Of course, it may be violated to the same extent that baryon number is expected to be violated.) Thus, the b,τ system is assumed to be distinct from the lighter four quarks and four leptons. There is, in particular, no mixing between b and d or s. The original GIM structure is left intact. An additional mechanism must be invoked to mediate b decay, which is not present in the $SU(3) \times SU(2) \times U(1)$ gauge theory.

One possibility is that there is an additional SU(2) gauge interaction whose effects we have not yet encountered. It could mediate such decays of b as these

$$b \rightarrow \tau^+ + (e^- \text{ or } \mu^-) + (d \text{ or } s).$$

All decays of b would result in the production of a pair of leptons, including a τ^+ or its neutral partner. There are other possibilities as well, which predict equally bizarre decay schemes for b-matter. How the b quark decays is not yet known, but it soon will be.

The new SU(2) gauge theory is called upon to explain CP violation as well as b decay. In order to fit experiment, three additional massive neutral vector bosons must exist, and they cannot be too heavy. One of them can be produced in $e^+ e^-$ annihilation, in addition to the expected Z^0. Our model is rife with experimental predictions, for example: a second Z^0, a heavier version of b and of τ, the production of τ b in e p collisions, and the existence of heavy neutral unstable leptons which may be produced and detected in $e^+ e^-$ or in νp collisions.

This is not the place to describe our views in detail. They are very speculative and probably false. The point I wish to make is simply that it is too early to convince ourselves that we know the future of particle physics. There are too many points at which the conventional picture may be wrong or incomplete. The $SU(3) \times SU(2) \times U(1)$ gauge theory with three families is certainly a good beginning, not to accept but to attack, extend, and exploit. We are far from the end.

ACKNOWLEDGEMENTS

I wish to thank the Nobel Foundation for granting me the greatest honor to which a scientist may aspire. There are many without whom my work would never have been. Let me thank my scientific collaborators, especially James Bjorken, Sidney Coleman, Alvaro De Rújula, Howard Georgi, John Iliopoulos, and Luciano Maiani; the Niels Bohr Institute and Harvard University for their hospitality while my research on the electroweak interaction was done: Julian Schwinger for teaching me how to do scientific research in the first place; the Public School System of New York City, Cornell University, and Harvard University for my formal education; my high-school friends Gary Feinberg and Steven Weinberg for making me learn too much too soon of what I might otherwise have never learned at all; my parents and my two brothers for always encouraging a child's dream to be a scientist. Finally, I wish to thank my wife and my children for the warmth of their love.

REFERENCES

1. Yang, C. N. and Mills, R., Phys. Rev. *96,* 191 (1954). Also, Shaw, R., unpublished.
2. Weinberg, S., Phys. Rev. Letters *19,* 1264 (1967); Salam, A., in *Elementary Particle Physics* (ed. Svartholm, N.; Almqvist and Wiksell; Stockholm; 1968).
3. 't Hooft, G., Nuclear Physics *B 33,* 173 and *B 35,* 167 (1971); Lee, B. W., and Zinn-Justin, J., Phys. Rev. D *5,* pp. 3121–3160 (1972); 't Hooft, G., and Veltman M., Nuclear Physics *B 44,* 189 (1972).
4. Gross, D. J. and Wilczek, F., Phys. Rev. Lett. *30,* 1343 (1973); Politzer, H. D., Phys. Rev. Lett. *30,* 1346 (1973).
5. Sakurai, J. J., Annals of Physics *11,* 1 (1960).
6. Gell-Mann, M., and Ne'eman, Y., *The Eightfold Way* (Benjamin, W. A., New York, 1964).
7. Feynman, R., and Gell-Mann, M., Phys. Rev. *109,* 193 (1958); Marshak, R., and Sudarshan, E. C. G., Phys. Rev. *109,* 1860 (1958).
8. Bludman, S., Nuovo Cimento Ser. 10 *9,* 433 (1958).
9. Schwinger, J., Annals of Physics *2,* 407 (1958).
10. Glashow, S. L., Harvard University Thesis, p. 75 (1958).
11. Georgi, H., and Glashow, S. L., Phys. Rev. Letters *28,* 1494 (1972).
12. Salam, A., and Ward, J., Nuovo Cimento *11,* 568 (1959).
13. Salam, A., and Ward, J., Nuovo Cimento *19,* 165 (1961).
14. Glashow, S. L., Nuclear Physics *22,* 579 (1961).
15. Salam, A., and Ward, J., Physics Letters *13,* 168 (1964).
16. Feinberg, G., Phys. Rev. *110,* 1482 (1958).
17. Tzou Kuo-Hsien, Comptes rendus *245,* 289 (1957).
18. Glashow, S. L., Nuclear Physics *10,* 107 (1959).
19. Glashow, S. L., and Iliopoulos J., Phys. Rev. D *3,* 1043 (1971).
20. Many authors are involved with this work: Brout, R., Englert, F., Goldstone, J., Guralnik, G., Hagen, C., Higgs, P., Jona-Lasinio, G., Kibble, T., and Nambu, Y.
21. Hasert, F. J., *et al.,* Physics Letters *46 B,* 138 (1973) and Nuclear Physics *B 73,* 1 (1974). Benvenuti, A., *et al.,* Phys. Rev. Letters *32,* 800 (1974).
22. Prescott, C. Y., *et al.,* Phys. Lett. *B 77,* 347 (1978).
23. Gell-Mann, M., and Glashow, S. L., Annals of Physics *15,* 437 (1961).
24. Bjorken, J., and Glashow, S. L., Physics Letters *11,* 84 (1964).

25. Amati, D., *et al.,* Nuovo Cimento *34,* 1732 (A 64); Hara, Y. Phys. Rev. *134,* B 701 (1964); Okun, L. B., Phys. Lett. *12,* 250 (1964); Maki, Z., and Ohnuki, Y., Progs. Theor. Phys. *32,* 144 (1964); Nauenberg, M., (unpublished); Teplitz, V., and Tarjanne, P., Phys. Rev. Lett. *11,* 447 (1963).
26. Glashow, S. L., Iliopoulos, J., and Maiani, L., Phys. Rev. D *2,* 1285 (1970).
27. Georgi, H., and Glashow, S. L., Phys. Rev. Letters *33,* 438 (1974).
28. Georgi, H., and Glashow, S. L., Harvard Preprint HUTP-79/A 053.

References

A number of books, review articles and papers are indicated here in an attempt to help graduate students willing to review the foundations of particle and field physics and continue the study of gauge field theories.

A. - SYMMETRIES AND FIELD EQUATIONS :

1. E.P. Wigner, On unitary representations of the inhomogeneous Lorentz group, Ann. Math. 40, 149 (1939)

2. V. Bargmann and E.P. Wigner, Group theoretical discussion of relativistic wave equations, Proc. Nat. Acad. Sci. (U.S.) 34, 211 (1946)

   These two papers are reproduced in the book :

3. F.J. Dyson, Symmetry groups in nuclear and particle physics, W.A. Benjamin, New York (1966)

4. A.S. Wightman, L'invariance dans la mécanique quantique relativiste in Relations de dispersion et particules élémentaires, Hermann, Paris (1960)

5. R.F. Streater and A.S. Wightman, Spin, statistics and all that, W.A. Benjamin, New York (1964)

6. R.M.F. Houtappel, H. van Dam and E.P. Wigner, The conceptual basis and use of the geometric invariance principles, Rev. Mod. Phys. 37, 595 (1965)

7. E.P. Wigner, Symmetry and conservation laws, Proc. Nat. Acad. Sci. (U.S.) 51, 956 (1964)

8. E.P. Wigner, Symmetries and reflections, Indiana University Press, Bloomington (1967)

9. J.J. Sakurai, Invariance principles and elementary particles, Princeton University Press (1964)

10. W.M. Gibson and B.R. Pollard, Symmetry principles in elementary particle physics,Cambridge University Press (1976)

11. Y. Takahashi, An introduction to field quantization, Pergamon Press, London (1969) and literature quoted there

12. I.M. Guelfand, R.A. Minlos and Z.Ya. Shapiro, Representations of the rotation and Lorentz group and their applications, Pergamon Press London (1963)

B. - FIELD THEORY, PARTICULARLY ELECTRODYNAMICS :

13. R.P. Feynman, Quantum electrodynamics, W.A. Benjamin, New York (1962)

14. J.D. Bjorken and S.D. Drell, Relativistic quantum mechanics, Mc Graw Hill, New York (1964)

15. J.D. Bjorken and S.D. Drell, Relativistic quantum fields, Mc Graw Hill, New York (1965)

16. V. Berestetski, E. Lifchitz and L. Pitayevski, Théorie quantique relativiste I, Ed. Mir, Moscow (1972)

17. E. Lifchitz and L. Pitayevski, Théorie quantique relativiste II, Ed. Mir, Moscow (1972)

18. S.S. Schweber, An introduction to relativistic quantum field theory, Harper & Row, New York (1961)

19. N.N. Bogoliubov and D.V. Shirkov, Introduction to the theory of quantized fields, Sohn Wiley New York (1959)

20. G't Hooft and M. Veltman, Diagrammar, CERN 73-9, Geneva (1973)

21. J.S. Bell, Experimental quantum field theory, CERN 77-18, Geneva (1977)

22. E. Brezin, J. Iliopoulos, C. Izytkson and R. Stora, Théorie des champs in Ecole d'Eté de Physique des Particules (edited by R.A. Salmeron), Institut National de Physique Nucléaire et de Physique des Particules (IN2P3) Paris (1973)

23. J. Leite Lopes, Introducción a la electrodinamica cuantica, Trillas, Mexico (1977)

. - SOLITONS, INSTANTONS AND TOPOLOGICAL QUANTUM NUMBERS :

24. Sidney Coleman, Classical lumps and their quantum descendants, Proc. 1975 Ettore Majornana School of Subnuclear Physics, Academic Press

25. R. Rajaraman, Non-perturbative classical methods in quantum field theory (a pedagogical review), Phys. Reports 21 , N° 5 (1975)

26. J. Arafune, P.G.O. Freund, C.J. Goebel, J. Math. Phys. 16, 433 (1975)

27. J.L. Gervais, A. Jevicki, B. Sakita, Phys. Rev. D12, 1038 (1975)

28. H. Nielsen and P. Olesen, Nucl. Phys. B61, 45 (1973)

29. G 't Hoft, Nucl. Phys. B79, 276 (1974)

30. A.M. Polyakov, JETP Lett. 20 , 194 (1974)

31. S. Coleman, Phys. Rev. D11, 2088 (1975)

32. R. Jackiw, Rev. Mod. Phys. 49, 681 (1977)

33. W. Marciano and H. Pagels, Quantum Chromodynamics, Phys. Reports 36C, N° 3 (1978)

34. C.J. Callan, A review of instanton physics, Proc. 19 th Intern. Conference High Energy Physics, Tokyo (1978)

35. S. Sciuto, Topics on Instantons, CERN preprint (1977)

36. C.G. Bollini, J.J. Giambiagi and J. Tiomno, Singular potentials and analytic regularisation in classical Yang-Mills equations, J. Math. Phys. 20, 1967 (1979)

37. G.H. Thomas, Introductory lectures on fibre bundles and topology for physicists, Rev. del Nuovo Cimento 3, N° 4 (1980)

38. N. Steenrod, The topology of fibre bundles, Princeton Univ. Press (1970)

D. - BASIC PAPERS ON WEAK INTERACTIONS :

39. E. Fermi, Nuovo Cimento 11, 1 (1934)

40. G. Gamow and E. Teller, Phys. Rev. 49, 895 (1936)

41. T.D. Lee and C.N. Yang, Phys. Rev. 104, 254 (1956)

42. A. Salam, Nuovo Cimento 5, 29 (1957)

43. L.D. Landau, Journ. Exp. Theor. Phys. (USSR) 32, 405 and 407 (1957)

44. T.D. Lee and C.N. Yang, Phys. Rev. 105, 1671 (1957)

45. C.G. Sudarshan and R.E. Marshak, Proc. Padua Conference on Mesons v - 14 (1957)

46. R.P. Feynman and M. Gell-Mann, Phys. Rev. 109, 193 (1958)

These and other papers are reproduced in :

47. P.K. Kabir, The development of weak interaction theory, Gordon and Breach, New York (1962)

E. - CLASSIFICATION OF STRONGLY INTERACTING PARTICLES, THEIR WEAK AND ELECTROMAGNETIC INTERACTIONS ; QUARKS :

48. M. Gell-Mann and Y. Néeman, The eightfold way, W.A. Benjamin (1964). This book reproduces the original articles by Gell-Mann and other authors including the paper of Cabibbo on the mixing angle θ_c

49. S.L. Adler and R.F Dashen, Current algebras and applications to particle physics, W.A. Benjamin (1968) (reproduces several original papers)

50. H.J. Lipkin, Lie groups for pedestrians, North-Holland, Amsterdam (1965)

51. J.J. Kokedee, The quark model, W.A. Benjamin, New York (1969)

52. H. Joos, Proc. 11th Scottish Univers. Summer School in Physics, (Ed. by J. Cumming, H. Osborn) Academic Press London (1971)

53. N. Cabibbo, M. Della Negra, G. Girardi, Ph. Heusse, M. Jacob, L. Maiani, F. Pierre, F. Richard and M. Veltmann, Les quarks in Ecole d'Eté de Physique des Particules (edited by R.A. Salmeron), IN2P3, Paris (1976)

54. O.W. Greenberg, Ann. Rev. Nucl. Sci. 28, 327 (1978)

Also :

55. D. Amati, H. Bacry, J. Nuyts and J. Prentki, Nuovo Cimento X34, 1732 (1964) (SU(4) ; see also Ref. 83)

56. M.K. Gaillard, B.W. Lee and J.L. Rosner, Rev. Mod. Phys. 47, 277 (1975) (Search for charmed quark evidence)

57. H.J. Lipkin, A quasinuclear colored quark model for hadrons, Fermilab Conf. 78/73-THY, Fermi National Accelerator Laboratory, Batavia (1978)

58. J. Ellis, Charm, après-charm and beyond, Hadron structure and lepton-hadron interactions (Cargèse 1977) Plenum Press, New York (1979)

F. - PARTON MODEL :

59. R.P. Feynman, Phys. Rev. Lett. 23, 1415 (1969)

60. J.D. Bjorken and E.A. Paschos, Phys. Rev. 185, 1975 (1969)

61. R.P. Feynman, Photon-Hadron Interactions, W.A. Benjamin, New York (1972)

62. J. Kogut and L. Susskind, Phys. Rep. 8C, 76 (1973)

63. E.A. Paschos and L. Wolfenstein, Phys. Rev. D7, 91 (1973)

64. G. Altarelli, B. Diu, F.M. Renard, P. Söding, J. Tran-Than Van, Interactions photon-hadrons, in Ecole d'Eté de Physique des Particules (edited by R.A. Salmeron) IN2P3, Paris (1974)

65. T.M. Yan, Ann. Rev. Nucl. Sci. 26, 199 (1976)

66. R.P. Feynman, Quark jets in VII Kaysersberg International Symposium on multiparticle dynamics (edited by R. Arnold, J.P. Gerber and P. Schübelin), Centre de Recherches Nucléaires, Strasbourg (1977)

67. R.D. Field and R.P. Feynman, Phys. Rev. D15, 2590 (1977)

68. R.P. Feynman, R.D. Field and G.C. Fox, Nucl. Phys. B128, 1 (1977)

G. - FURTHER READING IN PARTICLE PHYSICS :

69. R. Omnès, Introduction à l'étude des particules élémentaires, Ediscience, Paris (1970)

70. L.B. Okun, Weak interactions of elementary particles, Pergamon Press, London (1965)

71. R.E. Marshak, Riazzudin and C.P. Ryan, Theory of weak interactions in particle physics, Wiley - Interscience, New York (1969)

72. J. Bernstein, Elementary particles and their currents, W.H. Freeman, San Francisco (1968)

73. N. Cabibbo, J. Iliopoulos, J. Leite Lopes, L. Maiani, P. Musset, C. Rubia and R. Turlay, Interactions faibles, Ecole d'Eté de Physique des Particules (edited by R.A. Salmeron), IN2P3, Paris (1974)

74. G. Altarelli, J.E. Augustin, S. Glashow and L. Michel, Symétries et particules nouvelles, Ecole d'Eté de Physique des Particules (edited by R.A. Salmeron) IN2P3, Paris (1975)

75. C.A. Savoy, Théorie de interactions faibles, Département de Physique, Université de Genève (1976)

76. C.A. Garcia Canal, M.B. Gay Ducati and J.A. Martins Simões, Notes on inelastic scattering, Série de Cours et Conférences sur la physique des hautes énergies N° 15, Centre de Recherches Nucléaires, Strasbourg (1979)

77. D. Bachin, Weak interactions, Sussex University Press (1977)

78. S. Wojcicki, Weak decays, SLAC-PUB-2232, California Institute of Technology (1978)

H. - SOME OF THE BASIC PAPERS ON THE HIGGS MECHANISM :

79. J. Goldstone, Nuovo Cimento 19, 154 (1961)

80. P.W. Higgs, Phys. Lett. 12, 132 (1964) ; Phys. Rev. 145, 1156 (1966)

81. F. Englert and R. Brout, Phys. Rev. Lett. 13, 321 (1964)

82. G.S. Guralnik, C.R. Hagen and T.W. Kibble, Phys. Rev. Lett. 13, 585 (1965)

83. T.W. Kibble, Phys. Rev. 155, 1554 (1967)

I. - SOME OF THE EARLY PAPERS ON NEUTRAL CURRENTS AND THE EQUALITY $g_W = e$:

84. J. Leite Lopes, Nucl. Phys. 8, 234 (1958)

85. S.A. Bludman, Nuovo Cimento 9, 433 (1958)

86. A. Salam and J.C. Ward, Nuovo Cimento 11, 568 (1959)

87. T.D. Lee and C.N. Yang, Phys. Rev. 119, 1410 (1960)

88. T.D. Lee, Phys. Rev. Lett. 26, 801 (1971)

89. J. Schecht and Y. Ueda, Phys. Rev. D2, 736 (1970)

90. J. Leite Lopes, Nucl. Phys. B38, 555 (1972)

J. - SOME BASIC PAPERS ON GAUGE THEORIES :

91. C.N. Yang and R.L. Mills, Phys. Rev. 96, 191 (1954)

92. R. Utiyama, Phys. Rev. 101, 1597 (1956)

93. J. Schwinger, Ann. Phys. 2, 407 (1957)

94. S. Glashow, Nucl. Phys. 22, 579 (1961)

95. A. Salam and J. Ward, Phys. Lett. 13, 168 (1964)

96. S. Weinberg, Phys. Rev. 19, 1264 (1967)

97. A. Salam in Nobel Symposium ed. by N. Svartholm (1968)

98. S. Glashow, J. Iliopoulos and L. Maiani, Phys. Rev. D2, 1285 (1970)

99. S. Weinberg, Phys. Rev. Lett. 27, 1688 (1971)

100. G't Hooft, Nucl. Phys. B33, 173 (1971)

101. J. Prentki and B. Zumino, Nucl. Phys. B47, 99 (1972)

102. C.N. Yang, Phys. Rev. Lett. 33, 445 (1974) ; T.T. Wu and C.N. Yang, Phys. Rev. D12, 3843 (1975) ; Phys. Rev. D12, 3845 (1975)

103. C.N. Yang, Magnetic monopoles, fiber bundles and gauges fields in Five decades of weak interactions, New York Acad. of Sciences, New York (1977) ; Gauge fields, Proc. 6 th Hawaii Topical Conference in Particle Physics, University of Hawaii (1975) ; C. N. Yang, Theory of magnetic monopoles, Proc. 19 th Intern. High Energy Conference, Tokyo (1978) ; P. Goddard and Di Olive, Rep. Prog. Phys. 41, 1357 (1978)

For gauge fixing conditions see

104. E.S. Abers and B.W. Lee, Phys. Reports 9C, N°1 (1973)

105. G. Becchi, A. Rouet and R. Stora, Phys. Lett. 52B, 344 (1974)

106. V.N. Gribov, Nucl. Phys. B139, 1 (1978)

Also :

107. Ch. Ragiadakos and R.S. Viswanathan, Phys. Rev. D20, 1369 (1979) ; Phys. Lett. 86B, 288 (1979)

Also E. de Rafael, Ref. 127.

For gauge fixing conditions in gravitational theory see Weinberg, Ref. 174.

108. M. Kobayashi and K. Maskawa, Progr. Theor. Phys. 49, 652 (1973) ; J. Ellis, M.K. Gaillard and D. Nanopoulos, Nucl. Phys. B109, 213 (1976) ; L. Wolfenstein, Mixing angles, mass matrix and CP violation in the Kobayashi-Maskawa model, preprint COO-3006-124, Carnegie-Mellon University, Pittsburgh (1979)

K. - REVIEW ARTICLES ON GAUGE THEORIES :

109. L. O'Raifeartaigh, Rep. Prog. Phys. 42, 159 (1979)

110. J. Bernstein, Rev. Mod. Phys. 46, 7 (1974)

111. S. Weinberg, Rev. Mod. Phys. 46, 225 (1974)

112. M.A. Bag and A. Sirlin, Ann. Rev. Nucl. Sci. 24, 3791 (1974)

113. J. Leite Lopes, Le modèle des champs de jauge unifiés in Physical Reality and Mathematical description (edited by C.P. Enz and J. Mehra), D. Reidel, Dordrecht (1974)

114. S. Coleman, Proc. 1973 Ettore Majorana School of Subnuclear Physics, Academic Press (1975)

115. L. Maiani, Proc. 1976 CERN School of Physics, CERN 76-20, Geneva (1976)

116. J. Iliopoulos, Proc. 1977 CERN-JINR School of Physics, CERN 77-18, Geneva (1977)

117. R.M. Barnett, Implications of experiment on gauge theories, SLAC-PUB-1961, Stanford Linear Accelerator Center (1977)

118. G. Altarelli, The standard model of the weak interactions and its problems in Ecole d'Eté de Physique des Particules (edited by R.A. Salmeron) IN2P3, Paris (1977)

119. C.A. Savoy, L'unification des forces faibles et électromagnétiques dans la théorie de Salam-Weinberg, Congrès Soc. Française de Phys., Poitiers, preprint UGVA-DPT 1977/08-144, Univ. de Genève (1977)

120. M. Böhm and H. Joos, Eichtheorien der Schwachen, elektromagnetischen und starken wechselwirkung, DESY 78-27, Hamburg (1978)(and ref. therein)

121. C. Itzykson, Proc. 1978 CERN School of Physics, CERN 78-10, Geneva (1978)

122. S. Weinberg, Weak interactions in Proc. 19th Intern. Conf. High Energy Phys., Tokyo (1978) ; S. Weinberg, Nobel Lecture, Rev. Mod. Phys. 52 , 515 (1980) ; A. Salam, Nobel Lecture, Rev. Mod. Phys. 52, 525 (1980) ; S. Glashow, Nobel Lecture, Rev. Mod. Phys. 52, 539 (1980)

123. M. Levy, An introduction to gauge theories, Hadron Structure and lepton-hadron interactions, Plenum Press, New York (1979) ; J.J. Giambiagi, Teoria de campos de "gauge", Univ. Federal de São Carlos (1978)

L. - CHROMODYNAMICS :

124. C. Itzykson, Confinement , espoirs et problèmes in Ecole d'Eté de Physique des particules (edited by R.A. Salmeron) IN2P3, Paris (1977)

125. J. Swieca, Phys. Rev. D13, 312 (1976) ;
R.D. Field, Dynamics of high energy reactions, Proc. 19th Internat. Conf. High Energy Phys., Tokyo (1978)

126. G. Altarelli, Partons in quantum chromodynamics in Ecole d'Eté de Physique des Particules (edited by R.A. Salmeron) IN2P3, Paris (1977)

127. E. de Rafael, Quantum chromodynamics as a theoretical framework of the hadronic interactions in Ecole d'Eté de Physique des Particules (edited by C. Ghesquiere), IN2P3, Paris (1978) (and references therein)

128. B. Sakita, Quantum chromodynamics and related problems in Proc. 19 th International Conference High Energy Phys., Tokyo (1978)

129. M. Abud and P. Leal Ferreira, Rev. Bras. Fis. 9, 817 (1979)

130. H.J. Rothe, K.D. Rothe and J.A. Swieca, Screening versus confinement, Phys. Rev. D (to be published)

131. D.J. Gross, A theory of the strong interactions, Hadron Structure and lepton-hadron interactions (Cargèse 1977) Plenum Press, New York (1979)

132. H. Joos, Introduction to quark confinement in QCD, in Quarks and leptons as fundamental particles, Springer, Wien (1979)

133. Ch. Llewellyn Smith, Topics in quantum chromodynamics, Quantum flavordynamics, Quamtum chromodynamics and unified theories, Nato Advanced Study Institutes Vol. 54, Plenum Press, New York (1980) ; S.D. Ellis, Jets and Quantum chromodynamics, ibid. ; R.D. Field, Perturbative Quantum chromodynamics, ibid.

M. - HEAVY LEPTONS :

134. Y.S. Tsai, Phys. Rev. D4, 2821 (1971)

135. H. Georgi and S.L. Glashow, Phys. Rev. Lett. 28, 1494 (1972) (gauge model without neutral vector bosons)

Cf. also Prentki and Zumino, ref. 86.

136. J.D. Bjorken and C.H. Llewellyn-Smith, Phys. Rev. D7, 887 (1973)

137. J. Leite Lopes and Ch. Ragiadakos, Lett. Nuovo Cimento 16, 261 (1976)

138. M.L. Perl, Review of heavy lepton production in Proc. 1977 International Symposium on lepton and photon interactions at high energies, Hamburg (1977)

139. Ch. Ragiadakos, Les leptons lourds : étude théorique des implications de leur existence, Ph. D. thesis, Série de Cours et Conférences sur la physique des hautes énergies, Strasbourg (1978)

N. - NEUTRINO PHYSICS :

140. L.M. Sehgal, Phenomenology of neutrino reactions, Report ANL-HEP-PR-75-45, Argonne Nat. Laboratory (1975) ; S.M. Bilenky and B. Pontecorvo, Phys. Rep. 41, 225 (1978).

141. B.C. Barish, A review of neutrino physics at Fermi Lab, CALT 68-535, California Institute of Technology (1975)

142. B. Pontecorvo, Lepton charges and lepton mixing in Proc. European Conf. on Particle Phys., Budapest (1977)

143. P. Musset and J.P. Vialle, Phys. Rep. 39C, 1 (1978)

144. B.C. Barish, Phys. Rep. 39C, 280 (1978)

145. K. Tittel, Neutrino reactions I, Proc. 19 th International Conf. on High Energy Phys. 863, Tokyo (1978)

146. C. Baltay, Neutrino interactions II, Proc. 19 th International Conf. on High Energy Phys., 882, Tokyo (1978)

147. J. Steinberger, Some recent experimental results in high-energy neutrino physics in Hadron Structure and lepton-hadron interactions (Cargèse 1977), Plenum Press, New York (1979)

O. - SPECULATIONS ON LEPTON STRUCTURE :

148. J. Leite Lopes, Quarks for hadrons and leptons, Rev. Bras. Fis. 5 37 (1975)

149. J. Leite Lopes and N. Fleury, A model for leptons, Lett. Nuovo Cimento 19, 7 (1977)

150. S. Barshay and J. Leite Lopes, Non-conservation of muon number in a model of lepton structure, Phys. Lett. 68B, 174 (1977)

151. J. Leite Lopes and J.A. Martins Simões, The muon magnetic moment in a model of lepton structure, Rev. Bras. Fis. 8, 621 (1978)

152. M.B. Gay Ducati, J. Leite Lopes and J.A. Martins Simões, On new possible lepton interactions, Lett. Nuovo Cimento 24, 432 (1979)

153. Y. Ne'eman, Primitive particle model, Phys. Lett. 82B, 69 (1979)

154. H. Harari, A shematic model of quarks and leptons, Phys. Lett. 86B, 83 (1979)

155. M.A. Shupe, A composite model of leptons and quarks, Phys. Lett. 86B, 87 (1979)

156. M.A. Shupe, Intrinsic parity of quark and lepton constituents, Phys. Lett. (to be published)

157. R. Casalbuoni and R. Gatto, Unified description of quarks and leptons, Phys. Lett. 88B, 306 (1979)

158. R. Casalbuoni and R. Gatto, Unified theories for quarks and leptons based on Clifford algebras, Phys. Lett. 90B, 81 (1980)

159. J.G. Taylor, Spin 3/2 quarks in deep inelastic scattering, Phys. Lett 90B, 143 (1980)

160. J.G. Taylor, A model of composite quarks and leptons, preprint, King's College, London (1979)

161. W. Krulikowski, Primordial quantum chromodynamics, Phys. Lett. 90B, 241 (1980)

162. J. Leite Lopes, J.A. Martins Simões and D. Spehler, Production and decay properties of possible spin 3/2 leptons, Phys. Lett. 94B, 367 (1980)

P . - GRAND UNIFICATION :

163. H. Georgi and S.L. Glashow, Phys. Rev. 32, 438 (1974)

164. H. Georgi, H. Quinn and S. Weinberg, Phys. Rev. Lett. 33, 451 (1974)

165. C. Jarlskog, Gauge theories in New phenomena in lepton and hadron physics, Plenum Press, New York (1979)

166. J. Ellis, SU(5), Ref. TH.2723-CERN, Geneva (1979)

167. M.K. Gaillard and L. Maiani, New quarks and leptons, LAPP-TH-09, LAPP Annecy (1979)

168. J.F. Donnoghue, Proton lifetime and branching ratios in SU(5), Preprint Center for Theoretical Physics, MIT, Cambridge Mass. (1979)

169. C. Jarlskog and F.J. Yndurain , Nucl. Phys. B149, 29 (1979)

170. J. Ellis, M.K. Gaillard and D.V. Nanopoulos, On the effective lagrangean for baryon decay, preprint LAPP-TH-06 and Ref. TH. 2749-CERN, Geneva (1979)

171. J.C. Pati and A. Salam, Phys. Rev. D8, 1240 (1973) ; Phys. Rev. Lett. 31, 661 (1973) ; Phys. Rev. D10, 275 (1974)

172. J.C. Pati, Unification : its implications for present and future high-energy experimentation, Proc. 19 th International Conf. High Energy Phys., Tokyo (1978)

. - SOME REFERENCES ON GRAVITATION AND SUPERGRAVITATION

173. R.P. Feynman, Lectures on gravitation, Preprint, California Institute of Technology, Pasadena (1963)

174. S. Weinberg, Gravitation and cosmology, J. Wiley, New York (1972) (and the references in this book)

175. C.W. Misner, K.S. Thorne and J.A. Wheeler, Gravitation, W.H. Freeman, San Francisco (1973) (with abundant references); R.J. Adler, M.J. Bazin and M. Schiffer, Introduction to general relativity, Mc Graw-Hill, New York (1975)

176. H.C. Ohanian, Gravitation and space-time, W.W. Norton, New York (1976) (and the literature quoted)

177. G't Hoft, Quantum gravity in Trends in Elementary Particle Theory, Springer-Verlab, Heidelberg (1975)

178. H. Fleming, M. Novello, C.G. Oliveira, P.P. Srivastava et al in Proc. II Escola de Cosmologia e Gravitação (edited by M. Novello), Centro Brasileiro de Pesquisas Fisicas, Rio de Janeiro (1980)

The enormous value of the lepto-quark boson masses for grand unification and the fact that the proton lifetime must be higher than the age of the universe point out to the necessity of a superunification theory of gravitational forces with the strong, electromagnetic and weak interactions as well as of cosmological considerations. Supergravity as a gauge theory of supersymmetry is regarded as providing the route to such a super unified model. See the following review papers and the literature they quote :

179. B. Zumino, Supersymmetry and supergravity, Proc. European Conf. Particle Phys., Budapest (1977)

180. D.Z. Freedman and P. Van Nieuwenhuizen, Supergravity and the unification of the laws of physics, Sci. American, 126, February (1978)

181. D.Z. Freedman, Review of super-symmetry and supergravity, Proc. 19 th International Conf. High Energy Phys., Tokyo (1978)

182. A. Salam, Unification, superunification and new theoretical ideas, Proc. 19 th International Conf. High Energy Phys., Tokyo (1978)

Also :

183. J. Ellis, M.K. Gaillard and D.V. Nanopoulos, The smoothness of the universe, Phys. Lett. 90B, 253 (1980)

184. F. Englert, The creation of the universe as the break down of a grand unified symmetry, preprint, Université Libre de Bruxelles (1980)

Renormalization :

We have not studied renormalization in the present book.

The renormalizability of gauge field theories is the condition which gave them credibility. Although not a natural law -but rather a procedure required by the present mathematical status of field theories- renormalization is needed in order to transform away, when possible, the divergences which arise in the theory, by appropriate redefinition of the fields, propagators and physical parameters. It thus allows the calculation of observable quantities to any order in perturbation theory.

The fundamental work of Dyson, Salam, Weinberg, Ward and others in electrodynamics are quoted in Refs. 14, 15, 18, 19.

See also K. Hepp, Théorie de la renormalisation, Springer-Verlag, Heidelberg (1969).

The basic papers by 't Hooft, Bollini and Giambiagi, 't Hooft and Veltman, B.W. Lee, Fadeew and Popov, Taylor and Slavnov, Adler and also Bouchiat, Iliopoulos and Meyer among others, are quoted in Ref. 104 as well as in the excellent recent books :

185. J.C. Taylor, Gauge theories of weak interactions, Cambridge University Press (1976)

186. C. Nash, Relativistic quantum fields, Academic Press, New York (1978)

Index